TA 365 COM

Jens Blauert (Editor)

Communication Acoustics

Jens Blauert (Editor)

Communication Acoustics

With 146 Figures

 Springer

Prof. em. Dr.-Ing. Dr. techn. h.c. Jens Blauert
Institute of Communication Acoustics
Ruhr-University Bochum
44780 Bochum
Germany

ISBN-10   3-540-22162-X **Springer Berlin Heidelberg New York**
ISBN-13   978-3-540-22162-3 **Springer Berlin Heidelberg New York**

Library of Congress Control Number: 2005922819

Springer is a part of Springer Science+Business Media

springeronline.com

© Springer-Verlag Berlin Heidelberg 2005
Printed in The Netherlands

The use of general descriptive names, registered names, trademarks, etc. in this publication does not imply, even in the absence of a specific statement, that such names are exempt from the relevant protective laws and regulations and therefore free for general use.

Typesetting: Data conversion by the authors
Final processing: PTP-Berlin Protago-TEX-Production GmbH, Germany
Cover-Design: deblik, Berlin
Printed on acid-free paper   62/3141/Yu - 5 4 3 2 1 0

Preface

Communication Acoustics deals with those areas of acoustics which relate to the modern communication and information sciences and technologies.

Due to the advent of digital signal processing and recording in acoustics, these areas have encountered an enormous upswing during the last four decades, whereby psychological and physiological knowledge is extensively exploited for technical applications. Communication-acoustical components can be found in various systems in communication technology, as either stand-alone systems or embedded components.

Since a comprehensive handbook on *Communication Acoustics* is not yet available, this volume tries to fill the gap by providing a collection of review articles covering some of the most relevant areas of the field.

The articles are written with the goal in mind of providing students with comprehensive introductions and offer numerous references to relevant literature. Besides its usefulness as a textbook, this book will also be a source of valuable information for those who want to improve or refresh their knowledge in the field of *Communication Acoustics*.

The team of authors consists of the following scientist and technologists,

> *Jens Blauert, Jonas Braasch, Hugo Fastl, Volkmar Hamacher,*
> *Dorte Hammershøi, Ulrich Heute, Inga Holube, Herbert Hudde,*
> *Ute Jekosch, Georg Klump, Armin Kohlrausch, Arild Lacroix,*
> *Henrik Møller, Sebastian Möller, John N. Mourjopoulos,*
> *Pedro Novo,* and *Steven van der Par.*

Marc Hanisch assisted substantially in compiling the individual manuscripts. Each chapter has been reviewed by three other members of the team plus by at least two external reviewers.

Due to its interdisciplinary character, the book is bound to attract readers from many different areas, such as acousticians, audio engineers, audiologists, biologists, cognitive scientists, communication engineers, communication scientists, computer scientists, electronic engineers, hearing-aid users, hi-fi enthusiasts, linguists, musicians, musicologists, oto-laryngologists, physicists, physiologists, psycho-acousticians, psychologists, quality experts, sound designers, speech technologists, and tonmeisters.

Jens Blauert, editor

Contents

List of Contributors

Jens Blauert
Ruhr-University Bochum
Institute of Communication
Acoustics
Bochum, Germany
and
Aalborg University
Department of Acoustics
Aalborg, Denmark

Jonas Braasch
McGill University (CIRMMT)
Department of Music Theory
Montréal, Canada

Hugo Fastl
Technical University of Munich
Technical-Acoustics Group
Department of
Human-Machine-Communication
Munich, Germany

Volkmar Hamacher
Siemens Audiologische Technik
GmbH
Erlangen, Germany

Dorte Hammershøi
Aalborg University
Department of Acoustics
Aalborg, Denmark

Ulrich Heute
Christian-Albrecht University
Institute for Circuit
and System Theory
Faculty of Engineering
Kiel, Germany

Inga Holube
University of
Applied Sciences Oldenburg
Institute of Hearing Technology
and Audiology
Oldenburg, Germany

Herbert Hudde
Ruhr-University Bochum
Institute of
Communication Acoustics
Bochum, Germany

Ute Jekosch
School of Architecture
Rensselaer Polytechnic Institute
Troy NY, USA

Georg Klump
Oldenburg University
Zoophysiology and Behaviour Group
Oldenburg, Germany

Armin Kohlrausch
Philips Research Laboratories
Eindhoven, The Netherlands
and

Technical University Eindhoven
Department of
Technology Management
Eindhoven, The Netherlands

Arild Lacroix
Goethe-University Frankfurt
Institute of Applied Physics
Frankfurt am Main, Germany

Henrik Møller
Aalborg University
Department of Acoustics
Aalborg, Denmark

Sebastian Möller
Ruhr-University Bochum
Institute of
Communication Acoustics
Bochum, Germany

John N. Mourjopoulos
University of Patras
Audiogroup,
Wire Communications Laboratory
Electrical & Computer Engineering
Department
Patras, Greece

Pedro Novo
Ruhr-University Bochum
Institute of
Communication Acoustics
Bochum, Germany

Steven van de Par
Philips Research Laboratories
Eindhoven, The Netherlands

1 Analysis and Synthesis of Auditory Scenes

Jens Blauert[1,2]

[1] Institute of Communication Acoustics, Ruhr-University Bochum, Bochum
[2] Department of Acoustics, Aalborg University, Aalborg

Summary. Audition is the most important human sensory modality in inter-individual communication. Consequently, acoustics has always dealt with communication. Yet recently, due to the high amount of computing power available, communication-acoustical systems become increasingly complex and sophisticated. In particular, they become more intelligent and knowledgeable. This trend will be discussed in this chapter by taking two complementary research streams as examples which have been pursued at the Institute of Communication Acoustics at Bochum during the past three decades, namely, (a) analysis of auditory scenes with the goal of arriving at parametric representations and, complementary, (b) synthesis of auditory scenes from parametric representations. The discussion is based on two software systems which have been developed for research purposes, namely, a binaural-analysis system and an auditory-virtual-environment generator – both of which will be roughly explained. It is, however, not the aim of the chapter to introduce scientific or technological details, but rather to bring a trend to the fore which may well coin the profile of communication acoustics in the years to come.

1.1 Introduction

This section will identify three paramount milestones in the evolution of Communication Acoustics. The first one had been prompted by the invention of the vacuum triode – which made amplification of weak electrical signals feasible. The second one is linked to the advent of computers in acoustics and, thus, to the introduction of digital processing of acoustical signals. The third milestone is rather an ongoing current process and can be associated with a general trend in the information and communication technologies, namely, that technological systems are being furnished with increasing amounts of built-in intelligence and explicit knowledge. Placardingly one could say: "Communication Acoustics becomes cognitive!" It is essentially this third milestone that will be further explained and justified in this chapter.

1.1.1 Some Reminiscing

Acoustics is a latinized term stemming from the Greek verb $\alpha\kappa o\acute{u}\epsilon\iota\nu$ [akúɪn], which means "to hear". Accordingly, already at the brink of the 20th century, it was widely accepted that acoustics is inherently a multi-disciplinary science with physical phenomena, "acoustic events", at the one hand and perceptual phenomena, "auditory events", on the other one. Relevant scientific books like the ones of *Lord Rayleigh* [61] and *Von Helmholtz* [28] provide evidence for this view. Further, already at that time, some inventions had been made which where fundamental for later technological applications, e. g., for television [53], telephone [62], magnetic sound recording [59] and sound on film [64]. Nevertheless, it was only after 1910 that all these technological ideas prompted what one might call "killer applications" in today's marketing terms, namely, applications such as radio broadcast, sound-film cinema, public-address systems with loudspeakers, and the world-wide telephone network. This became possible since the vacuum triode had been invented [11] and amplification of "weak currents" had become feasible.

Consequently, in the years after about 1920, acoustics enjoyed a dramatic upswing. Besides pure physics and auditory perception, electrical engineering became part of the game and a new name for the field was created, namely, "electro-acoustics". Many research facilities were established and capable academic teachers were recruited, in Germany, e. g., Profs. *H. Barkhausen, W. Reichardt, E. Meyer, L. Cremer,* and *V. Aschoff.*

A further, substantial boost was given to Communication Acoustics by the advent of computers in the field. The Bell Laboratories, Summit NJ, were a prominent laboratory for pioneer work in this respect and the lecture of *M. Schroeder* [66] in 1968 opened the eyes of many in the field. Digital signal processing became an indispensable component of Communication Acoustics. In fact, almost all modern applications of it would be unthinkable without digital signal processing – see, e. g., [31, 44, 52] in this volume. Communication Acoustics has thus established itself as a relevant subfield of modern technology. This becomes particularly evident since various technologies originating from it can be regarded as essential "enabling technologies" for many advanced applications. Such enabling technologies are, for instance, speech technology, binaural technology, perceptual coding, audio engineering, surround sound, and auditory virtual environments.

Since a few years, a new trend becomes visible which requires thorough analysis, as it will most likely shape the scope of Communication Acoustics in the years to come. The trend reflects the fact that modern information and communication systems are about to contain more and more built-in intelligence and knowledge. In fact, this general trend applies particularly to Communication Acoustics where remarkable progress has already been made in this regard, e. g., in speech technology – think of advanced speech recognizers and dialog systems as examples. New algorithms will be developed along these lines and will allow for many sophisticated novel applications. Thus, there is ample evidence that the field of Communication Acoustics will enjoy further growth and attain even higher impact on technology and society.

1.1.2 The Classical Paradigm of Communication Technology

The traditional topic of communication technology is the transmission of signals, carrying information across space and time. A transmission system is usually characterized by the following elements, connected in cascade,

- an information source,
- an encoder,
- a transmission channel,
- a decoder and
- an information sink.

The channel may contain memory and add linear and/or nonlinear distortions and/or noise.

A comparable general architecture is generally applied for transmission in Communication Acoustics in particular – frequently called "audio" transmission. Here, the goal is to transmit acoustical signals from one point in space and/or time to another one such that the auditory perceptions at both instances correspond to each other. The most demanding task in this context is "authentic" reproduction, such that the listeners at the play-back end hear the same that they would hear at the recording end of the transmission chain.

One possible approach to arrive at authentic reproduction is given by the binaural technique [25], which tries to reproduce the sound signals at the entrance to the two ears of a listener authentically, thus providing all acoustic cues deemed sufficient for perceptual authenticity. The schematic of such a binaural transmission system is given in Fig. 1.1. The input signals to the system are either picked up at the ear canals of real humans or stem from microphones mounted at a head replica – a so-called artificial head or dummy head. Playback is accomplished by headphones here, yet, loudspeaker techniques would also be available for this purpose.

It is obvious that equalization is necessary in order to reproduce the ear signals authentically, e. g., to correct for distortions as imposed by microphones and headphones. Further, if intermediate storage of the recordings is desired, there must be memory in the system. In fact, we see in almost all

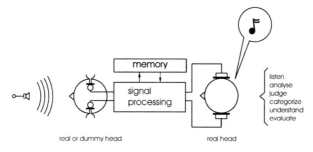

Fig. 1.1. Schematic of a binaural transmission system

modern audio-transmission systems that a significant amount of processing power is installed. Even if one looks at some standard audio equipment, such as a mixing console, one will very likely find a "computer" inside, these days.

1.1.3 Splitting Up the Classical Transmission Chain

Looking at Fig. 1.1, please consider the role of the listeners at the receiving end. The listeners – i. e. the information sinks in terms of information theory – do things like listening to and analyzing their auditory events, describing and categorizing them, evaluating them and judging upon them. Instructed well, they are even able to provide quantitative description of what they hear, thereby rendering parametric representations of their auditory scenes.

Having computational power at hand in any case, the question comes up of whether such listener tasks could be performed by computers as well. The answer is yes, and this is indeed already the case for various tasks, e. g., in speech recognition and sound-quality evaluation. The function of such systems is, in more general terms, called "auditory-scene analysis", ASA. Figure 1.2 a shows a block diagram accordingly.

What is feasible at the analysis side can, of course, also be attempted at the synthesis side. Here the issue is to generate acoustic input signals to the listeners' ears in such a way that the listeners perceive auditory events in a pre-described way, e. g., auditory scenes where the various components are parameter controlled. Such auditory scenes are called "auditory virtual environments", AVE, and a system which generates them is called a virtual-environment generator. Figure 1.2 b gives a block-diagram to illustrate the idea.

An ASA system and an AVE system, when connected in cascade, can form a parametric transmission system. The ASA system would analyze auditory scenes and represent them as sets of parametric descriptors. These parameter sets would then be transmitted to the synthesis side and the AVE system would re-synthesize the auditory scenes that were/are present at the recording end. Parametric transmission system are already widely used in Communication Acoustics. As an example for such a system with quite some sophistication, but within reach of today's technology, one may think of the following.

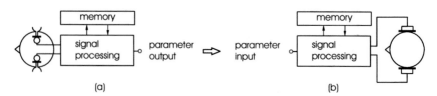

Fig. 1.2. Schematic as derived by separation of the diagram shown in Fig. 1.1. (**a**) System for "auditory-scene analysis", ASA. (**b**) System for the generation of "auditory virtual environments", AVE

A speech-recognition system produces a graphemic, i. e. ortho-graphic, transcript of what is said at the recording end – a simple example of ASA. The grapheme string is then transmitted to a speech-synthesis system which resynthesizes spoken speech – a simple example of an AVE generator. By the way, the transmission rate for the grapheme string can be as low as 50 bit/s for running speech at normal speed. Studies have shown that this roughly corresponds to the bit-rate that the brain can handle consciously – see [43] for interesting early work on this topic.

The following two sections will deal with ASA systems and AVE systems separately. It will, however, become evident that both kinds of systems share a common trend, namely, the more sophisticated they become, the more built-in intelligence and knowledge they require.

1.2 Analysis of Auditory Scenes

There is substantial technological demand for auditory-scene analysis. Hence, this field is currently an area of pronounced interest in international research. Important application areas are, among others, systems for the identification and localization of sound sources, especially in acoustically adverse environments such as multi-source, noisy or reverberant situations – for example, for acoustically-based surveillance and/or navigation. Further, systems to separate and "de-colourate" concurrent sound sources – so-called "cocktail-party processors" – which are, e. g., needed as front-ends for hearing aids or robust speech recognizers. Also, for the modelling of auditory recognition and assessment tasks, it is often advisable or even indispensable to start with an auditory-scene analysis, for example, in systems for analysis in architectural acoustics or in systems for quality assessment of speech and product sounds. In this context, so-called "content filters" are worth mentioning. These filters gain in relevance with respect to the tasks of automatic archiving and retrieval of audio-visual program material. There, they are used to analyze and code the contents of this material – compare the MPEG-7 coding as proposed by ISO/IEC [48].

As far as the systems for ASA make use of human auditory signal processing as a prototype, their structure follows more or less an architecture as given in Fig. 1.3. Yet, it has to be noted at this point that there are also approaches which are not directly motivated by biological analogies, e. g., electronically-controlled microphone arrays or blind source-separation algorithms. The following discussion is restricted to the fundamental ideas of binaural signal processing – see also [13].

Binaural systems have two front ports which take the signals from the left and right ear of a human or a dummy head as an input. Relevant signal-processing stages are as follows. After some sloppy band-pass filtering which simulates the middle ear, the two ear signals are fed into the inner-ear models,

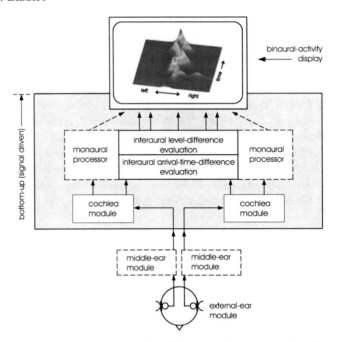

Fig. 1.3. Basic architecture of a model of binaural signal processing. Note that the processing is strictly signal-driven, i. e. bottom-up oriented

i. e. *cochlea* models. Here, the following operations are performed. The signals are first decomposed into ear-adequate spectral components, so-called critical-band components. Then it may follow what one would call "automatic volume control" in consumer electronics, namely a compression-like process. Finally, the band-pass filtered and eventually compressed signals are converted into signals which represent the neural activity as generated by the inner ear. One could call this process a special kind of A/D conversion. This is performed for each of the critical band components and in each of the two cochlea models, i. e. for the left and right ear, separately. The set of output signals differs in its character due to the model concept. It may, for example, be a set of series of discrete impulses, i. e. neural-spike series with variable rate, or signals which describe the rate of these spike series as a function of time.

In a next step, the two sets of cochlea-output signals are fed into a module which accepts signals from both ears. This "binaural" module analyzes the inter-aural arrival-time differences, i. e. differences between the left-ear and right-ear signals in each critical band. The algorithm used for these computation is usually based on estimating the inter-aural cross-correlation functions in each spectral region, e. g., from inter-aural coincidences as a function of the inter-aural arrival-time differences [32]. Multiple signals from uncorrelated sound sources will give rise to multiple peaks in the cross-correlation function. These peaks can be enhanced by contrast-enhancing procedures, such as the so-called "contra-lateral inhibition" [47]. The information rendered by the positions and forms of the peaks of the correlation functions is needed to identify individual sound sources and their lateral positions in space, among other features. In addition to the analysis of inter-aural arrival-time differences, an analysis of inter-aural level differences is usually performed as well. The results of this analysis can, e.g., be used to modify the results of the cross-correlation analysis in a meaningful way, such as to allow for a more precise sound-source localization. Of course, it has to be considered too that we can also hear with one ear only, so additional monaural modules are often added to the models.

From the processes as described above, a three-dimensional time-varying output finally results, its dimensions being running time, intensity and lateral position. Graphical representation of this output leads to what is called a running "binaural-activity map" – see Fig. 1.3, upper part. A number of tasks within ASA can be performed based on this, so far strictly bottom-up, processing and the resulting series of binaural-activity maps. Two examples will be discussed in the following subsections.

1.2.1 Quality Recognition and Assessment

In Fig. 1.4 we show, as an example, binaural-activity maps which have been generated by feeding the binaural model with binaural impulse responses which have been recorded in concert halls with the aid of an artificial head – so-called "binaural room responses". The maps indicate the direct sound and the individual reflections as they arrive at the ears of the dummy head. By inspecting and interpreting these maps, experienced experts of room acoustics are able to make prediction on the acoustic quality of the halls, i.e. on their suitability for specific kinds of musical performances. The left map with its

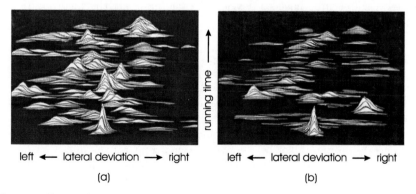

left ◀— lateral deviation —▶ right   left ◀— lateral deviation —▶ right

(a)   (b)

Fig. 1.4. Binaural-activity map of the binaural impulse responses of concert halls for (**a**) classic–romantic music and (**b**) modern symphonic and electro-acoustically enhanced music

dense distribution of early lateral reflections would be rated as belonging to a hall which is well suited for music of the classical-romantic genre, as in this hall a high amount of auditory spaciousness will be provided, accompanied by clearness and transparency. In fact, this map stems from a hall which is considered to be among the three best ones in the world for this kind of music, namely, the *Großer Musikvereinssaal* in Vienna. The map at the right shows only sparse early lateral reflections and the respective hall, consequently, sounds less spacious. This is what may be preferred for modern symphonic music and also for electro-acoustically enhanced performances.

We have chosen this example to demonstrate that binaural-activity maps can be used as a basis for the assessment and evaluation of the auditory quality of systems. Yet, the system, so far, only provides the basis for the judgement of experts. The experts have to use their knowledge and expertise to interpret the map. In fact, quality judgment at large is a very complex process, as becomes clear from the following definition of sound quality [34], translated from German.

"Sound quality is the result of an assessment of all features and feature values of a sound sample under examination which are recognized and nameable in the context of judging upon their suitability to meet all recognized and nameable features and feature values of individual expectations and/or social demands and/or pragmatic necessaries."

In other words, quality judgement starts out from a set of recognized and nameable features, the so-called "character" of the sound sample, which are then compared to a "reference set" of features. Quality might be defined as the distance between these two sets. In any case, what the experts do when assessing quality, requires expertise and knowledge. This is certainly more than can be modelled with a mere bottom-up, signals-driven algorithm. It has thus been realized in Communication Acoustics that for complex recognition task, such as "quality recognition" which we have taken as an example here, model architectures have to developed that allow for the incorporation of explicit knowledge.

1.2.2 "Cocktail-Party Processing"

To support the above statement, a further example will be discussed in the following, based on an auditory phenomenon which is known as the "cocktail-party effect" in the field. The term cocktail-party effect denotes the fact that human listeners with healthy binaural-hearing capabilities are able to concentrate on one talker in a crowd of concurrent talkers and discriminate the speech of this talker from the rest. Also, binaural hearing is able to suppress noise, reverberance and sound colouration to a certain extent.

Effort to model the cocktail-party effect may also start from binaural-activity maps as rendered by analysis systems as shown in Fig. 1.3. In Fig. 1.5 such a map is given which illustrates a case of two concurrent talkers. The

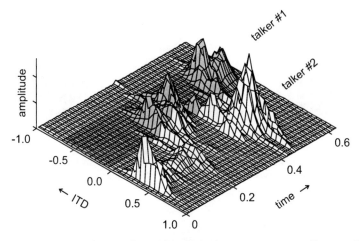

Fig. 1.5. Instantaneous binaural-activity map of two concurrent talkers at different azimuthal directions [60]

elevations and peaks in the map depict the spatial, temporal and spectral positions at which each of the two talkers is active. Suitable filter algorithms, working on this information, are able to enhance those chunks of the signals where the desired talker is active and suppress those where the undesired one speaks. In the figure, this is symbolized by shading the area of the activity of the undesired talker. Obviously, the effectiveness of such signal-processing depends on how reliably the point of activity of the desired talker can be identified. In the map, one can clearly see areas where the discrimination may be problematic.

Nevertheless, cocktail-party processors are very effective as long as they deal with a limited number of talkers – up to five – under an-echoic conditions and with low interfering noise. In these cases, some systems achieve an enhancement of the signal/noise ratio for the desired talker of up to 23 dB. This is more than humans can accomplish under these condition.

Yet, as soon as substantial amounts of reverberation and/or noise are added to the scene, today's instrumental cocktail-party processors run into problems and perform worse than humans. Figure 1.6 illustrates this situation. The left panel (a) shows the binaural-activity map of a section of an-echoic speech. The middle panel (b) illustrates what happens when wall reflections, i. e. reverberation, are introduced. It becomes clear that, if concurrent talkers were present in such a reverberant acoustic environment, it would become extremely difficult to identify the contributions of a particular talker. There are algorithms available which reduce the influence of the reflections [1, 24, 60]. The results of such an algorithm can be inspected in the right panel (c). Yet, it turns out that the de-reverberation algorithms do not remove all detrimental effects of the reflections, for instance, spectral modification are left which are perceivable as "colouration". These spectral

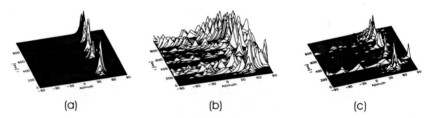

(a)   (b)   (c)

Fig. 1.6. Instantaneous binaural-activity map of a talker in (**a**) an-echoic environment (**b**) reverberant environment. In (**c**) the effect of the reverberation has partly been removed by signal processing

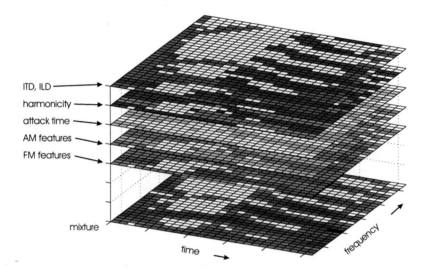

Fig. 1.7. Activity maps which consider different features of the ear-input signals to identify contributions of a desires sound source. The maps may be combined efficiently by means of a fuzzy temporal-cluster algorithm – see *bottom* plane [45]

modifications cause that the intelligibility of the processed speech hardly increases, although the "dryer" speech may be preferred in terms of quality.

Further algorithms to be applied on top of the binaural processing as schematically presented in Fig. 1.3 are, e. g., algorithms for de-colouration [14], or such which take further signals features into account in addition to inter-aural time differences, ITDs, and inter-aural level differences, ILDs, to identify the contribution of a target talker, namely, harmonicity, attack times, AM features and FM features. Figure 1.7 illustrates activity maps generated this way. They can, e.g., be combined in an optimum way by means of a fuzzy temporal-cluster analysis [45].

Nevertheless, it must be stated at this point that today's attempts to master the cocktail-party problem only lead to limited results as soon as acoustically adverse condition come into play. It is obvious that pure bottom-up signals processing has arrived at its limits. Better segregation would require more intelligence and knowledge.

1.2.3 Paradigmatic Consequences

Two application areas of auditory-scene analysis have been presented above, both based on binaural-activity maps, namely,

- sound-quality assessment,
- segregation of concurrent talkers.

Both application areas can be seen as examples to bring a general problem of auditory-scene analysis to the fore, namely, that there is the need to make the analysis systems more intelligent and more knowledgeable.

A sub-area of Communication Acoustics where this situation has been recognized already some years ago, is the field of speech recognition. Consequently, modern speech recognizers incorporate knowledge such as domain knowledge, semantic networks, language models, word models, grammatical, syntactic, phono-tactic and phonetic models, being represented in the form of rules, fuzzy logics, transition probabilities, look-up tables, dictionaries, and so on.

For more general auditory-scene analysis, knowledge to be incorporated may concern items such as spatial, temporal and spectral sound-source characteristics, information on the kind of signals, e.g., music, speech, noise, data on the room geometry, interaction rules between the acoustic/auditory objects – just to name a few. Further, information stemming from other sensory modalities, e.g., visual, tactile, olfactory, gustatory and/or proprioceptive cues may be available. Prior knowledge, e.g., the case history, would often be useful. The more the system knows about the contents of the auditory scene, the better analysis and recognition tasks can be accomplished – see [18] for pioneer work in this context.

To deal with such knowledge and to apply it effectively, strictly bottom-up processes are inadequate. Auditory-scene analysis has thus obviously reached a point where novel algorithms are required. For this reason more recent approaches provide modules on top of the binaural-activity-display level which work on a top-down basis, i.e. a hypothesis-driven rather than a bottom-up, signal-driven basis. In this way, it becomes possible to include knowledge-based processing into the structure. Signal-processing architectures of such a kind have already proved to be successful in instrumental speech recognition. One possible architecture for the complete system is depicted in Fig. 1.8.

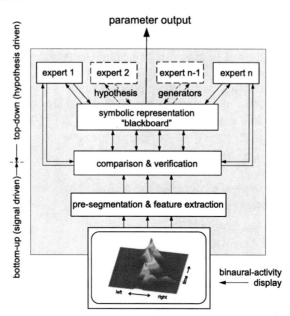

Fig. 1.8. Amendment to the schematic of Fig. 1.3, adding hypothesis-driven, i. e. top-down oriented, procedures in order to apply the knowledge implemented in the system

The running series of binaural-activity map is input to a categorization and segmentation process which produces a – usually error-infected – symbolic representation. The symbolic representation is then put on a so-called "black-board" module, where it can be inspected by different knowledge-based expert modules. These expert modules are software modules which contain and apply explicit knowledge in specific fields, relevant to the analysis tasks to be performed. The expert modules, under control of a coordination module, generate hypotheses with the aim of arriving at plausible interpretations of the series of binaural-activity maps to the end of producing a meaningful identification and analysis of the auditory scene. The individual hypotheses are evaluated step by step, eventually modified, and finally accepted or rejected. Once a plausible parametric representation of the auditory scene has been obtained in this way, any further processing and utilization depends on the specific application tasks at issue.

1.3 Synthesis of Auditory Scenes

Parametric synthesis of auditory scenes is currently of even higher technological relevance than their computational analysis, particularly when the listeners can act in and interact with the synthesized scenes. These interactive auditory scenes are often called "auditory virtual environments", AVEs.

Since auditory virtual environments, as virtual environments in general, are artificial, namely, are generated by computers, they rest on parametric representations of scenes. There exist description languages already which allow virtual worlds to be defined and specified in a formal, parametric way. The representation may, as of today, include spatial, but even content-related aspects. MPEG coding, as mentioned above, plays a role in this regard.

With the use of parametric coding it becomes possible that users which actually reside in different locations, displace themselves perceptually into a common virtual room, where they may confer together, i. e. tele-conferencing, or even jointly exercise a mechanical task, tele-operation. Further, one may enter a virtual environment to inspect it or visit objects in it, e. g., virtual museum or virtual tourism. Since an entrance to virtual spaces can be provided via the internet, manifold applications can be imagined. Virtual environments can further be superimposed on real environments – "augmented reality", "joint reality" – e. g., to assist navigation or provide other on-line support.

In the following, a number of realistic applications for AVEs are listed. The list is based on projects which the Bochum Institute has been involved in. Such projects have been, e. g., auditory displays for pilots of civil aircraft, AVEs for the acoustic design and evaluation of space for musical and oral performances, for individual interactive movie sound, and for tele-conferencing. Further, virtual sound studios, listening and control rooms, musical practicing rooms, and systems to generate artificial sound effects, especially so-called "spatializers". In addition, there is the auditory representation in simulators of all kinds of vehicles, e. g., aircraft, passenger cars, trucks, train, motorcycles. Also, AVEs are in use for the archiving of cultural heritage, for training, e. g., police and fire-fighter training, for rehabilitation purposes, e. g., motoric training, and as an interactive interface to the world-wide web, e. g., an internet kiosk.

Last, but not least, virtual environments play a pronounced role in scientific research. This is mainly due to the fact that they allow for flexible and economic presentation of complex experimental scenarios. Scenarios can be modified and changes can be performed without any physical effort. Research in areas like psycho-physics, psychology, usability, product-sound design and assessment is about to take advantage of this possibility. The use of virtual-environment techniques in these areas is currently dramatically increasing. For example, at our institute at Bochum, we use a generator for virtual auditory scenes to study the auditory precedence effect.

Details of the AVE technology are discussed at a different place in this volume [54]. Here we only give a broad overview, as needed to provide further evidence of a general trend in Communication Acoustics, namely, that communication-acoustical systems become more and more intelligent and knowledgeable.

1.3.1 Binaural Room Simulation

The basic ideas for parametric synthesis of auditory scenes originate from the sixties. There are mainly two roots. Firstly, it was proposed to generate auditory events at specific positions in space – so-called "virtual sound sources" – by modifying sound signals in the same way as they would be modified when impinging upon the human head and passing through the external ears, i. e. the *pinnae* and the ear canals, before reaching the ear drums [2]. Secondly, sound-field models which were based on geometrical acoustics, particularly on ray tracing, where amended in such a way that they now could render room impulse responses [42].

By bringing the two ends together, the impulse responses of the rooms, calculated for a specific room with a specific sound source at a specific position, could be convolved with impulse responses of a listener's ears, such rendering ear signals which correspond to those in a real sound field – see e. g. [6,25,46,54] for details. Listeners who are exposed to these signals have an auditory scene which is equivalent to the one that they would perceive when sitting in a correspondent real space with a sound source emitting signals – such as speech or music.

Figure 1.9 provides an overview of such a system. The processing steps are roughly as follows. Starting from a geometric description of the space to be modelled – including all surrounding surfaces with their absorption coefficients and the position and directional characteristics of the sound source – the sound-field model determines a cloud of virtual sound sources which characterizes the sound field as seen from the listener's position. The impulse response, as generated by these virtual sources, is then direction-specifically weighted with the impulse response of the listener's ear, which have to be measured beforehand. In such a way, a set of binaural impulse responses is finally rendered. These, in turn, are then convolved with "dry", i. e. an-echoic, speech or music – the resulting signals being delivered to the listener's ears.

This simulation technique, known as "binaural room simulation", can be considered state of the art these days. Commercial systems are available from the shelf. They are, for example, used in the planning process of spaces for musical performances – e. g., concert halls – as they enable listening into a hall when it is still on the construction board as a mere computer model. The term "auralization" is used for the process of making the signals from the computational models audible in this way. Typically, binaural room-simulation systems track many thousands of individual reflections and, additionally, provide for a stochastic reverberation tail. Further, the techniques used today for this purpose are no longer restricted to geometrical approximation, but may include explicit solutions of the wave equation by numerical methods – e. g., by using boundary-elements or finite elements. Diffraction and dispersion can thus be accounted for to a certain extent.

However, the listeners to such a system, although experiencing auditory scenes like in real halls, are easily able to discriminate that their auditory

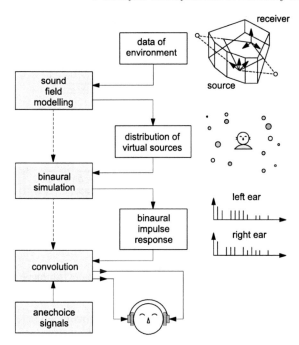

Fig. 1.9. Schematic of a system for binaural room simulation, after [46]

percepts do not stem from a real hall. For example, if they moved their heads about, their auditory scene would move accordingly, instead of staying spatially constant. Nevertheless, we have mentioned binaural room simulation here, as it forms the basis for AVE generators – as shall be explained in the next subsection.

1.3.2 Virtual-Environment Generation

The crucial step from binaural room simulation to virtual-environments is taken when the system becomes interactive, i. e. actions of the user can modify the auditory percept. For example, the lack of room constancy in binaural room simulation can be overcome in the following way [10]. At any instance the position at the listeners head is determined – so-called "head tracking" – and the position data are transmitted to the simulation system. Based on this information, the system can modify the ear signals for the listener accordingly.

The important issue from a system's point-of-view is the following. As soon as interaction is involved, real-time signal processing becomes indispensable. The system reaction must happen within perceptually plausible time spans – for the auditory representation within roughly 50 ms. Further, the refresh rate for the generated scene must be high enough for the perceptual scene neither to jolt nor to flicker. To this end the refresh rate has to

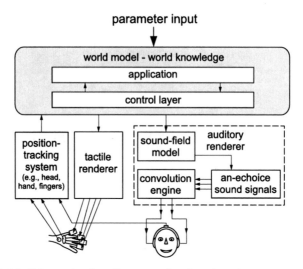

Fig. 1.10. Schematic of auditory-tactile virtual-environment generator

be above 30 times per second for moderately moving objects. Further, for objects moving fast, *Doppler* shifts may have to be taken into consideration and to be modelled.

Due to the interactivity requirements, the architecture of AVE generators systems differs from that of static-simulation systems. Instead of "authentic" auditory scenes, the aim is now to create perceptually "plausible" ones because, among other reasons, interactive perceptual authenticity would be out of the reach of today's technology in most cases. The developer of such a system obviously needs a very detailed knowledge on human sensory perception, since it has to be decided at every instant which attributes of the signals are perceptually relevant and, thus, have to be presented accurately and instantly. Less relevant attributes can be calculated later or even be omitted.

In Fig. 1.10 the architecture of an AVE generator is schematically depicted. It can currently be observed that virtual-environment generators become more and more multi-modal. In a close alliance with the auditory modality, tactile, visual and proprioceptive information is presented, eventually, among such for further senses. To clarify this fact, i. e. that AVE generators are usually components of multi-modal virtual-environments, i. e. embedded systems, the figure shows an auditory–tactile generator as a bi-modal example. Yet, our discussion focuses solely on the auditory component. The figure depicts a system where the acoustic signals are presented via headphones. In principle, loudspeaker presentation would be possible as well. Current multi-channel loudspeaker reproduction systems as used in consumer electronics, e. g., movie theatres, home theatres, TV, CD, DVD, radio, can indeed be seen as a pre-step to virtual environments. However, what is still missing in this context, as a rule, is interactivity.

The example system in Fig. 1.10 houses, as its core, a "world model". This is basically a data bank which contains detailed descriptions of all objects which are to exist in the virtual environments. In one layer of the world model, termed application, rules are listed which regulate the interaction of the virtual objects with respect to the specific applications intended. Further, a central-control layer collects the reactions of the subjects which use the virtual-environment system interactively and prompts the system to render appropriate responses. It goes without saying that the world model, including the application and central-control layers, have to be geared with a considerable amount of knowledge and intelligence.

In the system shown in the figure, head, hand, and finger positions of the user are continuously monitored. As explained above, the head positions are of relevance as the signals have to be adapted constantly for the subject to perceive a spatial perspective which stays spatially still. By moving hands and fingers, the subjects can influence the virtual environment, for example, move sound sources about. Those system components that generate the signals which are finally presented to the subjects via actors, i.e. headphones for the auditory modality, are called "renderers". The most important component of the auditory renderer is the sound-field model. This is the module which creates the set of binaural impulse responses, based on binaural room-simulation principles.

In many applications of virtual environments it is aimed at exposing the subjects to a virtual situation such that they feel perceptively "present" in it. This is especially important if scenarios are to be created in which the subjects are supposed to act intuitively, as they would do in a respective real environment. Human–system interfaces which base on the principle of virtual environments have the potency of simplifying human–system interaction considerably. One may think of tele-operation systems, design systems and dialog systems in this context, also of computer games. The effort involved in creating perceptual presence is task-depending and depending on user requirements. For example, for vehicle simulators the perceptual requirements are far less stringent than for virtual control rooms for sound engineers. Generally, the virtual environment must appear sufficiently "plausible" to the listener in order to provide perceptual presence.

Often users of AVEs use their own voices in the virtual environment e.g., in the case of tele-conferencing systems. In this context, it is worthwhile mentioning that the perception of one's own voice in virtual realities is an important issue. If the own voice does not sound natural, the perception of actually being in the virtual environment can hardly be achieved. After careful analysis and simulation of the voice-sound propagation through the air and through the skull, this task could recently be mastered [56]. Natural perception of ones own voice in AVEs can now be provided.

Modern speech technology offers various components which can be integrated into auditory virtual environments. Examples are systems for instrumental speech synthesis and recognition. By utilization of these, human-system interaction can be performed via voice signals. The system, then, reacts with artificial voice signals which the user can understand. Systems for spoken human–machine dialog and human–machine interaction have already been demonstrated more than once, yet often still in a quite rudimentary form – see [50] for a discussion.

There is no doubt that any further evolution of auditory virtual environments, as of virtual environments in general, strongly depends on our capability of supplying the AVE generator – in particular the world model in it – with sufficient intelligence and knowledge, such as to enable it to provide high perceptual plausibility and appropriate guidance to its users.

1.4 Discussion and Conclusions

It has been one goal of this contribution to show that the specific branch of acoustics which relates to the communication and information technologies, has experienced a dramatic evolution during the last decades. From electro-acoustics, which had been formed as a symbiosis of electrical engineering and acoustics, Communication Acoustics has finally evolved, with signal processing and computer science joining in. A clear shift from hardware to software activities can be observed throughout the field.

With auditory-scene analysis, ASA, taken as one example of typical activity in Communication Acoustic, it was shown that a major research aim in this field is the development of algorithms which analyze real auditory scenes in order to extract a parametric representation. Some human capabilities in analysis and recognition can already be mimicked or even be surpassed. As to the parametric-synthesis side, AVE, i. e. the second example of Communication-Acoustics activity discussed in this chapter, there is a clear tendency to multi-modal interactive virtual environments. Perceptual plausibility and appropriate user guidance are prominent research and development goals, among other reasons, to provide users with a sense of presence in these environments and to enable them to interact intuitively with them.

To achieve these goals, the systems become increasingly intelligent and knowledgeable. Consequently, in the case of ASA and AVE, there are components in the system which could well be addressed their "brains" – see Fig. 1.11. Evidently, the tendency of increasing built-in system intelligence and knowledge does not only hold for Communication Acoustics, but for the communication and information technologies at large. As a consequence, communication acoustical analyzers and synthesizers are often found as "embedded" components of more complex systems – which, by the way, makes it very hard or even impossible to evaluate their performance in isolation.

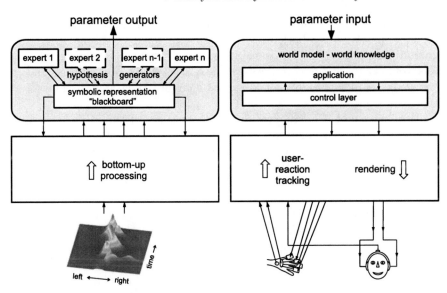

Fig. 1.11. Synopsis of an ASA and an AVE system – schematic. The *shaded* areas denote the blocks where knowledge and intelligence of the system are implemented, i. e. where the "brains" of the systems reside

In summing up and looking back into the past of Communication Acoustics, the field has dealt with the physical and perceptual issues of auditory communication for a long time. With the introduction of digital signal processing, it became possible to model some of these issues, e. g., the acoustical behaviour of the external ears and some sub-cortical biological signal processing in the auditory system. The algorithms used were, as a rule, strictly signal driven, i. e., bottom up.

This strategy was, however, not sufficient for more complex task. This became at first obvious in speech technology, were the task of speech recognition required hypotheses-driven, top-down processing procedures, and consequently, the modelling of functions which are located more centrally in the human nervous system. For example, linguistic knowledge was imported and applied to computational speech recognition. In term of computational algorithms, this means that pure signals processing had to be supplemented by processing of symbols and, ultimately, semantic content. To actually implement these issues, advanced software technology had to be employed.

As regards the human system, Communication Acoustics has now arrived at a point where functions of the central nervous system, i. e. brain functions, are about to be modelled. This applies to multi-modal interaction [39] and, in particularly, to cognitive functions. As we all know, the behaviour of human beings is not guided directly by the acoustical signals that we provide them with, e. g., like a reflex, but rather by the "meaning" which is transferred via these signals. Consequently, an important demand of advanced communication technologies is to gain more information on how human beings interpret acoustic signals on the background of their individual situational and background knowledge, i.e. what the signals actually "mean" to them – see [33], this volume, in this context. As we have worked out above, this kind of knowledge is indispensable for advanced computational analysis as well as synthesis.

Without doubt, Communication Acoustics is interdisciplinary in essence. Physicists, electrical engineers, signal-processing specialists, software technologist, audio-engineers, architects, psycho-acousticians, psychologists, computer linguist, biologists, physiologists, medical experts, audiologists, musicologists, and linguists have a long tradition of working together in it, among others experts.

Currently, we are about to experience that the view of Communication Acoustics becomes even wider. When it comes to the human being as a source and sink of information, the whole organism from sensory periphery up to central cognitive functions is now taken into consideration. Scientific fields like the cognitive sciences, brain research and artificial intelligence will, thus, increasingly contribute to our field. The author is positive that this development can only add to the attractiveness of Communication Acoustics, as the human being is, after all, the most interesting object of scientific research.

Acknowledgement

This chapter is, to a large extent, based on R&D which has been performed by the author's PhD students – more than 50 ones over the years. Most are given reference to in the following, but the author wants to thank all of them for most creative cooperation. In the field of speech technology, dissertations relevant in the context of this chapter are [4, 9, 12, 22, 30, 35–38, 41, 49, 63]. For binaural analysis, the respective list is [8, 13–16, 21, 24, 26, 45, 47, 57, 65, 67, 68, 70]. Sound and speech quality are dealt with in [3, 27, 29, 34, 50, 51], including two habilitations. Auditory virtual environments is the topic of [17, 19, 20, 23, 40, 46, 55, 56, 58, 69, 71]. Thanks are further due to the *Mogens-Balslevs Fond* for supporting a substantial part of the time spent on compiling this chapter and editing this book at large.

References

1. Allen J B, Berkley D A, Blauert J (1977), Multimicrophone signal-processing technique to remove room reverberation from speech signals. J Acoust Soc Amer 62:912–915
2. Bauer B B (1961) Stereophonic earphones and binaural loudspeakers. J Audio Engr Soc 9:148–151
3. Bednarzyk M (1999) Qualitätsbeurteilung der Geräusche industrieller Produkte: Der Stand der Forschung, abgehandelt am Beispiel der KFZ-Innenraumgeräusche (Quality assessment of the sounds of industrial products:

the state of research discussed by taking the interior sounds of cars as an example). Doct diss, Ruhr-Univ Bochum, VDI-Verlag, Düsseldorf

4. Belhoula K (1996) Ein regelbasiertes Verfahren zur maschinellen Graphem-nach-Phonem-Umsetzung von Eigennamen in der Sprachsynthese (A rule-based approach to automatic grapheme-to-phoneme conversion of proper names for speech synthesis. Doct diss, Ruhr-Univ Bochum, VDI-Verlag, Düsseldorf

5. Blauert J (1999) Binaural auditory models. Proc 18th DANAVOX Symp, Scanticon, Kolding

6. Blauert J (1996) Spatial hearing: the psychophysics of human sound localization. 2nd rev Engl ed, MIT Press, Cambridge MA

7. Blauert J (1999) Binaural auditory models: architectural considerations. Proc 18[th]Danavox Symp 189–206. Scanticon, Kolding

8. Bodden M (1992) Binaurale Signalverarbeitung: Modellierung der Richtungserkennung und des Cocktail-Party-Effektes (Binaural signal processing: modelling the recognition of direction and the cocktail-party effect). Doct diss, Ruhr-Univ Bochum, VDI-Verlag, Düsseldorf

9. Böhm A (1992) Maschinelle Sprachausgabe von deutschem und englischem Text (Speech synthesis for German and English text). Doct diss, Ruhr-Univ Bochum, Shaker, Aachen

10. Börger G, Blauert J, Laws P (1977) Stereophone Kopfhörerwiedergabe mit Steuerung bestimmter Übertragungsfaktoren durch Kopfdrehbewegungen (Stereophonic headphone reproduction with variations of specific transfer factors by means of head rotations. Acustica 39:22–26

11. Bosch B (2001) Lee de Forest – "Vater des Radios" (Lee de Forest – "father of radio"). Funk Gesch 24:5–22 and 24:57–73

12. Braas J (1981) Ein digitales Leitungsmodell als Hilfsmittel zur Sprachsynthese (A digital transmission-line model as a tool for speech synthesis), Doct diss, Ruhr-Univ Bochum, Bochum

13. Braasch J (2001) Auditory localization and detection in multiple-sound source scenarios. Doct diss, Ruhr-Univ Bochum, VDI-Verlag, Düsseldorf

14. Brüggen M (2001) Klangverfärbungen durch Rückwürfe und ihre auditive und instrumentelle Kompensation (Sound colouration due to reflections and its auditory and instrumental compensation). Doct diss, Ruhr-Univ Bochum, dissertation.de Verlag im Internet, Berlin

15. Buchholz J (2002) A computational auditory masking model based on signal-dependent compression. Doct diss, Ruhr-Univ Bochum, Shaker, Aachen

16. Djelani Th (2001) Psychoakustische Untersuchungen und Modellierungsansätze zur Aufbauphase des auditiven Präzedenzeffektes (Psycho-acoustic experiments and modelling approaches regarding the built-up of the auditory presedence effect). Doct diss, Ruhr-Univ Bochum, Shaker, Aachen

17. Dürrer B (2001) Untersuchungen zum Entwurf von Auditory Displays (Investigations into the design of auditory displays). Doct diss, Ruhr-Univ Bochum, dissertation.de Verlag im Internet, Berlin

18. Ellis D P W (1999) Using knowledge to organize sound: the prediction-driven approach to computational auditory scene analysis and its application to speech/nonspeech mixtures. Speech Comm 27:281–298

19. Els H (1986) Ein Messsystem für die akustische Modelltechnik (A measuring system for acoustical scale models). Doct diss, Ruhr-Univ Bochum, Wirtschaftsverlag NW, Bremerhaven

20. Farag H (2004) The psycho-acoustics of sound-source occlusion. Doct diss, Alexandria Univ, Alexandria
21. Gaik W (1990) Untersuchungen zur binauralen Verarbeitung kopfbezogener Signale (Investigations regarding the binaural processing of head-related signals). Doct diss, Ruhr-Univ Bochum, VDI-Verlag, Düsseldorf
22. Geravanchizadeh M (2001). Spectral voice conversion based on locally linear transformation rules for vocal-tract characteristics. Doct diss, Ruhr-Univ Bochum, Shaker, Aachen
23. Giron F (1997) Investigations about the directional characteristics of sound sources. Doct diss, Ruhr-Univ Bochum, Shaker, Aachen
24. Grabke, J (1997) Ein Beitrag zum Richtungshören in akustisch ungünstigen Umgebungen (A contribution to directional hearing in acoustically adverse environments). Doct diss, Ruhr-Univ Bochum, VDI-Verlag, Düsseldorf
25. Hammershøi D, Møller H (2005) Binaural technique: basic methods for recording, synthesis and reproduction, Chap 9 this vol
26. Hartung K (1998) Modellalgorithmen zum Richtungshören, basierend auf den Ergebnissen psychoakustischer und neurophysiologischer Experimente mit virtuellen Schallquellen (Model algorithms for directional hearing, based on the results of psycho-acoustic and neuro-physiologic experiments with virtual sound sources). Doct diss, Ruhr-Univ Bochum, Shaker, Aachen
27. Hegehofer Th (1998) Ein Analysemodell zur rechnerbasierten Durchführung von auditiven Sprachqualitätsmessverfahren und seine Realisierung (An analysis model for computer-aided execution of auditory speech-quality measuring procedures and its implementation). Doct diss, Ruhr-Univ Bochum, VDI-Verlag, Düsseldorf
28. von Helmholtz H (1863) Die Lehre von den Tonempfindungen als physiologische Grundlage für die Theorie der Musik (Sensation of tone als a physiological basis for the theory of music) Vieweg und Sohn, Braunschweig
29. Hempel Th (2001) Untersuchungen zur Korrelation auditiver und instrumenteller Messergebnisse für die Bewertung von Fahrzeuginnenraumgeräuschen als Grundlage eines Beitrages zur Klassifikation von Hörereignissen (Studies on the correlation of results of auditory and instrumental measurements regarding the assessment of interior car sounds as a basis for a contribution to the classification of sound events). Doct diss, Techn Univ Berlin, Herbert-Utz-Verlag, München
30. Henrich P (1988) Sprachenidentifizierung zur automatischen Graphem-zu-Phonem-Umsetzung von Fremdwörtern in einem deutschsprachigen Vorleseautomaten (Language identification for automatic graphem-to-phonem conversion in a German-speaking automatic machine for reading aloud). Doct diss, Ruhr-Univ Bochum, Wissenschaft und Kunst, Herne
31. Heute U (2005) Speech and audio coding – aiming at high quality and low data rates. Chap 14 this vol
32. Jeffress L A (1957) A place theory of sound localization. J Comp Physiol Psych 61:468–486
33. Jekosch U (2005) Assigning meanings to sounds – semiotics in the context of product-sound design. Chap 8 this volume
34. Jekosch U (2000). Sprache hören und beurteilen: Ein Ansatz zur Grundlegung der Sprachqualitätsbeurteilung (Perception and appraisal of speech: laying the foundations of speech-quality assessment and evaluation) Inaugural diss (habilitation) Univ Duisburg–Essen, Essen

35. Jekosch U (1989). Maschinelle Phonem-Graphem-Umsetzung für unbegrenzten deutschen Wortschatz (Instrumental phonem-to-grapheme conversion for unlimited German vocabular). Doct diss, Univ Diusburg–Essen, Verlag für Wissenschaft und Kunst, Herne

36. Kesselheim, Michael (1990). Computergestützte Konstruktion grosser Wortklassensysteme (Computer-aided construction of extensive word-class systems). Doct diss, Ruhr-Univ Bochum, Universitätsverlag Dr. N. Brockmeyer, Bochum

37. Knohl L (1996) Prosodiegesteuerte Sprecher- und Umweltadaption in einer Mehrsprecher-Architektur (Prosody-driven speaker and environment adaptation in a multi-speaker speech-recognition architecture). Doct diss, Ruhr-Univ Bochum, VDI-Verlag, Düsseldorf

38. Köster S (2002). Modellierung von Sprechweisen für widrige Kommunikationsbedingungen mit Anwendungen auf die Sprachsynthese (Modelling of speaking styles for adverse communication conditions and applications to speech synthesis). Doct diss, Ruhr-Univ Bochum, Shaker, Aachen

39. Kohlrausch A, van der Par S (2005) Audio-visual interactions in the context of multi-media applications. Chap 5 this vol

40. Korany N (2000) A model for the simulation of sound fields in enclosures: integrating the geometrical and the radiant approaches. Doct diss, Alexandria Univ, Alexandria

41. Kraft V (1996) Verkettung natürlichsprachlicher Bausteine zur Sprachsynthese: Anforderungen, Techniken und Evaluierung (Concatenation of elements stemming from natural speech for speech synthesis: requirements, techniques and evaluation). Doct diss, Ruhr-Univ Bochum, VDI-Verlag, Düsseldorf

42. Krokstadt A, Strøm S, Sørsdahl S (1968) Calculating the acoustical room response by use of a ray-traycing technique. J Sound Vibr 8:118–125

43. Küpfmüller K (1959) Informationsverarbeitung durch den Menschen (Information processing in humans). Nachrichtentechn. Z. 12/68–74

44. Lacroix A (2005) Speech production – acoustics, models and applications. Chap 13 this vol

45. Lehn K (2000) Unscharfe zeitliche Clusteranalyse von monauralen und interauralen Merkmalen als Modell der auditiven Szenenanalyse (Fuzzy temporal-cluster analysis of monaural and interaural cues as a model of auditory scene analysis). Doct diss, Ruhr-Univ Bochum, VDI-Verlag, Düsseldorf

46. Lehnert H (1992) Binaurale Raumsimulation: Ein Computermodell zur Erzeugung virtueller auditiver Umgebungen (A computer model for the generation of auditory viural environments). Doct diss, Ruhr-Univ Bochum, Shaker, Aachen

47. Lindemann W (1985). Die Erweiterung eines Kreuzkorrelationsmodells der binauralen Signalverarbeitung durch kontralaterale Inhibitionsmechanismen (Amendment of a cross-correlation model of binaural signal processing with contra-lateral inhibition mechanisms). Doct diss, Ruhr-Univ Bochum, Bochum

48. MPEG-7-document: http://www.chiariglione.org/mpeg /standards/mpeg-7/mpeg7-htm. Accessed Jan 2005

49. Mersdorf J (2000) Sprecherspezifische Parametrisierung von Sprachgrundfrequenzverläufen: Analyse, Synthese und Evaluation (Speaker-specific parametrizing of F_0 contours: analysis, synthesis and evaluation) Doct diss, Ruhr-Univ Bochum, Shaker, Aachen

50. Möller S (2005) Quality of telephone-based spoken dialogue systems. Inaugural diss (habilitation) Ruhr-Univ Bochum, Bochum. Springer New York

51. Möller S (1999) Assessment and prediction of speech quality in telecommunications. Doct diss Ruhr-Univ Bochum. Kluwer New York
52. Mourjopoulos J N (2005) The evolution of digital audio technology. Chap 12 this vol
53. Nipkow P (1884) Elektrisches Teleskop. German Pat #30105
54. Novo P (2005) Auditory virtual environments. Chap 11 this vol
55. Pellegrini R (2001). A virtual reference listening room as an application of auditory virtual environments. Doct diss, Ruhr-Univ Bochum, dissertation.de Verlag im Internet, Berlin
56. Pörschmann Ch (2001) Eigenwahrnehmung der Stimme in auditiven virtuellen Umgebungen (Perception of one's own voice in auditory virtual environments). Doct diss, Ruhr-Univ Bochum, VDI-Verlag Düsseldorf
57. Pösselt Ch (1986) Einfluss von Knochenschall auf die Schalldämmung von Gehörschützern (Effect of ear-occlusion on the bone-conduction pathways in man and its influence on the sound attenuation performance of ear protectors). Doct diss, Ruhr-Univ Bochum, Wirtschaftverlag NW, Bremerhaven
58. Pompetzki W (1993). Psychoakustische Verifikation von Computermodellen zur binauralen Raumsimulation (Psycho-acoustic verification of computer models for binaural room simulation). Doct diss, Ruhr-Univ Bochum, Shaker, Aachen
59. Poulsen V (1898) Telegraphone, Danish Pat, cited after US Pat #8961 (1899)
60. Rateitschek K (1998) Ein binauraler Signalverarbeitungsansatz zur robusten maschinellen Spracherkennung in lärmerfüllter Umgebung (A binaural signal-processing approach for robust speech recognition in noisy environments). Doct diss, Ruhr-Univ Bochum, VDI-Verlag, Düsseldorf
61. Lord Rayleigh J W (1896) The theory of sound. 2nd ed, McMillan, London,
62. Reis P (1867) Publ demonstr of the "Telephone" at Frankfurt. After http://de.wikipedia.org/wiki/Philipp_Reis. Accessed Febr 2005
63. Rühl, H-W (1984) Sprachsynthese nach Regeln für unbeschränkten deutschen Text (Rule-based speech synthesis for unlimited vocabulary). Doct diss, Ruhr-Univ Bochum, Bochum
64. Ruhmer E (1901) The Photographone, Scientif Amer July 20, 1901. After http://www.fsfl.se/backspegel/ruhmer.html. Acessed Febr 2005
65. Schlichthärle D (1980) Modelle des Hörens – mit Anwendung auf die Hörbarkeit von Laufzeitverzerrungen (Models of hearing with application to the audibility of arrival-time differences). Doct diss, Ruhr-Univ Bochum, Bochum
66. Schroeder M R (1968) Computers in acoustics: symbiosis of an old science and a new tool. Proc 6th Int Congr Acoust ICA'68, vol I:71–87, Tokyo
67. Schröter J (1983) Messung der Schalldämmung von Gehörschützern mit einem physikalischen Verfahren – Kunstkopfmethode (Measuring the insertion loss of hearing protectors with a physical technique – dummy-head method). Doct diss, Ruhr-Univ Bochum, Wirtschaftsverlag NW, Bremerhafen
68. Slatky H (1993) Algorithmen zur richtungsselektiven Verarbeitung von Schallsignalen eines binauralen "Cocktail-Party-Prozessors" (Algorithms for direction-selective processing of sound signals for a binaural "Cocktail-Party Processor". Doct diss, Ruhr-Univ Bochum, VDI-Verlag, Düsseldorf
69. Strauss H (2000) Simulation instationärer Schallfelder in auditiven virtuellen Umgebungen (Simulation of non-stationary sound fields for auditory virtual environments). Doct diss, Ruhr-Univ Bochum, VDI-Verlag Düsseldorf

70. Wolf S (1991). Lokalisation von Schallquellen in geschlossenen Räumen (Localization of sound sources in enclosed spaces). Doct diss, Ruhr-Univ Bochum, Bochum

71. Xiang N (1991) Mobile universal measuring system for the binaural room-acoustic-model technique. Doct diss, Ruhr-Univ Bochum, Wirtschaftsverlag NW, Bremerhaven

2 Evolutionary Adaptations for Auditory Communication

Georg Klump

Zoophysiology and Behaviour Group, Oldenburg University, Oldenburg

Summary. Many organisms have evolved efficient means for acoustic communication. Adaptations can be found concerning all components of the communication system: signal generation at the sender is optimised, signal characteristics are tailored to the transmission channel, and receivers have evolved elaborate mechanisms for segregating the signals from separate sources and for analysing signal characteristics. The acoustics of the environment often imposes similar demands on the mechanisms for auditory analysis in different animal species. Thus, mechanisms of auditory analysis show many similarities in different animal species ranging from insects to mammals. These similarities result either from convergent evolution of auditory systems that are selected to achieve a similar performance or they are the consequence of the preservation of structures in evolutionary history. Examples for both types of traits are provided that have evolved as adaptations for auditory communication.

2.1 Introduction

In many animal species, sound plays a vital role in communicating. For example, it is used to convey information about the location of individuals to coordinate their movements in the environment, to convey information about what an individual has perceived in its surroundings – such as a predator or suitable prey – or to identify individuals or determine whether an individual is a conspecific or a suitable mating partner. Each animal species has evolved a set of communication signals to transmit information. These signals are not limited to reflect the internal motivational state of an animal – e. g., [56] – but they can also be used as a reference to objects in the environment. The latter case is exemplified by alarm calls used by vervet monkeys, *Cercopithecus aethiops*, specifying the type of predator and eliciting a specific escape response – e. g., [73]. What the auditory system of animals and humans has to achieve in communicating is to detect and classify the various signals and to locate their source.

Solutions for achieving efficient communication have evolved concerning all components of the communication system comprised of the sender, the transmission channel and the receiver [74]). Evolutionary adaptations both on the sender and the receiver side will be described in this chapter that enable the animals to cope with the acoustic environment.

It is not the goal of this chapter to provide a comprehensive review of animal communication, which is an undertaking that requires a book by itself – see e. g., [3]. Here I want to provide exemplary evidence that similar solutions have been developed in evolution to solve general tasks in acoustic communication and that the basic functional principles employed in auditory systems are similar across such diverse groups of species as mammals and insects.

2.2 Sender Adaptations

Sending out a communication signal is costly to an animal. It requires energy to produce the signal and we observe many adaptations for efficiently broadcasting communication signals. In some animals, the energy demands for producing the signal can be quite high. Gray tree frogs, *Hyla versicolor*, for example, consume up to about 20 times more oxygen when calling, as compared to their resting metabolic rate [78]. This may be considerable more than they need for locomotion during constant exercise. To meet the high energy demands for calling, the tree frogs have evolved specialized muscles with many mitochondria for call production – for a review see [82]. Despite these adaptations, energy may be limiting the duration of calling behaviour that a frog can afford. It has been suggested that gray tree frogs may deplete their glycogen and lipid resources in the muscles so much while calling in a chorus during an evening that they have to stop [18]. Furthermore, for frogs of many species in which the males advertise their presence to the females searching for a mate by calling, it may not be possible to obtain enough food to replenish the energy used for the high rate of broadcasting communication signals. Thus, the males rely on stored resources, e. g., fat reserves, and they are no longer able to continue attracting partners with communication signals if the energy reserves are used up [82]. Consequently, some frogs have been found to adjust the energy expenditure when calling depending on the acoustic context. If calling by themselves without any competition, they produce less expensive signals than in a situation in which they compete with other males. Gray tree frog males, for example, produce signals at an increased duration and rate, costing more than double the energy, measured through their oxygen consumption, when hearing other males, than when they call without any acoustic competition – e. g., [83]. Similar to , birds have been found to increase the amplitude of their song output when confronted with a loud background noise [5]. In humans, this is known as the *Lombard* effect.

Accessory structures serve to enhance the sound radiation. For example, many frogs have vocal sacs that may serve to couple the sound output to the environment. Gray tree frogs convert about 2.4 % of their metabolic energy to sound energy [52]. Most of the energy loss in this species occurs in the muscles producing the energy for the airflow resulting in the vibrations of the vocal chords. The energy of the vocal chord vibrations appears to be transformed very efficiently into radiated acoustic energy – 44 % of the energy of the vibrations is broadcast, see [52]. In the Bornean tree-hole frog, *Metaphrynella sundana*, which calls in a cavity formed within a hollow tree trunk, it has been demonstrated that an individual adjusted the dominant frequency of its call to the resonant properties of the cavity [46].

In addition, frogs and other animals have evolved behavioural patterns to enhance the possibility of broadcasting sound without interference by conspecifics. Males of many frog species alternate their advertisement calls – see review [34] – thereby enhancing the possibility to broadcast a call that is not jammed by the signal of a competitor.

Besides the energy considerations, there are other costs that have shaped the signal structure, the mechanisms of signal production and the senders' broadcast behaviour. Most importantly, eavesdropping on the signals by undesired receivers is reduced by the sender – see review [51]. This can be achieved, for example, by broadcasting communication signals of a frequency and amplitude that is adapted to the desired transmission distance. High-frequency sounds are attenuated more in air than low-frequency sounds. Thus, by broadcasting high-frequency alarm signals birds can alert their neighbours without providing evidence of their presence to a distant predator – e.g., see [36].

2.3 Coping with the Properties of the Transmission Channel

The signals are modified on the path of transmission to the receiver – for a review see [84]. They are subject to frequency-dependent attenuation resulting in a change of the spectral characteristics. The attenuation also affects the contrast to the background noise as described by the signal-to-noise ratio. To be detectable, the signal must be sufficiently above the background noise. Also the temporal characteristics of the signal are modified on the path of transmission. Reflections occurring whenever the signal impinges on a surface will add a clutter of echoes that trail the elements of a signal. These echoes will fill up silent gaps between signal elements and wind or turbulence created by heating of the air over sun-exposed surfaces will modulate the envelope of a transmitted signal.

Whereas, on the one side, signals must remain recognizable despite the modifications on the path of transmission, on the other hand these changes can be used by the receiver to estimate the transmission distance of a signal [58]. Birds have been shown to fly further towards the source of a playback sound if it was more degraded – e.g., [57, 60] – indicating that they can extract information about the distance of the source using degradation cues.

2.4 Receiver Adaptations

To respond appropriately to a communication signal, the receiver has first to detect and then to classify the signal as being of a certain type, and the receiver usually benefits from being able to locate the signal source. These dual functions that its auditory system has to fulfil are often separated in two parallel processing pathways: a "what" pathway and a "where" pathway [66]. Parallel processing in the auditory system is also found within these pathways. Temporal and spectral features of a sound are often processed by separate neuron populations. For example, in the first station of the "where" pathway in the cochlear nucleus of birds and mammals, the auditory-nerve fibres provide input to neurons that have features optimized for a high temporal resolution and to neurons that have a lower temporal resolution and represent spectral information. These two pathways can separately process the interaural spectral differences and the interaural temporal differences – e. g., [79] – that are combined at the midbrain level in the computation of auditory space, as demonstrated in the barn owl [38]. By combining the multiple pathways at higher levels of the auditory processing system, complex stimulus characteristics can be encoded without much loss of information as would result from the shortcomings in the processing of stimulus features by non-specialized neurons.

Besides parallel processing mechanisms, we can find other common characteristics in the auditory systems of receivers that have evolved to match the demands posed by the acoustic environment. Often the different animals have developed quite comparable capabilities that either are the result of a common origin, i. e. have evolved in a common predecessor and were conserved when the diversity of animal groups formed, or are the result of independent evolutionary solutions due to similar requirements – i. e. are so-called convergent developments. Examples of both types of adaptations are given in the following paragraphs when discussing adaptations for performing different auditory processing tasks.

2.4.1 Auditory Non-Linearities and Gain-Control Mechanisms

The range of amplitudes of sound that the auditory system has to analyze may exceed 6 orders of magnitude, e. g., the sound pressure may vary by a factor of 1,000,000 between a soft whisper and the level of the music in a rock concert [55, 68]. To make it possible for the mechano-receptor cells to represent this variation in amplitude, given their own limited dynamic range, auditory sensory organs of insects as well as of have evolved non-linear transduction mechanisms and active mechanisms of gain-control – for reviews see [6, 15, 20, 49, 68]. At high sound-pressure levels, the animals' hearing organ exhibits a linear response. At low sound-pressure levels, however, the response of the animals' hearing organ is highly non-linear, showing a large amplification that appears to depend on active and energy consuming

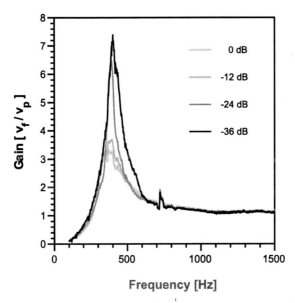

Fig. 2.1. The gain of the antennal "ear" of the mosquito *Toxorhynchites brevipalpis* in relation to frequency. The gain is expressed as the ratio of the vibration velocity of the antenna flagellum, v_f, and the particle velocity, v_p, of the acoustic stimulus. The different curves show data for four relative levels of the stimulus. The frequency response of the flagellum is narrowly tuned and shows an increasing gain at lower levels of the stimulus. Data kindly provided by *M. Göpfert* – for details see [15]

physiological processes. What constitutes the evidence of active mechanical processes both in the insect and in the vertebrate ear that contribute to the sensitivity and the frequency selectivity of the sound receiving organ? The first indicator is the change in the spectrum of vibration amplitude of structures of the sensory organ when stimulated with sounds of different level. This is nicely exemplified by the relation between the vibration velocity of the of the mosquito antenna, flagellum velocity, v_f, and the level of the sound applied when stimulating the antenna, represented by the particle velocity, v_p – for an example using the displacement as a measure of gain see [15]. The sensitivity of the antenna as expressed by the ratio v_f/v_p, i.e. the gain, is considerably increased with decreasing level of the sound impinging of the "mosquito ear" – Fig. 2.1.

At the same time, the frequency tuning is sharpened, thus providing the mosquito with a narrow auditory frequency filter tuned to about 400 Hz – which is at the frequency of the mosquito's wing-beats. In male mosquitoes the increase in sensitivity is much higher than in female mosquitoes, indicating that there has been a higher evolutionary pressure on the auditory sensitivity of males that compete for females and that will be able to acoustically survey a larger space around them if their sensitivity is improved. The in-

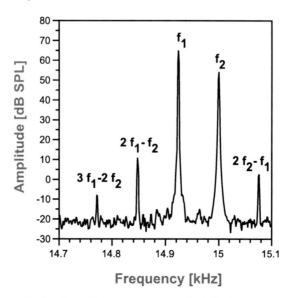

Fig. 2.2. Acoustic two-tone distortion measured at the tympanum of the ear of the grasshopper *Locusta migratoria*. The labels at the peaks indicate the frequencies, f_1, f_2, of the tones used for stimulation and the various distortion products – data kindly provided by *M. Kössl*

crease in the gain of the mosquito antenna is only found in vivo. Post-mortem, no changes in the gain with level are observed and the system becomes linear, indicating that the active processes depend on a functional metabolism. Very similar observations have been made in the mammalian cochlea – for reviews see [21, 62, 68, 69]. It exhibits a level-dependent gain that results in an increased sensitivity at low to intermediate levels of stimulation. Post mortem, the cochlea responds linearly to the sound input indicating that active processes are involved in the normal functioning of the living mammalian cochlea. The gain of the amplifier in the sound-receiving organs is adjusted to such a level that the positive feedback itself generates vibrations that become evident as acoustic emissions from the ear. Such emissions of sound are found in the insect "ear" , e. g., in the antenna of *Drosophila* or a mosquito, [17, 67] as well as in the ears of vertebrates – e.g. [41, 49, 65]. Acoustic distortion products, which are another hallmark of non-linear function of the human ear, can be measured in a wide range of sound-receiving organs of animals – e. g., [41–43, 49, 65]. Distortion-product emissions occurring in the ear of grasshoppers have been found to share many features with those in the vertebrate and the human ear [42, 43] – see Fig. 2.2.

As in the mammalian ear, cubic difference tones with frequency $2f_1 - f_2$, with $f_1 < f_2$ – where the frequencies f_1 and f_2 are the the primary tones used to generate the distortion – are the most prominent form of distortion in the grasshopper ear and they can be elicited at relatively low levels of the primary tones – for a review of mammalian data see [65, 68]. Similar to the results found in the study of vertebrate ears, the distortion-product emissions in the grasshopper are susceptible to hypoxia [43]. Although the different invertebrate and vertebrate species have morphologically diverse hearing organs exploiting different mechanical principles, among others such as particle-movement receivers, sound-pressure and sound-pressure-gradient receivers – for a review see [3] – the various species seem to share many characteristics of the molecular transduction mechanisms of mechano-receptors – see review [14]. Studies in the fly *Drosophila* indicate that mutations affecting the function of the transduction mechanism in mechano-sensory neurons of the "antennal ear" also affect the active processes obvious in the generation of spontaneous vibratory activity [17].

2.4.2 Frequency Selectivity

Starting in the inner ear, the auditory system of vertebrates selectively processes the different frequencies of a sound. This ability has been described as a bank of band-pass filters, i. e. the auditory filters – see review [70] – that comprise the auditory system. It results from the tuned response of the basilar membrane carrying the sensory cells, which shows a maximum vibration amplitude and velocity at a specific location of the cochlea depending on the frequency of the acoustic signal – for reviews see [68, 69]. The active processes in the sound-receiving organ as described above, e. g., in the cochlea, not only increase the sensitivity, but also lead to a sharpening of the frequency selectivity. In mammals, the properties of the vibrational pattern of the basilar membrane are sufficient to explain the auditory filter properties. In other vertebrates, such as in birds, or in amphibians the tuning of the sensory cells themselves may contribute substantially to the frequency selectivity of the ear [69]. The frequency selectivity of auditory-nerve fibres resembles the frequency selectivity of the basilar membrane, but the auditory nerve fibre response also indicates suppressive effects across frequencies that affect the tuning [69]. In the ascending auditory pathway the auditory filters are further sharpened by suppressive effects that are due to inhibitory interactions across frequencies. These contribute to generating a frequency tuning that is much less dependent on the overall level than the frequency tuning in the inner ear. Psychophysical and neurophysiological frequency tuning is matched. The tuning of at least a fraction of the neurons in the ascending auditory pathway is similar to the frequency tuning of auditory filters, as have been determined psychophysically, e. g., in the European starling – see Fig. 2.3.

In general, the frequency selectivity of animal ears is not very different from that of humans. Auditory-filter bandwidths, i. e. critical bands, measured psychophysically at 2 kHz in the European starling, e. g., [45], the gerbil [29], and in humans, e. g., [63], are all between about 12 to 15 % of the

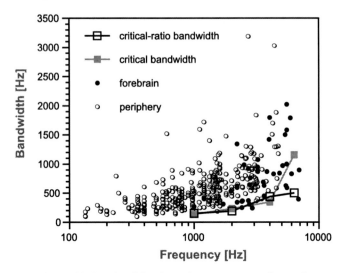

Fig. 2.3. The Q–10-dB bandwidth of single neurons in the auditory periphery, namely, auditory-nerve fibres and neurons in the *ganglion cochleare* – data from *Gleich* and *Manley*, see [45], and of multi-unit recordings in the auditory forebrain of the European starling – data from [61]. Data plotted in relation to psychophysical estimates of the starling's auditory-filter bandwidth, i.e. critical-ratio bandwidth and critical band – the latter corresponding to the 3-dB bandwidth. Given the V-shaped tuning curves, the neurons' bandwidth 3 dB above the most sensitive point of the tuning curve is about 2.5 times narrower than the Q–10-dB bandwidth. The most selective neurons thus show a sharpness of tuning that matches the psychophysical measures of frequency selectivity

center frequency of the filter. Auditory filter bandwidths have also been estimated by masking experiments in which a tone has to be detected in a wide-band noise. The auditory-filter bandwidth is then computed by assuming that at the detection threshold the energy of the noise in the filter is similar to the energy of the tone signal. This measure of frequency selectivity, called the "critical-ratio bandwidth" [11], has been determined in a wide range of vertebrates in which it shows a similar pattern – with some interesting exceptions. As indicated above, also this measure of the frequency selectivity of filters in the auditory system in general amounts to a constant fraction of the center frequency, namely, between about 10 and 40% – see review [9]. The exceptions from the rule can be explained as evolutionary adaptations. In the horseshoe bat, *Rhinolophus ferrumequinum*, for example, a sharp reduction of the critical masking ratio in the frequency range of the bat's echolocation call has been observed [48], indicating very-sharply tuned auditory filters. Further, a correlated increase, by an order of magnitude, in the sharpness of tuning of neurons in the bat's auditory periphery has been found [77]. The horseshoe bat uses calls that are basically 50-ms to

60-ms tones of a constant frequency of about 83 kHz, with a short frequency sweep at the beginning and the end. Using the sharply tuned filters at the call frequency, it can detect and classify fluttering insects by analyzing the amplitude modulations and the frequency shift in the echo that is generated by the moving insect wing and results from the *Doppler* effect – e. g., [81]. In this process, the horseshoe bat even adapts the frequency of its emitted calls to account for its own flying speed in such a way, that the echoes always fall within the restricted frequency range in which the auditory system is so narrowly tuned and able to detect small changes in frequency – an effect called *Doppler*-shift compensation [72].

2.4.3 Sound Localization

Receivers benefit from being able to pinpoint the location of sound sources. Good sound-localization abilities will for example reduce the time to find an adequate mating partner, e. g., in many frog species in which the female approaches calling males, or enable the receiver to find out whether a sender already intruded into the receiver's territory. In the latter case, the localization ability must not only include the direction from which the sender of a signal is heard, but also its distance – e. g., [58, 60]. Compared to humans, many animals face a large problem in sound localization that is due to the fact that these species have small heads and use communication sounds with a wavelength that can be much larger than their head diameter. A good example is provided by the gray tree frog, *Hyla versicolor*, in which males use advertisement calls with dominant frequencies of 1.1 and 2.2 kHz corresponding to a wavelength of 310 and 155 mm, respectively. Females, with a distance between their ears of about 15 mm, need to localize these calls. Two classes of cues, namely, inter-aural time differences, ITD, and inter-aural intensity differences, IID, allow the female frogs to estimate the direction from which a male called. If they would rely only on comparing the ITD and the IID of the sounds impinging on the outer surface of their left and right tympanum, they would experience a maximum ITD of 66 μs and a maximum IID of much less than 1 dB if the position of the sound source is 90° to the side – estimated by approximating the head by a sphere, see [44, 75] for ITDs and IIDs, respectively.

Although the inter-aural time and intensity differences predicted from the geometry of the frogs head are very small, gray tree frog females are able to localize a sound source broadcasting a synthetic advertisement call with frequency components at 1.1 and 2.2 kHz with an average horizontal accuracy of 19° jump error angle, i.e. the angle between the vector pointing to the sound source and the vector describing the leap of the frog, and an average vertical accuracy of 12° – see [24].

Measurements of the typanum vibrations of various frog species indicate that frogs have evolved a hearing organ that provides them with much larger

ITDs and IIDs than would be expected from the simple calculations of maximum inter-aural cues shown above – e. g., [24, 26, 27]. In the grass frog *Rana temporaria*, a species that has a slightly larger head than the gray tree frog, ITDs of about 660 μs and IIDs of about 7 dB can be estimated from such measurements at the tympanum for the dominant frequency of the grass frog's advertisement call of 400 Hz and the sound source located 90° to the side [24, 27]. This large difference cannot be explained by the distance between the frog's ears. It is most likely due to the ear acting as a pressure-gradient receiver – see review [7]. Sound impinges not only from the outside onto the tympanum, but also from the inside getting into the middle-ear cavity by means of the tympanum of the other ear, the nares, and possibly the skin overlaying the lungs – e. g., [8]. The ITDs and IIDs observed at the level of the tympanum vibration alone are not sufficient to explain the large phase shift indicated by the action potentials of the auditory nerve. This had lead to discussing of an additional non-tympanic input to the ear providing for increased inter-aural differences in time [26, 27].

The pressure-gradient system appears to operate at low frequencies in small birds as well – see review [30]. In birds, the two middle ears are connected by an air-filled space that allows sound transmission. There is sufficient crosstalk between the two middle ears at frequencies below 2 kHz to enhance the ITD substantially and somewhat less the IID – e. g., [47]. At frequencies of 4 kHz and above, measurements so far have revealed an attenuation of the sound transmitted from one middle ear to the other that renders the function of the birds' ears as a sound-pressure gradient system at high frequencies unlikely. At these frequencies, the ears of the bird species in which sound transmission has been studied in detail so far can best be described as being independent pressure receivers.

In mammals there is no constantly open air-filled space connecting the middle ears. Sound travelling from one ear through the head to the other ear by bone conduction is attenuated by more than 40 dB – e. g., [1]. Therefore, the ITDs available to the mammalian auditory system and to that of other vertebrate species in which the transfer function of the inter-aural connection indicates a high attenuation, e. g., > 20 dB, are in general determined by the dimensions of their heads and the spatial separation of their ears. In a small mammals, like the gerbil with a separation of the ears between 2 and 3 cm, the maximum ITD is in the range of about 90 to 130 μs. IIDs in mammals are generated only by the sound shadow provided by the head and the directional properties of the pinnae. IIDs of a useful amount are only generated, if the diameter of the head is in the range of a quarter of the wavelength or larger [75]. In the case of the gerbil, the computed shadowing effect of the head will amount to about 1 dB at 2 kHz and the value will grow with increasing frequency.

The limits to sound localization are determined by the limits of detecting ITDs or IIDs. In animal species that have not evolved pressure-gradient receivers enhancing the inter-aural differences in time and amplitude, and that have much smaller heads than humans, the localization accuracy is expected to be not as good as in humans. Humans are able to localize a sound source presenting low-frequency tones, $< 1\,\text{kHz}$, in the horizontal plane in the free sound field with an accuracy of about $10°$ – e.g., [76]. However, in animal species such as the *Tyto alba*, accurate localization ability is of great importance. Barn owls can catch prey in total darkness relying on their hearing only. Despite a much smaller head, the barn owl has a localization accuracy for tones, 4 to 8 kHz, of between $5°$ and about $10°$ in azimuth and elevation [39]. Its localization accuracy for broadband-noise signals is better than that of humans. The barn owl uses ITDs to compute the azimuth location of the sound and IIDs to compute its elevation [37, 53]. It has been calculated from the experiments measuring the localization accuracy that barn owls can resolve ITDs of $8\,\mu s$ [37]. Other species appear to be nearly as accurate if the ITD threshold is computed from the minimum angle that can be resolved. For example, great tits, *Parus major*, can discriminate sound sources that are separated by an angle of $20°$. The time difference deduced from this minimum resolvable azimuth angle amounts to $18\,\mu s$ [35]. This is close to the ITD threshold in humans who have been shown to be able to perceive that a noise of 300 ms duration is lateralized if the signals in the headphones differ by about $10\,\mu s$ – see [85] for a review. Studies in humans have demonstrated lateralizationof the auditory event if an IID of between 0.5 and 1 dB is presented [85]. We know little about the IID threshold in animals. If we assess IID thresholds using a similar method as we used in the estimation of ITD threshold, i.e. estimating the IID that corresponds to the minimum angle that can be discriminated), small animals probably are also make use of small IIDs in the order of 1 dB.

2.5 Auditory Scene Analysis

So far it was demonstrated that the basic mechanisms used in the analysis of communication sounds from one source are similar in humans, other vertebrates, and even in at least some invertebrates. In the context of communicating in the natural environment, however, multiple sound sources are normally active at the same time. For example, in a bird morning chorus, individuals of many species broadcast their songs at the same time such that considerable overlap occurs. In this situation the bird's auditory system has to be able to analyze the sound from each of the sources separately to respond appropriately to its conspecifics. In analogy to the identification of objects in visual scenes, *Bregman* [4] has used the term "auditory-scene analysis" for the task of grouping the sounds from each source into one percept, and the term "auditory stream" is used for the sounds that are attributed to one specific source. Auditory streaming has been reported not only to occur in humans – see review [4]. We have evidence that auditory-stream segregation also occurs in animal communication.

In an experiment studying the perception in European starlings, *Sturnus vulgaris*, the following was observed [23]. Mixed-species song stimuli were presented which were constructed from the songs of two different species – two out of the collection of songs of starling, nightingale, mocking bird, and brown thrasher. The birds were then trained to peck two different keys if the mixed stimuli contained starling song or did not contain starling song, respectively. Once the birds achieved a good discrimination with an overall set of 30 mixed-species stimuli, the actual test began. Being presented with novel stimuli composed of songs of the singers of the four species that were different from the training songs, the starlings readily transferred the discrimination to the previously unknown stimulus set. This transfer indicates that they had not learned individual stimulus exemplars, but were using some general feature common to the novel and the training stimuli. When the starlings were presented with probe stimuli composed only of the song of a single species during sessions in which they were discriminating mixed-species songs, they still classified the isolated songs according to the discrimination learned with the mixture. When being presented with isolated starling songs, they were classifying these as accurate as starling songs in the mixture. This suggests that the birds perform auditory-stream segregation when analyzing the mixed-species songs. In other words, they are able to parse the single species songs and apply the learned characteristics of each species' song to discriminating the isolated songs. The starlings' discrimination was not severely impaired if a bird dawn chorus was added to the stimuli during the discrimination, indicating that these were processed separately from the background noise – the latter being composed from signals of multiple sources.

Evidence that the animals' hearing system is able to perform auditory-scene analysis also comes from studies using less complex artificial stimuli, revealing the mechanisms underlying auditory scene analysis – for reviews see [10, 22]. It has been has shown in psycho-physical experiments [80] that humans assign sequential tones of similar frequency presented in rapid succession to one auditory stream. If the frequency separation is increased, the tones are assigned to separate streams. European starlings appear to have a similar percept in segregating tone sequences [50]. The birds were first trained to discriminate a series of tones with an isochronous rhythm from a series of tones with a galloping rhythm. Then the starlings were presented with two inter-leaved series of tones of two different frequencies that – according to [80] – in humans would lead to the perception of a galloping rhythm in the case of a small frequency separation, 1 half-tone, as they would be perceived as one auditory stream, and lead to the perception of two inter-leaved tone series, i.e. two auditory streams with an isochronous rhythm, in the case of a large frequency separation, i.e. 9 half-tones. In the case of the large frequency separation the starlings reported more likely an isochronous rhythm, and in the case of a small frequency separation the starlings reported more likely a galloping rhythm. This is discussed as evidence [50] that starlings experience auditory streaming in a similar way as humans do. A study of the responses of neurons in the auditory forebrain area of the starling, which is analogous to the mammalian primary auditory cortex, suggests that suppressive interactions between frequencies and forward masking contribute to the segregation of tone series into separate auditory streams [2]. However, these effects fail to explain some of the psychophysical observation of humans in the context of auditory streaming, thus pointing to additional processing mechanisms that so far have not been revealed.

In humans, the grouping of components of a sound to one auditory stream is also supported by common amplitude modulation of the different frequency components and their synchronous onset and offset – see review [4]. Research on animal perception suggests similar auditory-grouping phenomena. In a study exploiting the natural communication behaviour, mice have been shown to recognize a sound only if a synchronous or nearly synchronous – time differences < 30 ms – onset and offset allows them to integrate the frequency components of the sound [13]. Comparable to humans, European starlings [33] and gerbils [28] show an improved ability to detect a tone in a background noise if the frequency components of the background noise exhibit coherent, namely, positively correlated, amplitude modulation – Fig. 2.4. This effect is known as "co-modulation masking release", CRM – for a review see [54] – and has been attributed both to within-channel mechanism, i.e. involving one auditory filter channel, and to across-channel mechanisms, i.e. involving two or more separated auditory filter channels. In the case of within-channel mechanisms, the change in the temporal pattern of excitation is thought to provide the cue to the detection of the signal. In the case of across-channel mechanisms, interactions between the channels that are similarly involved in auditory grouping phenomena may contribute to the improved signal detection in the co-modulated noise background. Both psycho-acoustic studies in humans and in European starlings have demonstrated that auditory grouping resulting from synchronous onset and offset of components of a sound affects co-modulation masking release [19,31]. This suggests that at least some of the underlying mechanisms of both phenomena are shared. CMR has also been demonstrated on the neuronal level. It has been demonstrated in the mammalian primary auditory cortex for cat [59], gerbil [12]), in the analogue brain area of a bird, i.e. European starling [32]. In the guinea pig, *Cavia porcellus*, response patterns that could give rise to CMR are found already on the level of the dorsal cochlear nucleus [64].

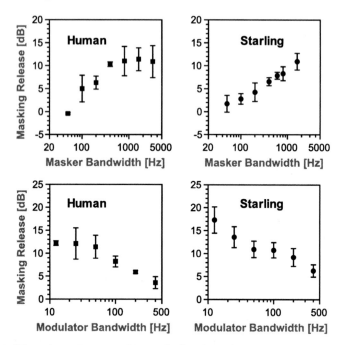

Fig. 2.4. The release from masking calculated as the threshold difference for the detection of 2-kHz tones in noise of a certain bandwidth, centered at the tone frequency, and amplitude-modulated noise of the same center frequency, bandwidth and overall level. The modulator was a low-pass noise which was multiplied with the carrier, thus resulting in a temporally coherent modulation at all frequencies. A positive masking release in these psychophysical experiments indicates better signal detection in the amplitude-modulated noise. Human data from [71], starling data from [33]. In the graphs at the *top row*, the modulator was a low-pass noise with a cut-off frequency of 50 Hz and the overall bandwidth of the masker was varied. In the graphs at the *bottom row*, the overall bandwidth of the masker was 1600 Hz and the cut-off frequency of the noise modulator was varied – low modulator bandwidth being equivalent to slow amplitude modulation. In both humans and starlings the coherent modulation of the masker allows for a better separation of the signal from the background noise

2.6 Concluding Remarks

The examples in this chapter have demonstrated that the animals' and the human auditory system have evolved similar auditory-processing capabilities. Even the underlying mechanisms may have common roots going as far as having a comparable protein machinery involved in the construction of mechano-receptors from nematodes to [14]. The active processes in the receptor organ are another example in which we find very similar mechanisms in species as diverse as the fruit fly *Drosophila melanogaster* – e. g., [16,17] – and some mammals including the human – e. g., see review [62]. The wide systematic distribution of these mechanisms in the sound-receiving organ suggests that they may have old phylogenetic roots – see also [49]. On the other hand, the chapter has also demonstrated a number of similarities between humans, other mammals and birds on the perceptual level. These similarities extend from the processing of simple stimulus features, e. g., as provided by frequency filtering in the auditory system, to the processing of auditory objects in the complex acoustic environment providing the opportunity to study the neural basis of auditory perception. Given the flexibility that the nervous system has when establishing its neuronal circuitry, it is well possible that similar physical characteristics of the environment have led to adaptations in auditory processing mechanisms that have emerged independently as a result of convergent evolution or developmental processes.

Acknowledgement

M. Göpfert and *M. Kössl* provided data for Fig. 2.1 and Fig. 2.2, respectively. The studies on auditory perception of the European starling by the author have been funded by the Deutsche Forschungsgemeinschaft, DFG.

References

1. Arnold S, Burkard R (2000) Studies of interaural attenuation to investigate the validity of a dichotic difference tone response recorded from the inferior colliculus in the chinchilla. J Acoust Soc Amer 107:1541–1547
2. Bee M A, Klump G M (2004) Primitive auditory stream segregation: a neurophysiological study in the songbird forebrain. J Neurophysiol 92:1088-1104
3. Bradbury J W, Vehrencamp S L (1998) Principles of animal communication. Sinauer, Sunderland MA.
4. Bregman A S (1990) Auditory scene analysis: The perceptual organization of sound. MIT Press, Cambridge MA.
5. Cynx J, Lewis R, Tavel B, Tse H (1998) Amplitude regulation of vocalizations in noise by a songbird, *Taeniopygia guttata*. Anim Behav 56:107–113
6. Dallos P (1992) The active cochlea. J Neurosci 12:4575–4585
7. Eggermont J J (1988) Mechanisms of sound localization in anurans. In: Fritsch B, Ryan M J, Wilczynski W, Hetherington T E (eds) The evolution of the amphibian auditory system, 307–336
8. Ehret G, Tautz J., Schmitz B, Narins P N (1990) Hearing through the lungs: lung-eardrum transmission of sound in the frog *Eleutherodactylus coqui*. Naturwiss 77:192–194
9. Fay R R (1992) Structure and function in sound discrimination among vertebrates. In: Webster D B, Fay R R, Popper A N (eds) The evolutionary biology of hearing. Springer, New York, 229–263
10. Feng A S, Ratnam R (2000) Neuronal basis of hearing in real world situations. Annu Rev Psychol 51:699–725
11. Fletcher H (1940) Auditory patterns. Rev Mod Phys 12:47–65

12. Foeller E, Klump G M, Kössl M (2001) Neural correlates of comodulation masking release in the auditory cortex of the gerbil. Abstr Assoc Res Otolaryngol 24:50–51

13. Geissler D B, Ehret G (2003) Time-critical integration of formants for perception of communication calls in mice. Proc Nat Acad Sci. 99:9021–9025

14. Gillespie P G, Walker R G (2001) Molecular basis of mechanosensory transduction. Nature 413:194–202

15. Göpfert M C, Robert D (2001) Active auditory mechanics in mosquitoes. Proc R Soc Lond. B 268:333–339

16. Göpfert M C, Robert D (2002) The mechanical basis of *Drosophila* audition. J Exp Biol. 205:1199–1208

17. Göpfert M C, Robert D (2003) Motion generation by Drosophila mechanosensory neurons. Proc Natl Acad Sci USA 100:5514-5519

18. Grafe U (1997) Use of metabolic substrates in the gray treefrog *Hyla versicolor*: implications for calling behavior. Copeia, 356–362

19. Grose J H, Hall J W (1993) Comodulation masking release: Is Comodulation sufficient?. J Acoust Soc Amer 93:2896–2902

20. Hudspeth A J (1997) Mechanical amplification of stimuli by hair cells. Curr Opin Neurobiol 7:480–486

21. Hudde H (2005) A functional view on the human peripheral hearing organ. Chap 3 this vol

22. Hulse S H (2002) Auditory scene analysis in animal communication. Adv Study Behav 31:163–200

23. Hulse S H, MacDougall-Shackleton S A, Wisniewsky A B (1997) Auditory scene analysis by songbirds: stream segregation of birdsong by European starlings (*Sturnus vulgaris*). J Comp Psychol 111:3–13

24. Jørgensen M B (1991) Comparative studies of the biophysics of directional hearing in anurans. J Comp Physiol. A 169:591–598

25. Jørgensen M B, Gerhardt H C (1991) Directional hearing in the gray tree frog *Hyla versicolor*: eardrum vibration and phonotaxis. J Comp Physiol A 169:177–183

26. Jørgensen M B, Christensen-Dalsgaard J (1997a) Directionality of auditory nerve fiber responses to pure tone stimuli in the grassfrog, *Rana temporaria*. I. Spike rate responses. J Comp Physiol. A 180:493–502

27. Jørgensen M B, Christensen-Dalsgaard J (1997b) Directionality of auditory nerve fiber responses to pure tone stimuli in the grassfrog, *Rana temporaria*. II. Spike timing. J Comp Physio. A 180:503–511

28. Kittel M, Klump G M, Wagner E (2000) Hearing in the gerbil (*Meriones unguiculatus*): Comodulation masking release. Zoology 103:Suppl III 68

29. Kittel M, Klump G M, Wagner E (2002) An estimate of the auditory-filter bandwidth in the Mongolian gerbil. Hear Res 164:69–76

30. Klump G M (2000) Sound localization in birds. In: Dooling R J,, Fay R R, A.N. Popper A N (eds) Comparative hearing: birds and reptiles, Springer, New York, 249–307

31. Klump G M, Hofer S B, Blohm B, Langemann U (2003) Auditory grouping and CMR: psychophysics and physiology. In: Elsner N, Zimmermann H (eds) The neurosciences from basic research to therapy . Thieme, Stuttgart, 461

32. Klump G M, Nieder A (2001) Release from masking in fluctuating background noise in a songbird's auditory forebrain. Neuroreport 12:1825–1829

33. Klump G M., Langemann U (1995) Comodulation masking release in a songbird. Hear. Res. 87:157–164
34. Klump G M, Gerhardt H C (1992) Mechanisms and function of call-timing in male-male interactions in frogs. In: McGregor P K (ed) Playback and studies of animal communication: problems and prospects. Plenum Press, London–New York, 153–174
35. Klump G M, Windt W, Curio E (1986) The great tit's (*Parus major*) auditory resolution in azimuth. Comp Physiol 158:383–390
36. Klump G M, Shalter M D (1984) Acoustic behaviour of birds and mammals in the predator context. Z Tierpsychol 66:189–226
37. Knudsen E I (1980) Sound localization in birds. In:A.N. Popper A N, Fay RR (eds) Comparative studies of hearing in vertebrates. Springer, New York, 289–332
38. Knudsen E I, Konishi M (1978) A neural map of auditory space in the owl. Science 200:795–797
39. Knudsen E I, Konishi M (1979) Mechanisms of sound localization in the barn owl (*Tyto alba*). J Comp. Physiol 133:13–21
40. Konishi M (1973) How the owl tracks its prey. Amer Sci 61:414–424
41. Köppl C (1995) Otoacoustic emissions as an indicator for active cochlear mechanics: a primitive property of vertebrate hearing organs. In: Manley G A, Klump G M, Köppl C, Fastl H, Oeckinghaus H (eds) Advances in hearing research 207–218
42. Kössl M, Boyan G S (1998) Otoacoustic emissions from a Nnonvertebrate ear. Naturwiss. 85:124–127
43. Kössl M, Boyan G S (1998) Acoustic distortion products from the ear of a grasshopper. J Acoust Soc Amer 104:326–335
44. Kuhn G F (1977) Model for the interaural time differences in the horizontal plane. J Acoust Soc Amer 62:157–167
45. Langemann U, Klump G M, Dooling R J (1995) Critical bands and critical ratio bandwidth in the European starling. Hear Res 84:167-176
46. Lardner B, bin Lakim M (2002) Tree-hole frogs exploit resonance effects. Nature 420:475
47. Larsen O N, Dooling R J, Ryals B M (1997) Roles of intracranial air pressure in bird audition. In: Lewis E R, Long G R, Lyon R F, Narins P M, Steele C R (eds) Diversity in auditory mechanics, World Scientific Pub, Singapore, 253–259
48. Long G (1977) Masked auditory thresholds from the bat, *Rhinolophus ferrumequinum*. J Comp Physiol A 116:247–255
49. Manley G A (2001) Evidence for an active process and a cochlear amplifier in non-mammals. J Neurophysiol. 86:541–549
50. MacDougall-Shackleton S A, Hulse S H, Gentner T Q, White W (1998) Auditory scene analysis by European starlings (*Sturnus vulgaris*): Perceptual segregation of tone sequences. J Acoust. Soc Amer. 103:3581–3587
51. McGregor P K, Dabelsteen T (1996) Communication networks. In: Kroodsma D E, Miller E H (eds) Ecology and evolution of acoustic communication in birds. Cornell University Press, Ithaca NY, 409–425
52. McLister J D (2001) Physical factors affecting the cost and efficiency of sound producing in the treefrog Hyla versicolor. J Exp Biol 204:69–80
53. Moiseff A (1989) Bi-coordinate sound localization by the barn owl. J Comp Physiol A 164:637–644

54. Moore B C J (1992) Across-channel processes in auditory masking. J Acoust Soc Jpn. 13:25–37

55. Moore B C J (2003) An introduction to the psychology of hearing. Academic Press, London

56. Morton E S (1977) On the occurrence and significance of motivation-structural rules in some bird and mammal sounds. Amer Nat 111:855–869

57. Naguib M, Klump G M, Hillmann E, Griebmann B, Teige T (2000) Assessment of auditory distance in a territorial songbird: accurate feat or rule of thumb. Anim Behav 59:715–721

58. Naguib M, Wiley H R (2001) Estimating the distance to a source of sound: mechanisms and adaptations for long-range communication. Anim Behav 62:825–837

59. Nelken I, Rotman Y, Yosef O B (1999) Responses of auditory-cortex neurons to structural features of natural sound. Nature 397:154–157

60. Nelson B S, Stoddard P K (1998) Accuracy of auditory distance and azimuth perception by a passerine bird in natural habitat. Anim Behav 56:467–477

61. Nieder A, Klump G M (1999) Adjustable frequency selectivity of auditory forebrain neurons recorded in freely moving songbird via radiotelemetry. Hear Res 127:41–54

62. Nobili R, Mammano F, Ashmore J (1998) How well do we understand the cochlea? Trends Neurosci. 21:159–167

63. Patterson R D, Moore B J C (1986) Auditory filters and excitation patterns as representations of frequency resolution. In: Moore B J C (ed) Frequency selectivity in hearing, Academic Press, London, 123–177

64. Pressnitzer D, Meddis R, Delahaye R, Winter I M (2001) Physiological correlates of comodulation masking release in the mammalian ventral cochlear nucleus. J Neurosci. 21:6377–6386

65. Probst R, Lonsbury-Martin B L, Martin G K (1991) A review of otoacoustic emissions. J Acoust Soc Amer 89:2027–2067

66. Rauschecker J P, Tian B (2000) Mechanisms and streams for processing of "what" and "where" in auditory cortex. Proc Natl Acad Sci USA 97:11800–11806

67. Robert D, Göpfert M C (2002) Novel schemes for hearing and orientation in insects. Curr Opin Neurobiol 12:715–720

68. Robles L, Ruggero M A (2001) Mechanics of the mammalian cochlea. Phys Rev 81:1305–1352

69. Ruggero M A, Narayan S S, Temchin A N, Recio A (2000) Mechanical basis of frequency tuning and neural excitation at the base of the cochlea: comparison of basilar-membrane vibrations and auditory-nerve-fiber responses in chinchilla. Proc Natl Acad Sci USA 97:11744–11750

70. Scharf B (1970) Critical bands. In: Tobias J V (ed) Foundations of modern auditory theory, Academic Press, London, 159–202

71. Schooneveldt G P, Moore B C J (1989) Comodulation masking release (CMR) as a function of masker bandwidth, modulator bandwidth, and signal duration. J Acoust Soc Amer 85:273–281

72. Schuller G, Beuter K, Schnitzler H-U (1974) Response to frequency-shifted artificial echoes in the bat, *Rhinolophus ferrumequinum*. J Comp Physiol 89:275–286

73. Seyfarth R M, Cheney D L, Marler P (1980) Vervet monkey alarm calls: evidence for predator classification and semantic communication. Science 210:801–803

74. Shannon C E, Weaver W (1949) The mathematical theory of communication. paperback ed 1963, Univ. Illinois Press, Champaign IL

75. Shaw E A G (1974) The external ear. In: Keidel W E, Neff W (eds), Handbook of sensory physiology, Vol V: Auditory System, Springer, Berlin, 455–490

76. Stevens S S, Newman E B (1936) The localization of actual sources of sound. Am J Psychol 48:297–306

77. Suga N, Neuweiler G, Möller J (1976) Peripheral auditory tuning for fine frequency analysis by the CF-FM bat, Rhinolophus ferrumequinum. J Comp Physiol 106:111–125

78. Taigen T L, Wells K D (1985) Energetics of vocalization by an anuran amphibian (Hyla versicolor). J Comp Physiol B 155:163–170

79. Takahashi T T, Moiseff A, Konishi M (1984) Time and intensity cues are processed independently in the auditory system of the owl. J Neurosci 4:1781–1786

80. Van Noorden L P A S (1975) Temporal coherence in the perception of tone sequences. Doct diss, Eindhoven Univ of Technol, Eindhoven

81. Von der Emde G, Schnitzler H-U (1990) Classification of insects by echolocating greater horseshoe bats. J Comp Physiol A 167:423–430

82. Wells K D (2001) The eregetics of calling in frogs. In: Ryan M J (ed) Anuran communication. Smithsonian Inst. Press, Washington, 45–60

83. Wells K D, Taigen T L (1986) The effects of social interactions on calling energetics in the Gray treefrog (Hyla versicolor). Behav Ecol Sociobiol 25:8–18

84. Wiley R H, Richards D G (1982) Adaptation for acoustic communication in birds: sound transmission and signal detection. In: Kroodsma D E, Miller E H (eds) Acoustic communication in birds, vol 1 , Academic Press, New York, 131–181

85. Yost W A, Hafter E R (1987) Lateralization. In: Yost W A, Gourevitch G (eds) Directional Hearing, Springer, New York, 49–84

3 A Functional View on the Peripheral Human Hearing Organ

Herbert Hudde

Institute of Communication Acoustics, Ruhr-University Bochum, Bochum

Summary. The human hearing organ is a signal processor par excellence. Its amazing abilities are often described in terms of psycho-acoustic models. However, in this chapter the focus is laid on the physical background, particularly on the physics of the peripheral hearing organ. The peripheral system can be looked at as a signal conditioner and preprocessor which stimulates the central nervous system. It comprises acoustic, mechanic, hydro-acoustic, and electric components which, in total, realize a sensitive receiver and high-resolution spectral analyzer. For daily life it is extremely important that the hearing organ can also work under adverse conditions. This includes the need for general robustness and low sensitivity with respect to varying external and internal working conditions. In the hearing organ several strategies are found which noticeably differ from technical solutions.

3.1 Introduction

It goes without saying that the remarkable abilities of the human hearing organ require very complex anatomical structures and features. Thus a survey of the hearing organ in a single chapter would necessarily be very superficial. To avoid this, the general survey is held short but certain components of the hearing organ are considered in more depth. The term "functional view", as announced in the title essentially denotes an engineering point of view.

Often the function of the hearing organ is regarded in perspective of its amazing signal processing facilities. In such a case signal processing in the central parts, i. e. in the brain stem and in the brain, come to the fore. Due to the highly complex nature of the central structures a physical understanding of the relevant processes is particularly difficult at this point. Therefore psycho-acoustic models have to be used – which often express quantitative findings by rather simplified signal-processing blocks. Yet, for many applications the perception of sound can be sufficiently accurately modelled in this way. From the viewpoint of signal processing the underlying physical principles are often of little interest. On the contrary, a too detailed description of very complex systems is even undesired in such cases, because computing times are unnecessarily increased. For instance, it is often entirely sufficient to describe the middle ear as a simple band-pass filter without taking into account any details of its structure, especially when the signal analysis as performed by the hearing organ is the main item of interest.

In this chapter, however, the focus is put on the physical mechanisms as occurring in the hearing organ. This means a total change of perspectives. The focus is necessarily shifted to the periphery where the structures are much simpler. In fact, the objective of this chapter is to improve our understanding of why the hearing organ is "designed" just as it is. This includes aspects like the robustness of the auditory perception with respect to external influences, e. g., varying static pressure. We will see that the "design principles" as found in biological systems are fairly different from technical design principles. Features which are seemingly disadvantageous if only frequency responses are considered, emerge as benefits if looked at from a more holistic point of view.

This chapter is predominantly directed to readers who already have some basic knowledge of the hearing organ. Yet, the discussions should be understandable also for readers without such previous knowledge, as necessary details of the considered structures are always given. As announced above, we start with an extremely compact survey of the hearing organ in order to make the reader acquainted with basic facts.

Auditory signal processing already starts outside the head. The external sound field has to couple into the ear canals. The relative positions of the two ear canals and the sound source(s) lead to a coupling that is strongly dependent on frequency, except at very low frequencies, i. e. below some hundred Hertz. In this contexts not only the two *pinnae* but also the whole head have an important functional role, which is best described as a spatial filtering process. This – linear – filtering is usually quantified in terms of so-called head-related transfer functions, HRTFs, which can also be interpreted as the directivity characteristics of the two ears.

The ear canals are short tubes of up to 3 cm in length which act as acoustic transmission lines. Due to their small lateral dimensions only the fundamental-mode wave can propagate in them. This wave is characterized by pressure variations occurring only along the axis of the ear canal. In cross-sectional areas, perpendicular to the axis, the pressure is almost constant. This type of propagation reduces the auditorily relevant input to basically two "ear signals", as, e. g., being specified at the entrance to the ear canals or in front of the tympanic membranes, i.,e. eardrums, which terminate the ear canals inwards. The level difference and arrival-time differences of these two signals are the basis of binaural hearing, which enables a fairly accurate spatial localization of sound sources. The combination of pinna and ear canal is usually denoted as outer – external – ear. In addition to the sound conduction via the ear canals, called "air conduction" sound reaches the sensors in the inner ear also via the skull, "bone conduction" . However, under normal conditions, bone conduction is perceptually almost irrelevant as it results in levels in the inner ear that are at least 40 dB lower than those excited by air conduction.

The sound wave impinging on an eardrum generates vibrating forces which are transmitted to the inner ear via a chain of three ossicles, *malleus, incus,* and *stapes.* A schematic representation of the peripheral ear is shown in Fig. 3.3. The ossicles and their elastic suspensions are placed in the tympanic cavity. The last ossicle, the stapes, acts like a piston that transmits the vibrations to the fluids in the inner ear. Eardrum and ossicular chain constitute the middle ear. The main function of the middle ear is to implement a more effective transmission of sound energy from the air in the ear canal to the fluids in the inner ear.

The inner ear comprises two organs which are directly coupled by common fluids, the vestibular organ and the *cochlea.* Only the cochlea has auditory functions and is thus to be considered in this chapter. It contains a system of three fluid-filled ducts which, together, form the "cochlear duct", see Fig. 3.1. The three individual ducts are separated by membranes. The cochlear duct coils round from the base to the apex, thereby forming a kind of snail-like capsule, thus called cochlea. The main task of the cochlea is a spectral analysis of the input signals which are imposed by stapes vibration. This means that the input signal is decomposed into many parallel output signals, each of which represents a spectral component in a different frequency band. Altogether these components can cover the full auditory range. The frequency separation is performed by a very complex system comprising hydro-acoustic, mechanic and electric components.

The decisive element, at least according to classical theory, is the basilar membrane. This membrane separates two of the ducts and follows the two and a half windings of the cochlea from the base to the apex. The special function of spectral analysis is achieved by inhomogeneous tuning along this membrane which, at a given location, yields a maximum response for a characteristic frequency, called "best frequency". The vibrations at different points are sensed by many so-called inner hair cells, IHCs. These hair cells can be considered as the output ports of the cochlea. They are connected to endings of the hearing nerve via synaptic clefts. External ears, middle ears and cochleae are usually together referred to as the peripheral hearing organ.

The central hearing organ is part of the central nervous system. Here the auditory information is transmitted via neurons, i. e. nerve fibres,. The fibres conduct short electric pulses, called spikes. At this point the character of signals is no longer time-continuous, but rather time-discrete. The neurons are part of a huge network called auditory pathway. This network comprises several *nuclei* in the brain stem where neurons can interact. These nuclei can be regarded as relay stations where information from different fibres are jointly processed by means of specialized cells. This also includes interaction of fibres from both ears. The highest cerebral level is the auditory *cortex,* which is situated on the so-called superior temporal plane – fairly close to the surface of the brain, but hidden in a fissure.

The primary spikes of the auditory pathway are elicited by the inner hair cells. Each fibre represents a one-way connection. But there are not only ascending fibres running from the cochlea via several nuclei to the auditory cortex, but also descending ones. Fibres that transport pulses upwards to higher neural levels are called afferent, those that conduct information back from higher levels to lower ones, are called efferent. Actually, the cochlea is also innervated by efferent fibres. Yet, most of the efferent fibres are not connected to the above-mentioned inner hair cells, but to further cells which are called outer hair cells, OHCs . Many specialized cells in the auditory pathway contribute to the highly complex signal processing – which by far exceeds the performance of modern computers. There is some evidence that in the course of the analysis of signals in and from the auditory system, expert knowledge is utilized to recognize already known auditory patterns. A permanent process of interpretation seems to be an essential part of hearing.

In the following sections we shall now address mainly acoustic and mechanic aspects of the hearing organ. Readers interested in more details of the electric processes within the cochlea and on the neural level may consult, e. g., the books [12, 14] – for more findings see a collection of papers introduced by [8].

Understanding of the hearing organ with respect to its physical details is of relevance not only in fundamental research, but it is also a necessary prerequisite for a variety of technical and medical applications. This includes the optimization of technical devices related to hearing, e. g., earphones, hearing aids, hearing protectors, and the improvement of surgical procedures and equipment such as middle ear prostheses. Further, the development of electromechanical transducers to drive the ossicles, middle-ear implants requires a thorough knowledge of middle-ear mechanics.

In the further flow of this chapter, we shall now reverse the direction and start the functional view in the brain stem rather then tracking the auditory signals on its way from an outer sound field up to the central parts. The conditions found in the brain stem specify the tasks which have to be performed already in the periphery. This is indeed a general advantage of the reversed view, namely, the structural design and the conditions found in a more central system of higher complexity determine the requirements to be met by a preceding, less complex pre-processing system.

In this sense, the cochlea is the first component to be regarded. Subsequently, the boundary conditions of the inner ear determine the functions to be accomplished by middle ear and ear canal. Thus, in this chapter, only cochlea, middle ear and ear canal are dealt with, albeit not in equal depth.

In spite of many decades of research into the cochlea, there is still no generally accepted model which satisfactorily explains all the physical processes involved. On the other hand, there are so many suggestions and such a variety of data around that it is quite impossible to provide a fairly comprehensive representation in this chapter. Thus we shall restrict ourselves to a fairly short review of the cochlea under some particular aspects. The main focus is then led to the middle ear. In contrast to the situation with the cochlea, this author thinks that the design of the middle ear can be regarded as being essentially understood by now.

To complete the functional view on the hearing organ, the effect of skull vibrations on hearing will be regarded as well, namely, bone conduction, as already mentioned. Here, like with the cochlea, many questions cannot be answered at present. Particularly, the relative contributions of different pathways transmitting vibrations to the cochlea has not yet been clarified. We shall, in this context, offer preliminary results based on a finite-element model of the human skull. Nevertheless, quite some work will still be necessary to understand complex phenomena such as the perception of one's own in its physical details [13].

3.2 Cochlea

In up-to-date technical measuring systems signal analysis is usually predominantly performed in the digital domain, i.e. using time-discrete signal. Only the first stages, e.g., sensors and signal-conditioning amplifiers, deal with time-continuous signals. As the human brain is often considered as a kind of computer, a naive reader might expect that the conditions found in the human hearing organ are similar. Along such lines of thinking, the peripheral parts would be assumed to act as sensitive receivers of sound, with the central parts having the task of performing all the signal processing.

In contrast to such a view, measurements with electrodes placed in auditory structures of the brain clearly reveal that at least the spectral analysis is already performed in a previous stage: Contiguous frequency regions are already mapped into contiguous spatial sectors of the brain, so-called tonotopic representation. In fact, there is no doubt that the cochlea contains structures which perform frequency separation, i.e. spectral decomposition.

Of course the brain is not a computer in today's technological sense. In particular it is not able to deal with numbers, and consequently, such a thing like discrete *Fourier* transform, DFT, cannot be performed. On the other hand, the brain is very well suited to perform temporal analyses – an essential part of the auditory signal processing which is often interpreted as a kind of correlation. The computation, i.e. estimation, of correlations within and between signals complies with the signal processing properties at a neural level as time differences between spikes can easily be evaluated by specialized cells.

Some details of the cochlea are shown in Fig. 3.1. However, the primary goal of this short treatment is not to explain all the mechanisms occurring in an actually existing cochlea, but to consider possible elements that could realize a spectral analyzer on a mechanic basis.

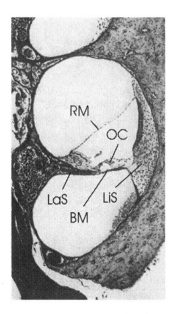

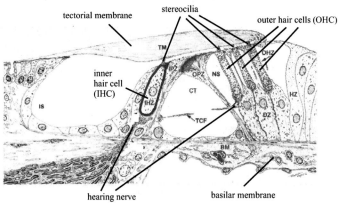

Fig. 3.1. The cochlea contains two and a half turns of a duct running from the base turn to the apex – not shown. **Top**: Cross section through the duct. It is separated into three fluid-filled scales by *Reissner*'s membrane, RM, and the basilar membrane, BM, which connects the osseous lamina spiralis, LaS, and the ligamentum spirale, LiS. Upon the basilar membrane rests a further structure, the organ of Corti, OC. **Bottom**: Details of the organ of *Corti* – not all of the abbreviations in this figure are to be explained here. The relevant elements are the inner and outer hair cells, IHCs and OHCs. The tectorial membrane, TM, has tight contact only to the tallest stereocilia, the sensory hairs on the top of the OHCs. The IHCs are mainly innervated by afferent fibers of the hearing nerve – both figures from [15]

The basic sensors in the cochlea that convert motion into electrical spikes are the inner hair cells, IHC, see Fig. 3.1 . An IHC is an oblong flask terminated at the top by a plate which carries a tuft of sensory hairs, commonly called *stereocilia*. Bending of the stereocilia causes changes of the electrical potential inside the hair-cell body, i.e. of the receptor potential. These changes elicit spikes, so called action potentials, in the afferent fibres of the hearing nerve via the synapses at the bottom of the hair cell. Thus, quite obviously, the IHCs are the primary sensors of the cochlea.

To obtain the frequency separation observed in psycho-acoustic and physiologic measurements a fixed relationship between certain frequencies and certain IHCs has to be established. It is worth questioning why such a sophisticated structure as shown in Fig. 3.1 is necessary at all. A much simpler frequency analyzer could, e.g., be constructed from an array of tuned hair cells.

Tuning could then be achieved by a variation of the bending stiffness of the stereocilia or by a variation of the dimensions and elasticity of the hair-cell body. Such an array of tuned hair cells would work akin to a tuned-reed frequency meter. Tuning of the hair cells themselves instead of tuning of a structure which drives non-tuned hair cells is found in many insects and also in lower vertebrates. Mammals are usually considered to have non-tuned hair cells. In this case not the hair cell itself but the location where it is placed must be associated with a certain "best frequency". According to classical theory only the position along the basilar membrane determines the best frequency, however, recent investigators [2] have suggested that also mammals do have tuned hair cells.

Before the question of tuning is followed up further, we consider general advantages of the mammalian cochlea design as a fluid-filled capsule. For one thing, the rigid capsule provides much better protection of the vulnerable elements inside compared to a structure that is more or less directly accessible by an outer sound wave via air-filled chambers. Whereas the capsule is surrounded by less robust spongeous material, the bounding walls of the cochlea are made of especially rigid bone. Further, the fluid allows a much more effective coupling of the incoming wave to the sensory hairs by means of a gelatinous layer on the top of stereocilia. Vibrations of the fluid are directly transmitted to this layer which, in turn, is closely coupled to the stereocilia. This hydro-acoustic coupling is less dependant upon on frequency and much more effective than air coupling. Several variations of this principle of coupling are found in various species [10]. In mammals the gelatinous layer is shaped as the tectorial membrane.

The driving mechanism is quite different for inner and outer hair cells. The stereocilia of the OHCs are in close contact with the tectorial membrane, whereas the stereocilia of the IHCs are driven by a flow in the narrow cleft between the organ of Corti and the tectorial membrane.

Fluids are also necessary to implement a kind of "battery" by means of different ionic concentrations in the different ducts. The organ of *Corti* is situated in the middle duct filled with endolymph which has a resting potential of about +80 mV. In contrast, the resting potential inside the hair cells is about -45 mV for IHCs and -70 mV for OHCs – all potentials measured against a neutral, e. g., a neck electrode. Therefore between the OHCs and the endolymph a voltage of 150 mV can be exploited to drive an electric current via ionic channels. Direct transduction between vibrational and electrical energy would certainly also be possible without a "battery" such as by means of piezo-electric materials, but the use of auxiliary energy appreciably extends the potency of the total design.

We now return to the question of tuning. According to classical theory tuning is mainly caused by the elastic properties of the basilar membrane in combination with the organ of *Corti* resting on it. The stiffness of the membrane is highest at the base and decreases continuously along the two windings to the apex. This can be interpreted as an array of tuned sections which are associated with certain resonant frequencies. However, one does not observe isolated responses at certain sections, but rather a transversal wave running along the basilar membrane from the base to the *apex*. Propagation in the reverse direction is strongly attenuated. Due to the stiffness distribution along the membrane the velocity of the travelling wave is highest at the base and decreases on its the way to the apex. The "highway" for this wave is represented in Fig. 3.2 on the left.

The conditions can be described by differential equations which comprise the impedance variation of the basilar membrane as a key feature. A more intuitive understanding can be obtained when considering the decelerating wave. The velocity of propagation vanishes at a certain location depending on frequency. Hence the wave is virtually stopped here, and several periods of a stimulating sinusoidal signal are superimposed. As a consequence the amplitudes become very high at this position. Behind this location no wave at the best frequency and at higher frequencies, can propagate and the wave energy is rapidly absorbed. Thus the locations associated with the highest frequencies within the auditory range are found near the base. Waves of lower frequencies can penetrate deeper, up to the apex which is reached by waves at frequencies below about 100 Hz. This, in short, explains the formation of the travelling wave and the location of maximum response at the best frequency, shown at the right side of Fig. 3.2.

The existence of travelling waves was experimentally found by *Von Békésy* in 1960 [1]. Although the majority of researchers considers the travelling wave as the primary effect in the cochlea, there are others who interpret the travelling wave only as an epiphenomenon. Actually, the number of OHCs which are mainly innervated by efferent fibres exceeds the number of IHCs by a factor of about 3.5. It is indeed very surprising to find more hair cells

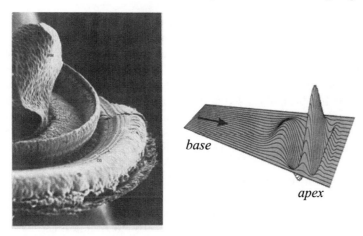

base

apex

Fig. 3.2. Left: View into an opened cochlea. Unavoidably the vulnerable membranes alter in shape and position during the course of preparation. Nevertheless the photo gives a good insight into the "highway". The organ of *Corti* rests on the basilar membrane which is disrupted on the *right side* in this figure. On the top of the *Corti* organ the rows of the stereocilia of the hair cells can be observed. Photo from [15]. **Right**: Travelling wave running from the base in the direction of the apex, represented on an un-wrapped basilar membrane. At a certain frequency-dependent location a maximum response arises

connected to efferent neurons than those connected to afferent neurons in a sensory organ.

There is no doubt that OHCs increase the sensitivity of the cochlea. Damaged OHCs in hearing impaired persons cause the so-called recruitment phenomenon. The threshold of hearing is then increased by 40 to 50 dB, whereas, at high sound pressure levels, SPLs, the sensitivity becomes fairly normal. Thus the OHCs are obviously mainly effective at low SPLs. Another important finding is that the sharp tuning found in an intact cochlea vanishes if the OHCs are damaged. Obviously, the OHC improve both the minimum audible pressure level and the frequency discrimination.

We know from experiments that OHCs change their shape, mainly their length, when excited via the stereocilia or by an electrical voltages [3, 22]. According to classical theory the vibrations of the OHC are fed back to the basilar membrane, which primarily excites the OHCs via the stereocilia. By means of this feed-back loop the basilar-membrane vibrations are amplified. The components which contribute to the feed-back loop are often referred to as "cochlear amplifier". The effect of the OHC saturates at about 40 dB above the hearing threshold. This explains the recruitment phenomenon just mentioned.

The term "cochlear amplifier" implies a fundamental concept, namely that of a the passively travelling wave which is somehow amplified by the

OHCs. As already mentioned, other theories have been proposed which interpret the OHCs as a self-contained system designed for low SPLs. In this case an additional tuning mechanism must exist, since we know that the frequency selectivity is best at low levels. The second tuned structure is conjectured to be established in the OHCs themselves and/or in the tectorial membrane. It has been discussed [2] to interpret the three rows of OHCs as a surface-acoustic-wave resonator, SAW resonator, for waves travelling on the tectorial membrane. The best frequency corresponds to the condition of half a wavelength occurring between two adjacent rows.

After years of research based on classical notions, new ideas on possible mechanisms are now about to stimulate the discussion on the functioning of the cochlea. A definite decision regarding the actually dominating mechanisms could be obtained from a model which is entirely physical without using any fictive elements. This model had to be able to simulate the behaviour of the cochlea in quantitative accordance with all reliable measurements. All models published so far need some generalized assumptions or unexplained feedback loops to generate the desired output.

However, the functioning of cochlear amplifier is not the only question which has not been answered definitely so far. Can we really be sure that the cochlea is actually a pure spectrum analyzer? Many reviewers stress the high frequency resolution found in the tuning curves of the cochlea measured at low SPLs. Yet, how does this comply with the psycho-acoustically measured result that the frequency resolution as finally perceived even increases slightly with sound intensity. When using speech-like sounds instead of sinusoids the effective bandwidths appears much wider than the narrow tuning curves found in mechanical and neural responses. Usually such unexplained features are supposed to be created in the central hearing organ. In [11] it is suggested that the cochlea could be immediately involved in these processes.

There is another important feature of the auditory signal processing that is widely agreed to be realized in the cochlea. Due to non-linearity in the inner hair cells, weaker portions of the spectral energy distribution of a sound can be "masked" by stronger portions of nearby frequency bands. In consequence some information contained in the original sound signals does not show up in the auditory percept, in other words: It is erased. This fact is being heavily exploited in modern audio coding. Thus not only spectral decomposition but also reduction of some irrelevance in auditory signals is an intrinsic task of the cochlea.

3.3 Ear Canal and Middle Ear

A functional understanding of the ear canal and, particularly, of the middle ear is the prominent goal of this chapter. Therefore this section is more detailed. It includes an analysis of the spatial vibrations of the middle ear. Further, it attempts to present reasons for the comparably sophisticated shape of the ossicular chain. As ear canal and middle ear work as a functional unit they are dealt with in common.

We have seen that there are good reasons to fill the cochlea with fluids. Yet, fluids also induce a big disadvantage, namely, the sound wave as propagating in the air about the head must somehow reach the fluids. The change in media goes with a corresponding change of the magnitude of the impedance. If an acoustic wave in air, specific impedance $Z'_{air} \approx 390\,Ns/m^3$, impinges on a boundary with a specific input impedance, Z'_{in}, reflections occur which are determined by the impedance ratio according to the reflectance magnitude

$$|r| = \left| \frac{Z'_{in}/Z'_{air} - 1}{Z'_{in}/Z'_{air} + 1} \right|. \tag{3.1}$$

Often the specific impedance of water is taken to estimate these reflections. It is about 3850 times higher than the specific impedance of air. This results in a reflection coefficient of 0.9995, which means that only a small fraction of the incident acoustic power on the order of $1 - |r|^2 \approx 10^{-3}$, corresponding to a reduction of 30 dB, would reach the fluids. In order to reduce the reflections, a device is needed that transforms the low impedances in air into the higher ones of the fluids. In fact, the main purpose of the middle ear is that of such a transformer.

However, it is certainly a considerable over-simplification to only consider the specific impedances. Actually, an initial sudden change of the impedance and corresponding reflections happen already the ear-canal entrance. Thus the ear canal has obviously to be included into the considerations. Furthermore, the simple two-media calculation at the boundary to the inner ear fluids is completely wrong as it assumes propagating waves in both media. But we know from the preceding paragraph that the fluids rather behave as a lumped element, at least up to several Kilohertz. Therefore the acoustic input impedance of the cochlea at the stapes footplate, Z_C, has to be taken into account. This impedance not only reflects the lymphatic fluids but also the complete vibrating system of the basilar membrane, the organ of *Corti*, the tectorial membrane and, in addition, the round-window membrane.

A still rather general, but more appropriate description of the underlying conditions is represented in Fig. 3.3 in terms of a circuit model. As usual, the sound pressures are treated in analogy to voltages and volume velocities are considered analogous to electrical currents [9]. The model does not imply any special assumptions about the ear canal and the middle ear except that simple one-dimensional treatment is applicable. This means that the vibrations at any points can be expressed by only one component. In a one-dimensional system the amplitudes and phases of vibrations change with frequency, but not their spatial directions. Indeed the modes of vibration of the eardrum and the ossicles do change considerably with frequency. Nevertheless the structure of the circuit of Fig. 3.3 is well suited to understand the basics of the sound transmission, at least with respect to input-output relationships. Actually,

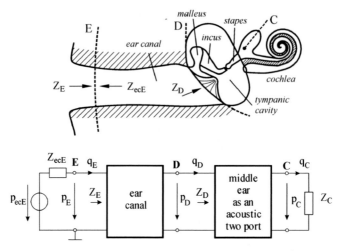

Fig. 3.3. Schematic diagram of the peripheral ear and a basic circuit structure of a corresponding acoustical model. The three ports of the circuit belong to the entrance to the ear canal, E, the eardrum, D, and the cochlea, C. According to the electro-acoustic analogies sound pressures, p, are treated like voltages and volume velocities, q, like electric currents

a one-dimensional model can also describe many details of the transmission fairly well. For substructures of a one-dimensional model of the ear canal and the middle ear, including parameter values, see [5].

If the reference plane, E, is chosen to lie deep enough in the ear canal, the sound wave is nearly reduced to the fundamental mode at this plane. This mode is the only one that can propagate in the ear canal because its tranversal dimensions remain smaller than half a wave length up to 20 kHz, which may be considered the upper limit of the auditory range. Thus, at E, we can define a one-dimensional port as used in system theory. At this port we observe a pressure, p_E, and a volume velocity, q_E, and, consequently, also an acoustic ear canal input impedance, Z_E. Because of the one-dimensionality at this position, the external sound field can be represented by an equivalent source acting at E. In the circuit a *Thevenin* source comprising an ideal pressure source, p_{ecE}, and an inner impedance, Z_{ecE}, is used.

At the end of the ear canal, port D, the conditions are complicated by the inclined position of the eardrum and its inhomogeneity. Near the tympanic membrane this leads to wave fronts which vary considerably in their spatial orientation with frequency – see Fig. 3.4. Hence, at a first sight, a one-dimensional description using a simple middle ear impedance, Z_D, seems impossible. But upon closer examination the variations turn out to be restricted to low frequencies. In this context 1150 Hz, underlying the mid panel of Fig. 3.4, is a low frequency as the pressure changes are small, i.e. 1,2%. For low frequencies the pressure near the drum is approximately constant and no

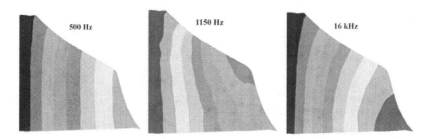

Fig. 3.4. Regions of equal sound pressure near the eardrum, to the *right* in all three figures, at different frequencies, calculated by a finite-element model of the ear canal. The boundaries between different *grey tones* can be interpreted as wave fronts. The variations of sound pressure strongly depend on frequency. Therefore the meaning of the *grey tones* in the three cases shown is very different. The *grey tones* represent sections of equal pressure variations in the following ranges, at 500 Hz ... 1.0–1.009, at 1150 Hz ... 1,0–0.988, at 16 kHz ... 1.0–9.96. The pressure at the beginning of the section is normalized to unity

transformation at all is needed. At higher frequencies the wave fronts are similar to the ones shown in the right panel of Fig. 3.4. Here the wave fronts are nearly perpendicular to the ear canal walls. Therefore the eardrum can be considered as a continuation of the ear-canal walls. This allows for the use of a simplified treatment as an acoustic transmission line with a lumped-element middle-ear impedance, Z_D, embedded at a reasonable location – see [5] for more details. Port C in Fig. 3.3 represents the entrance to the cochlea. More exactly, this port is defined by the sound pressure below the footplate, p_C, and the volume velocity of the footplate, q_C. Due to the small dimensions and the large sound velocity in fluids this port is truly one-dimensional. Only the velocity component perpendicular to the oval window contributes to the volume velocity.

On the basis of what has been discussed above the question of the impedances to be matched by a possible transformer can now be treated in a more realistic way. The well-known postulate to be met by the transformer, with the transformer ratio, t, is

$$Z_C = t^2 Z_{ecE}^* \tag{3.2}$$

where the asterisk means complex conjugation.

A look at the impedances to be matched – Fig. 3.5 – shows that (a) the transformer ratio had to vary with frequency in a well-defined way and (b) a real transformer can never provide an ideal match. A realistic transformer can only attempt to partly "fill the gap" across the different impedance magnitudes in the relevant frequency range. In particular, the very low real part of the external-ear radiation impedance at low frequencies cannot be compensated.

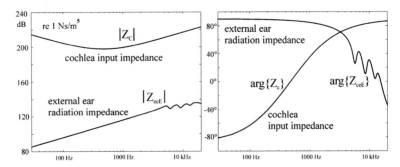

Fig. 3.5. Comparison of the external ear radiation impedance, Z_{ecE}, and the cochlea input impedance, Z_C – magnitudes, **left**, phases, **right**

The two impedances determine an upper limit of the gain that can be achieved by passive means. From this postulate follows that the maximum power as available from the external sound field at port E must reach the cochlea at port C. This statement can be mathematically expressed as follows:

$$\frac{1}{8}\frac{|p_{ecE}|}{\mathrm{Re}\{Z_{ecE}\}} = \frac{1}{2}|p_C|^2\mathrm{Re}\{Y_C\} = \frac{1}{2}|q_C|^2\mathrm{Re}\{Z_C\}. \tag{3.3}$$

It yields the optimum gain

$$\left|\frac{p_C}{p_{ecE}}\right| = \frac{1}{2\sqrt{\mathrm{Re}\{Z_{ecE}\} \times \mathrm{Re}\{Y_C\}}}, \tag{3.4}$$

which is only determined by the real part of the two impedances involved. Equation 3.4 serves as a reference in Fig. 3.6, showing the pressure gain provided by a normal middle ear with and without the contribution of the ear canal.

The normal middle ear does not approach the theoretical limit more closely than about 20 dB. Nevertheless, there is a broad frequency range

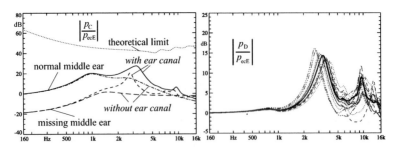

Fig. 3.6. Combined ear canal and middle ear gain, magnitude of pressure transfer function, **left**, ear canal resonances represented as magnitude of the pressure transfer functions from the outer ear canal source to the drum, **right**. Different lines belong to different ear canals

where the middle ear works fairly well indeed, at least from 800 Hz to 4 kHz. At low and high frequencies the gap becomes 40 dB and more. This is unavoidable due to elastic and inertial elements, but perhaps even desirable – at least at low frequencies where a higher sensitivity would increase the risk of an inner ear damage. Around 3 kHz the ear canal provides an essential fraction of the gain. This additional gain stems from the quarter-wavelength resonance. The contribution of the ear canal is depicted in the right panel of Fig. 3.6. Obviously the length of the ear canal is well adapted to the task of broadening the frequency range of high sensitivity towards the high frequency edge. By the way, the resulting frequency range of best sensitivity matches the frequency band of human speech communication.

As can also be seen from Fig. 3.6, the middle ear only provides about 20 dB improvement as compared to the case of a missing middle ear – provided that the round window is carefully protected against sound. At frequencies above 5 kHz the gain even reduces to 0 dB. Nevertheless a 20-dB gain at low and mid frequencies is obviously important enough to justify the fairly sophisticated design of the human middle ear – see Fig. 3.7.

A even better view of the shape of the ossicular chain and of the positions of the relevant ligaments suspending the ossicles can be obtained from Fig. 3.8. The important elastic elements are (a) the eardrum, (b) the posterior incudal ligament, (c) the anterior malleal ligament and (d) the annular ligament around the stapes footplate, which is not depicted in the figure.

The vibrations of the ossicular chain have been thoroughly investigated in [21]. This dissertation comprises the numerical treatment of the middle ear mechanics as well as the corresponding measurements. Most of the data have been measured as complex transfer functions relating three-dimensional

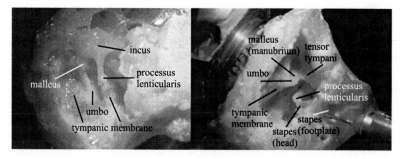

Fig. 3.7. Photos of the middle ear, taken from a temporal bone preparation – tympanic membrane, malleus and incus, **left**, a complete middle ear during a measurement stimulated by an acoustical measuring device, force response at the stapes footplate measured by a piezo-electric sensor, **rigth**. The manubrium of the malleus is firmly attached to the rear side of the drum. The drum is conically retracted. The umbo, i. e. navel, of this cone coincides with the end of the manubrium. The sharp bending of the long process of the incus, forming the processus lenticularis, provides a pusher that acts on the stapes head

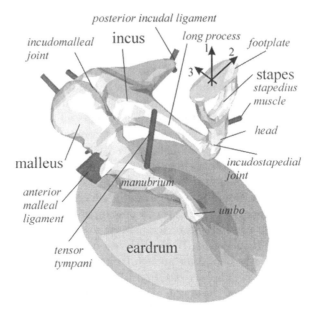

Fig. 3.8. Elements of the middle ear, i.e. eardrum or membrana tympani, the ossicles malleus, incus and stapes, the ligaments and two muscle tendons, musculus stapedius and tensor tympani. The posterior incudal and the anterior malleal ligament are stiff enough such as to establish a main axis of rotation at low frequencies. The axes 1, 2, and 3, see *reference frame*, are used to specify directions of vibration in the text

vibrations at many points on the three ossicles to a stimulating signal. The measurements were performed using complete temporal- bone preparations or isolated parts from it. The transfer functions were recorded under different conditions regarding stimulation, the components involved, and the mechanical loading applied. Taken as a whole, the data provide a broad base that has been used in [21] to match model parameters. The model consists of a finite-element description of the eardrum and a generalized circuit model describing the ossicles together with the joints and suspending ligaments.

The model given in [21] allows for the spatial vibrations occurring at several points on the ossicles to be calculated. Mainly two kinds of stimulation of the middle ear are provided, namely, (a) acoustical stimulation by a sound pressure applied at a certain distance from the drum and (b) mechanical stimulation induced by shaking the temporal bone, i.e., the tympanic cavity walls, in an arbitrary manner. Vibrations as calculated from the model are shown in Fig. 3.9. The lines with grey gradations represent the trajectories of several points on the ossicles during a cycle of vibration. Equal grey tones belong to equal time intervals in a cycle. The numbers in the figures denote effective magnitudes of the displacements in Nanometer at different locations.

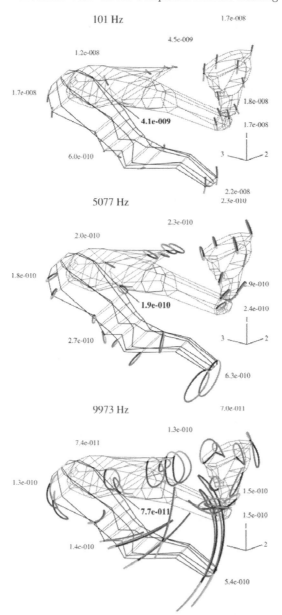

Fig. 3.9. Vibrations of the middle-ear ossicles at different frequencies. The middle ear is stimulated by an acoustic pressure of 1 Pascal, i. e. 94 dB SPL, at the ear drum. The numbers denote effective values of the displacement, in Meter, at different locations. The *bold numbers* belong to the center of gravity of malleus and incus being considered a rigid unit – from [21]

The bold numbers belong to the center of gravity of the malleus and incus considered as a unit.

At low frequencies, i. e. below a few hundred Hertz, the model confirms the old idea of a main axis of rotation which is established by the *anterior malleal* and the *posterior incudal ligament*. This has been clearly demonstrated by a video animation, but can also be recognized in the graphic representation in Fig. 3.9, upper panel. The main axis is given by a line connecting the anterior malleal and the posterior incudal ligament where little translatory motions occur. This is indicated by small trajectories and low values of effective displacements. A pure rotatory motion which occurs at a certain point is not visible in the graphic representation.

The middle panel of Fig. 3.9, showing the situation at about 5 kHz, illustrates that the main-axis ligaments are no longer able to fix to an axis of rotation. The vibrational mode depicted is characterized by a rocking motion. The majority of trajectories become approximately elliptic, i. e. the predominantly translatory motions as occurring at low frequencies are supplemented by strong rotatory ones. The fixation between malleus and incus at the *incudo-malleal* joint becomes a little weakened, but the ossicles remain essentially rigid. [1]

Dramatic changes occur at even higher frequencies. The ossicles, then, become increasingly de-coupled. The mode at about 10 kHz is particularly eye-catching because of the opposite-phase vibrations of the *manubrium* and the long process of the incus. Bending of both processes and the increased motility within the incudo-malleal joint govern this eigen-mode. Also the motion of the stapes is not at all piston-like any more. We observe a strong rotation about the short axis of the footplate, which does not contribute to the perception of hearing.

The irregular motion of the ossicles is certainly a consequence of the restricted ability of mother nature to design more perfect bearings. From an engineer's point of view the ossicular vibrations appears to be relatively ineffective in transmitting acoustic energy to the inner ear. But does this conclusion hold true, or may the imprecise suspension of the ossicular chain actually have some advantage? We shall discuss this question further down.

The middle ear transformer that we are seeking for is best recognized by considering the regular vibration about a fixed axis of rotation at low frequencies. Figure 3.8 shows that the *processus lenticularis* has about the same distance to the rotatory axis as a point of effective force transmission from the drum to the manubrium. Therefore a possible lever effect due to these two processes would have a lever ratio, L, of about unity and, therefore, has no effect on transmission. To be sure, the most relevant effects are provided by the pressure-to-force transduction at the eardrum and the force-to-pressure transduction at the stapes footplate. According to the substantially different

---

[1] Corresponding animations can be found at http://www.ruhr-uni-bochum.de /ika/research/audiology/animations/animations.html

transduction areas, A_D and A_F, the resulting transformer ratio of

$$T = L \times \frac{A_D}{A_F} \qquad (3.5)$$

essentially equals the area ratio.

Although at higher frequencies the regular motion discontinues, the transformer effect of the two areas remains partly effective. Due to the position and the suspension of the stapes, a considerable fraction of the motion remains perpendicular to the footplate. Also the eardrum has a preference to vibrate in the same direction as at low frequencies. However, the peripheral parts of the drum become more and more de-coupled at higher frequencies. Hence the sound-collecting area reduces considerably and, consequently, the effect of the middle ear on sound transmission almost vanishes at high frequencies. This is in agreement with the transfer functions shown in Fig. 3.6, left.

A comprehensive investigation of the middle-ear function must not only focus on maximizing the sound transmission. There are several other aspects to be considered, e. g., the following ones. (i) The middle ear must provide an overload protection as the inner ear contains very vulnerable structures. (ii) The middle ear is subject to static pressure differences. Small pressure changes should not have too much effect on the transmission of sound. This means that the general parameter sensitivity of the middle ear transmission should be kept low. (iii) Further, the middle ear acts as a bone conduction receiver, unavoidably. Consequently, it should be as insensitive as possible with regard to bone conduction.

A very simple form of overload protection is realized in *columella* middle ears of non-mammalian vertebrates. Here the piston driving the inner ear is essentially an almost straight beam-like columella which is coupled to the eardrum via a bony so-called extracolumella. Under overload conditions the elastic connection between columella and extra-columella is altered in a way which decreases the coupling to some extent. In human middle ears the overload protection is mainly implemented in the incudo-malleal joint, which is far more elaborated than the columella/extra-columella design. The normally almost stiff connection between the two ossicles involved breaks down at higher levels. This mechanism allows a much stronger decoupling than a simple columella design could do. The joint has a fairly large articular area with robust bony material on both sides. The necessity of such a robust joint probably explains the shape of the malleus and incus and the surprisingly large masses concentrated near the joint.

It is well known that at high frequencies large masses have a negative effect on sound transmission due to their inertia. However, the effect is kept low if the masses do not move translatorily but rotatorily about an axis through the center of gravity. In fact, mass has the lowest effect when moving rotatorily about the axis of the smallest moment of inertia. Yet, such a consideration can only provide a rough guideline in our context because the rule can only be applied to rigid bodies. However, the ossicular chain must certainly not be treated as a rigid body. Only at low frequencies the malleus and the incus can be assumed to form an approximately rigid body.

In contrast to the *incudo-malleal* joint the *incudo-stapedial* joint is not at all a rigid fixation, even at low frequencies. The joint actually implements a relatively stiff connection in the direction towards the oval window, but it is very compliant for lateral motions and particularly for torques. Hence the relative positions of the long incudal process and the stapes can vary considerably. In the case of sound transmission malleus and incus thus form a kind of pick-up device that transmits the force collected by the eardrum to a rigid but flexibly clamped piston, the stapes. Static forces are suppressed by the joint. The advantages of the elastic coupling between the pick-up and the piston will become clear in the following.

If malleus and incus can be treated as a widely rigid unit, it makes sense to consider the common center of gravity, which is done in all panels of Fig. 3.9 – actually we will see that at higher frequencies the notion of a rigid malleus and incus must be abandoned. The displacements given in Fig. 3.9 show that at 100 Hz the ratio of the displacements at the stapes footplate and the center of gravity is greater than 1.7×10^{-8}m$/4.1 \times 10^{-9}$m. Here the motion of the center of gravity is fairly small. Only the displacement at the anterior hammer ligament is even smaller, due to its stiffness. The main axis of rotation shown in Fig. 3.8 is close to the center of gravity, which means a low effective mass, but slightly shifted towards the stapes. However, it must be kept in mind that at such a low frequency inertial forces are irrelevant.

At 5000 Hz the common center of gravity of the malleus and incus moves almost as strongly as the stapes footplate does. At 10 kHz the displacement of the stapes footplate is even a little lower than that of the center of gravity of the malleus and incus. The point of minimal motion lies considerably closer to the stapes. These findings reveal a trend of shifting the rotatory axis into the direction of the stapes. Yet, at higher frequencies a fixed axis of rotation does no no longer exist at all. Even an instantaneous axis cannot describe the behaviour since, with increasing frequencies, bending of the ossicles and motions in the joints lead to strong deviations from simple rigid-body motions. With increasing frequency the system of the suspended and elastic ossicular chain passes through a series of complex vibrational patterns which come close to the free eigen-modes. This corresponds to vibrations of minimal energy. It is essentially the flexible incudo-stapedial joint that enables vibrations close to the eigen-modes.

As already stated, the resulting vibrations are fairly irregular and depend on individual details. Thus we should expect a high parameter sensitivity in contrast to the one postulated above. Moreover, the just mentioned minimal vibrational energy alone is no proof of an effective transmission. The transmission via the *processus lenticularis* to the stapes head is certainly a preferred path, but also other vibrational patterns not supporting this path of transmission are possible. The appearance of nearby eigen-modes yields large motions, but one link is missing to explain a good transmission in spite of the irregular motion, namely, it is necessary to force the motion into the desired direction of transmission.

It is amazingly simply to explain how the forces are being directed. Due to the almost parallel positions of the eardrum base-plane and the stapes footplate, a stimulation of the ossicular chain by sound via the eardrum produces a force directed perpendicular to the ear drum base and the stapes footplate. Therefore a strong motion in this direction is superimposed upon all the irregular motions. Thus the large motions in "useless" directions, which correspond to the irregular character of the eigen-modes, do not at all indicate ineffective transmission. It is rather that the appearance of almost free vibrations provides high motility and minimal vibrational energy. Under these conditions the transmission of a directed force via a preferred path is particularly effective.

The flexibility of the ossicular chain provides a further benefit which is very important for good functionality under real conditions as follows. The parameter sensitivity is low as the transmission of forces can react flexibly to changes of various kind. Of course, the aforementioned statement only refers to the general trend of transmission, not to the transmission at a certain frequency. Therefore individual frequency responses differ considerably. But changes due to varying static pressures mainly shift the eigen-frequencies. Thus the corresponding frequency responses are also shifted, but not structurally altered. In contrast, a system designed to vibrate in a well-defined simple manner could fail if the conditions were changed to much. For instance, an incudo-stapedial joint designed to move only in a certain way, would tend to be immobilized if the relative positions of incus and stapes changed too much.

Also the problem of realizing small bone-conduction sensitivity is solved by the flexible ossicular chain. Here we consider the effect of vibrations of the tympanic cavity walls on the ossicles as transmitted via the ligaments. The direct impact of skull vibrations to the inner ear will be investigated afterwards. The general irregularity of motion means that the direction of stimulation determines the direction of force transmission. Whereas the normal acoustical stimulation prefers the axis perpendicular to the stapes footplate, an undirected excitation via the skull is much less effective in pushing the stapes to the oval window. But the missing "directivity" is not the whole effect. Even in the case that the tympanic cavity is shaken in a direction perpendicular to the stapes footplate transmission of forces is not very high. This is shown in Fig. 3.10 where acoustical stimulation and stimulation by shaking the tympanic cavity are compared.

At first sight an important difference can be observed in Fig. 3.10. On the left, in the case of a sound pressure stimulating the middle ear, a rocking motion about a time-variant axis near the main axis forces the long process of the incus to push onto the stapes head. On the right the main relative

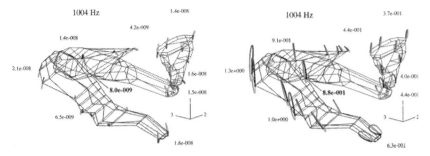

Fig. 3.10. Vibrations of the middle-ear ossicles at about 1000 Hz. On the **left** the ossicles are excited by a pressure at the drum. On the **right** the tympanic cavity is shaken perpendicular to the stapes footplate. As auditory perception follows the difference of the motions of the oval window and the stapes footplate, the relative motion is shown. The numbers denote effective values of the displacement, in Meter, at different locations. Again the *bold numbers* belong to the center of gravity of malleus and incus considered as a rigid unit

motion occurs at the heavy parts of malleus and incus. The vibrations of the tympanic cavity let the *malleus-incus* unit rather circle around the flexible incudo-stapedial joint. Therefore a comparably small force is transmitted to the stapes in this case. In other words, for non-acoustic stimulation the transmission via the processus lenticularis is no longer preferred, among other things due to the high stiffness of the annular ligament in the oval window. Instead a rotation about the incudo-stapedial joint becomes the preferred vibrational mode. This is a nice trick of nature which allows for robust and rather heavy ossicles, but nevertheless avoids large structure-borne forces acting on the inner ear fluids.

We have seen that the "design" of the middle ear makes a virtue of necessity in the following way. The lacking ability to create precise bearings and guideways is converted into a sophisticated usage of flexibility and mass distribution. This is what this author likes to call "biological design principle". This principle does not only provide for a good sound transmission, but simultaneously reduces the sensitivity for bone-conducted sound via the middle ear and for alterations in general.

3.4 Direct Bone Conduction to the Cochlea

As already mentioned in the introduction, acoustically excited structure-borne sound reaches the cochlea at a level which is at least 40 dB lower than that of the air-conducted sound. In most cases the bone conduction is undesired. For example, it limits the attenuation of hearing protectors or it can generate interfering signals when electro-acoustic equipment as earphones or hearing aids are used. But even the effect of bone conduction can be exploited. For example, it can be used diagnostically to discriminate between middle and inner ear damages and it is a prerequisite for bone-conduction hearing aids.

Of course, motions of the skull do not only have an impact on the auditory system via the ossicles, as examined in the preceding paragraph, but also have a direct effect on the cochlea. To estimate this contribution it would be necessary to know the types of vibrational modes which do actually occur in the temporal bone and, particularly, in the cochlea. Unfortunately it is hardly possible to study the details of bone conduction within the skull just by measurements. The only measurement which can be performed without intervention into the system under test is to record vibrations of the skull's surface by optical methods, e. g., [4]. Measurements of this kind reveal that the largest vibrations arise in the temporal bone.

But what does that mean for auditory perception? The latter surely depends on the vibrations of the internal cochlear structures. As of today, it is unclear whether relevant internal vibrations exist at all or whether the bony parts of the cochlea are so stiff as to vibrate as a whole, i. e. like a rigid body. To answer such questions computer models are necessary which simulate what cannot be measured directly. However, modelling the direct path of bone conduction to the inner ear is a difficult task as the skull is fairly intricate in structure and many of its mechanical parameters are not well known.

The modes of vibrations as possibly arising in the cochlea were already discussed in the fifties and sixties of last century. The research on bone conduction in this period came to a preliminary end with [20]. In this paper evidence for internal vibrations to actually occur is reported. Accordingly, the most relevant type of vibration to be expected would to have a torsional character. This means that the shape of the cochlea would change during a vibration cycle. As the perception of sound is the decisive criterion to judge on the relevance of different types of vibrations, the investigations must include at least the basilar membrane and its position within the scalae. According to [20] an asymmetry is necessary to generate an auditory perception, because otherwise the different forces acting on the basilar membrane would cancel out. The asymmetry is established by an unequal volume of the upper canal, as formed by *scala vestibuli* and *scala media*, and the lower canal, which is essentially the *scala tympani*.

Nowadays it is feasible to try a novel approach to study the vibrations, namely, by means of computer simulations. In our laboratory we use a model of a human skull [17] derived from computer-tomographic data as rendered by the "Visible-Human Project" of the National Library of Medicine, Bethesda ML, and from photos of anatomical sections of the inner ear which we have obtained from the ENT clinic of the University Hospital of Zürich. Mechanical data, i.e. density and *Young*'s modulus, were derived from grey shades of tomographic images and from own measurements using a complete skull as well as bony material from the skull, particularly from the inner ear. The

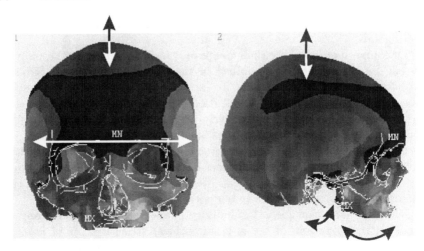

Fig. 3.11. Fundamental mode of skull vibrations. The *grey tones* represent regions of similar translatory-motion magnitudes. The *arrows* indicate the kind of motion – from [19]

geometric and mechanic data were used to construct a finite-element model, which can be analyzed by means of commercially available software.

In Fig. 3.11 the fundamental mode of a human skull – without brain – as computed by the model is depicted. Seen from ahead, the skull performs vibrations changing height and width in anti-phase, roughly keeping the enclosed volume constant. A view from the side shows additional rotatory motions of the upper jaw and the *processus styloideus* including its surrounding, also in anti-phase. At this lowest resonant frequency, which is below 2000 Hz, the temporal bone mainly vibrates translatorily. In addition bending occurs, noticeably at the processus styloideus.

As already mentioned, considering only the vibrations of the skull surface does not provide enough insight into the inner-ear vibrations. To obtain more information about the kind of vibrations it is necessary to consider the relative motions of points in the inner ear. Not the vibration of a single point, but the differences between the motions of different points would reflect changes in the shape of internal structures.

In a first approach [19], the vibrations of the complete vestibu-cochlear organ have been considered – using a frame, x, y, z, as shown in the insert at the bottom of Fig. 3.12. On each axis two points were chosen to ensure that the inner ear was fully enclosed. Negligible internal vibrations have been established at almost constant distances between the two points on an axis. Therefore the axial differences of vibrations are represented in Fig. 3.12. These differences indicate compressional vibrations. As the absolute values of the vibrations and their differences have no meaning here – only modes are considered – the differences have been normalized to the mean translatory

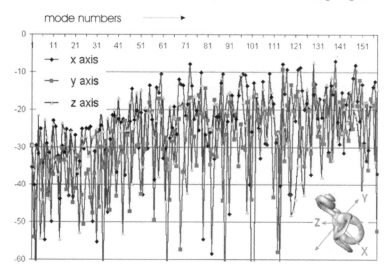

Fig. 3.12. Normalized compressional vibrations of the inner ear at its eigen frequencies – adapted from [19]. Instead of frequencies, mode numbers are given. The mode number 40 roughly corresponds to 5 kHz.

vibration amplitude across all six points considered. For the lower modes the compressional vibration is typically lower by 25 dB or more than the mean translatory vibration. Thus at frequencies up to 5 kHz the excitation of the inner ear fluids by translatory and rotatory motions clearly dominates. In fact, we see no indication of torsional vibrations as postulated in [20].

However, it is certainly not sufficient to only regard the magnitude of eigen-modes when estimating auditory perception. If there were a mechanism that would amplify compressional vibrations much more than translatory ones, the structural vibrations could play an important role in spite of their relative weakness. Therefore, in the future, the impact of inner-ear vibrations on the basilar membrane will be included in our model simulations in a more detailed way.

In a recent study from our laboratory [18] some preliminary results have already been published along these lines of thinking, which include the basilar membrane in a very simplified way. The simulations comprise shaking and compressing the complete inner ear in different directions and, for comparison, the usual acoustical stimulation. All the responses to the different kinds of excitations show the well-known travelling wave with almost equal locations of maximum sensitivity. This is in agreement with the surprising experimental finding that the auditory percept is widely independent of the manner and location of stimulation.

3.5 Conclusions

The peripheral hearing organ has been considered under functional aspects, whereby different parts have been examined at different depths, paying tribute to the authors own generic fields of research.

The human cochlea is primarily a spectrum analyzer. It is implemented as a fluid-filled very rigid capsule which provides a good protection of the vulnerable elements inside. The fluid in combination with the tectorial membrane improves the coupling of a sound wave to the stereociliae as compared to the case of an air-filled chamber. It also provides the prerequsites for realizing a "battery" which enables higher sensitivity, e. g., in comparison to piezo-electric sensors without auxiliary energy. Details of the tuning mechanisms of the cochlea are still discussed controversially but, without doubt, the basilar membrane is an important tuned structure involved. Yet, also the tectorial membrane or even the outer hair cells have been suggested to act as tuned elements, being responsible for the sharp tuning as supplied by the so-called "cochlear amplifier".

The ear canal has a protective function as well, namely to protect the middle ear against external impact. Its acoustic function is that of a transmission line. The quarter-wavelength resonance of the ear canal improves the over-all sensitivity of the ear in a frequency range from 2 kHz to 4 kHz, a band which is particularly relevant for human speech communication. The pressure gain of the ear canal in that frequency range is caused by the quarter-wavelength resonance of a transmission line terminated by a high impedance.

The middle ear has been investigated in larger depth than the other components. The usual interpretation of its function as an impedance transformer is confirmed, but only in a mid-frequency range of 800 Hz–4 kHz. The only relevant mechanism which implements the transformer ratio is the area ratio of the tympanic membrane and the stapes footplate. Only at low frequencies up to about 700 Hz the classical notion of a main axis through the two important ligaments of the malleus and incus can be maintained. The often used "lever ratio" is only applicable below this frequency and is near unity.

The ossicular chain vibrates in an surprisingly irregular manner, varying with frequency. It could be shown that the irregularity does not mean ineffective sound transmission. On the contrary, the irregular motions express an important feature of "biological design". Well-defined bearings and guideways of typical technical design are replaced by a design which is characterized by high flexibility. The most flexible component is the incudo-stapedial joint. The high flexibility ensures a low general parameter sensitivity and a low conduction of structure-borne sound via the middle ear, although sound transmission through the middle ear remains high.

Investigation of direct bone conduction to the cochlea requires a complete model of the human skull. Such a model has been developed. As a preliminary result it has been revealed that the internal structural vibrations of the cochlea seem to play a minor role in the reception of structure-borne sound.

This would mean that mainly translatory motions of the cochlea as a whole, which dominate in the temporal bone at lower frequencies, govern the experimentally found difference of some 40 dB between air and bone conduction.

References

1. v. Békésy G (1960) Experiments in hearing. McGraw-Hill, New York
2. Bell A, Fetcher N H (2001) The cochlear amplifier is a surface acoustic wave: "squirting" waves between rows of outer hair cell? J Acoust Soc Amer 116:1016–1024
3. Brownell W E, Bader C R, Betrand D, de Ribaupierre Y (1985) Evoked mechanical responses of isolated cochlear outer hair cells. Sci 227:194–196
4. Hoyer H E, Dörheide J (1983) A study of human head vibrations using time-averaged holography. J Neurosurg 58:729–733
5. Hudde H, Engel A (1998) Measuring and modelling basic properties of the human middle ear and ear canal. Part I: model structure and measuring techniques. ACUSTICA/acta acustica 84:720–738
6. Hudde H, Engel A (1998) Measuring and modelling basic properties of the human middle ear and ear canal. Part II: ear canal, middle ear cavities, eardrum, and ossicles. ACUSTICA/acta acustica 84:894–913
7. Hudde H, Engel A (1998) Measuring and modelling basic properties of the human middle ear and ear canal. Part III: eardrum impedances, transfer functions, and complete model. ACUSTICA/acta acustica 84:1091–1108
8. Hudspeth A J, Konishi M (2000) Auditory neuroscience, development, transduction and integration: Introduction to a collection of papers from a NAS colloquium. Proc Natl Acad Sci USA 97:11690–11691
9. Letens U (1988) Über die Interpretation von Impedanzmessungen im Gehörgang anhand von Mittelohr-Modellen (Interpretation of impedance measurement in the ear canal by means of middle-ear models). Doct diss Ruhr-Univ Bochum, Bochum
10. Manley G A (1990) Peripheral hearing mechanisms in reptiles and birds. Springer, Berlin
11. Nobili R, Mammano F, Ashmore J (1998) How well do we understand the cochlea? Trends Neurosci 21:159–167
12. Pickles J O (1988) An introduction to the physiology of hearing. Academic Press, London
13. Pörschmann C (2000) Influences of bone conduction and air conduction on the sound of one's own voice. ACUSTICA/Acta acustica 86:1038–1045
14. Purves D (ed) (2001) Neuroscience. Sinauer Ass, Sunderland MA
15. Reiss G, Walkowiak W, Zenner H P, Plinkert K, Lehnhardt E (1989) Das statoakustische Organ. Duphar Pharma, Hannover
16. Shera C A, Guinan jr. J J, Oxenham A J (2002) Revised estimates of human cochlear tuning from otoacoustic and behavioural measurements. Proc Natl Acad Sci USA 99:3318–3323
17. Taschke H, Baierl K, Hudde H (2000) Knochenschallleitung: Ein Modell des menschlichen Schädels zur Bestimmung von Schädelschwingungen (Bone conduction: a model of the human scull for determining scull vibrations). Fortschr. Akust, DAGA 2000, 250–251. Dtsch Ges Akust, Oldenburg,

18. Taschke H, Hülskemper M (2001) Knochenschall - Finite-Elemente-Berechnungen zur Innenohrkomponente (Bone conduction – finite-element calculations regarding the inner-ear component). Fortschr Akust, DAGA'01. Dtsch Ges Akust, Oldenburg. CD-ROM

19. Taschke H, Curdes Y (2002) Knochenschall - Schwingungsformen des menschlichen Schädels (Bone conduction – vibrbational modes of the human scull9. Fortschr Akust, DAGA'02, 62–63. Dtsch Ges Akust, Oldenburg,

20. Tonndorf J (1968) A new concept of bone conduction. Arch Otolaryng 87:49–54.

21. Weistenhöfer Ch (2002) Funktionale Analyse des menschlichen Mittelohres durch dreidimensionale Messung und Modellierung (Functional analysis of the human middle ear by menas of three-dimensional measurement and modelling). Doct diss, Ruhr-Univ Bochum, GCA-Verlag, Herdecke

22. Zenner H P, Zimmermann U, Gitter A H (1987) Fast motility of isolated mammalian auditory sensory cells. Biochem Biophys Res Commun 149:304–308

4 Modelling of Binaural Hearing

Jonas Braasch

CIRMMT, Department of Music Theory, McGill University, Montréal

Summary. In many everyday listening situations, humans benefit from having two ears. For more than a century, research has been conducted to understand which acoustic cues are resolved by the auditory system to localize sounds and to separate concurrent sounds. Since *Jeffress* proposed the first lateralization model in 1948, binaural models have become increasingly popular to aid in understanding the auditory system and to solve engineering tasks related to the localization and detection of acoustic signals. In the following chapter, a number of binaural models will be described – starting from the classical coincidence model to recent approaches which simulate human localization in three dimensions. The equalization-cancellation model will be also addressed, as a classical example to predict binaural detection experiments.

4.1 Introduction

The term "binaural hearing" refers to the mode of functioning of the auditory system of humans or animals in the context of tasks where the system benefits from having two ears. Such tasks are typically related to auditory localization, detection, or recognition. In the latter case, the task is not simply to detect a target sound within noise, but to recognize certain attributes of the target, e. g., to recognize spoken language or to analyze music.

To simulate binaural hearing, several models exist. Throughout literature, they are usually divided into physiologically- or psychologically-oriented approaches. While the first type of models aims at simulating the behaviour of neuronal cells in detail, the latter works on a more abstract phenomenological basis. A strict distinction between both types of models, however, does not exist. On the one hand, the number and accuracy of the simulated cells in physiological models is limited by the computational power and knowledge in this field. On the other hand, psychologically-based models have to simulate the auditory pathway accurately enough in order to predict a number of psycho-acoustical phenomena. This chapter mainly addresses psychologically-based models – with the main focus being put on localization models. A more general overview on binaural models, including physiological ones, can be found in [25]. To establish a binaural model, three prerequisites are typically required.

- The spatial cues which are analyzed by the auditory system have to be known – see Sect. 4.2.
- An algorithm for simulating the auditory analysis of at least one spatial cue has to be developed – see Sect. 4.3.
- A mechanism that estimates the sound-source positions from the results of the spatial-cue analysis has to be established – see Sect. 4.4.

In the following, after introducing various approaches to implementing localization models, detection models are addressed with a focus on models that are based on the "equalization-cancellation" algorithm, EC – see Sect. 4.5. In fact, a complete model, i. e. a model that is able to describe all types of binaural phenomena, does not exist until now. Yet, a number of promising approaches are available which focus on particular kinds of complex hearing tasks, e. g., on sound localization in the presence of concurring sound sources. In the last section, some examples for such approaches will be described – see Sect. 4.6.

4.2 Analysis of Binaural Cues by Humans

For many centuries, scientists have been trying to understand how the human auditory system is able to localize sound sources in space. As early as 1882, a review of different theories with respect to this topic has been written, listing what can be regarded the fundamental laws of binaural hearing up to the present day [98]. According to that review, the sensitivity of the auditory system to inter-aural phase differences had been found in 1877 by two researches independently from each other [97]. Their findings led to a theory according to which the arrival times of the sound wave emitted from a single source are usually not exactly the same at the left and right eardrums – due to the different path-lengths to both ears. This arrival-time difference between the left and right ear is called "inter-aural time difference", ITD. With a simple geometric model [50], see Fig. 4.1, it can be shown that the maximal ITD is measured when the sound wave arrives from the side along the axis which intersects both eardrums. In this case, the ITD can be estimated as the distance between the eardrums, ≈ 18 cm, divided by the speed of sound, ≈ 340 m/s, to a value of 529 μs.[1] However, larger ITDs than those are observed in nature. Because of shadowing effects of the head, the measured ITDs can be, depending on the head size, as large as 800 μs. A model which estimates the ITDs on the basis of the wave travelling around a sphere has been proposed in [100], which is still a good prediction in the high-frequency range. Later this model was modified to predict the ITD for all frequencies throughout the human hearing range [63].

---

[1] *Hornbostel & Wertheimer* estimated the distance between the two eardrums to 21 cm. This value, however, is too large, and nowadays it is common to use 18 cm instead

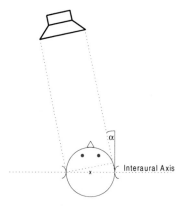

Fig. 4.1. Simple geometric model to estimate inter-aural time differences [50]

The existence of the head between both ears does not only determine the detour the travelling sound wave has to follow, but also causes attenuation of the sound wave at the contra-lateral eardrum, which leads to "inter-aural level differences", ILD[2] of both ear signals. Already in the end of the 19th century, a geometric model was established to estimate ILDs for various sound-source positions [92]. In contrast to the ITDs, the ILDs are strongly frequency dependent. In the low-frequency range, the human head is small in comparison to the wave length and, therefore, diffraction has only a minor effect on the sound wave. In the high-frequency range, however, the wave length is short as compared to the dimensions of the head, and much larger ILDs than in the low-frequency range can be observed. In this frequency region, the ILDs are not only determined by the shape of the head, but are also greatly influenced by the shape of the outer ears.

The frequency dependence of the ILDs led to the idea of the duplex theory, which claimes that ITDs are the dominant cue in the low-frequency range, while ILDs are more important than ITDs at high frequencies. *Lord Rayleigh* [87] showed both theoretically and in a psycho-acoustic experiment that the head is very effective in attenuating the sound at the contra-lateral ear for high, but not for low frequencies. For this reason, ILDs are considered too small in the low-frequency range to provide a reliable localization cue. For ITDs, it was concluded that the unequivocal relationship between the phase difference between both ear signals and auditory lateralization vanishes for high frequencies, where the path-length difference between both ears exceeds the wave length of signals stemming from a sideway sound source. Indeed, it was later shown for sinusoidal test signals [75] that our auditory system is not able to resolve the fine structure of signal frequencies above approximately 1.5 kHz.

---

[2] Note that inter-aural level differences, ILDs, are frequently referred to as inter-aural intensity differences, IIDs

Nowadays, however, the duplex theory is not seen in such a strict way anymore as it was originally proposed. It is now being realized that ITDs are important in the high-frequency range as well, as they can be evaluated through envelope fluctuations [27]. The analysis of the envelope instead of the fine structure of the signals helps to avoid phase ambiguities in the ITD/lateralization relationship. Furthermore, it has recently been revealed in detection experiments that our auditory system is principally equally sensitive to ITDs throughout the whole frequency range when so-called "transposed" signals are employed at high frequencies [81]. In this study, a transposed signal was created by half-wave rectifying a low-frequency sinusoidal tone, e. g., 125 Hz, and multiplying it with a sinusoidal tone of higher frequency, e. g., 2000 Hz. Assuming that the auditory system is analyzing the envelope of the composed signal to estimate ITDs, the actually-analyzed signal is more similar to low-frequency signals than to amplitude-modulated signals. It was further shown in lateralization experiments [68] that the influence of ITDs in the high-frequency range on the displacement of the auditory events could be increased by amplitude-modulating the stimuli.

The question of how ITDs and ILDs are combined in the auditory system to estimate the position of the sound source has been only partly answered so far. For a long period of time, it was assumed that ITDs and ILDs are evaluated separately in the auditory system, namely, ITDs in the *medial superior olive* MSO, as was first shown for dogs [38], and ILDs in the *lateral superior olive*, LSO. Yet, recent neuro-physiological findings have shown that the ITDs in the envelopes of modulated sounds, and even in low-frequency carriers, are also processed in the LSO [58,59]. In addition, the occurrence of the so-called "time-intensity-trading" effect revealed a high complexity of the processing in the auditory system already half a century ago [27,46]. The trading effect describes the phenomenon that the auditory event often evolves midway between the positions with ITD and ILD cues leading into opposite direction and, such, compensate for each other. It has been suggested [58] that the combined sensitivity of single neurons in the LSO to ITDs and ILDs offers an easy explanation for the time-intensity trading effect. However, the auditory events of the listeners become spatially diffuse or they even split up into more than one auditory event when the ITDs and ILDs differ too much from the "natural" combinations of ITDs and ILDs as observed in free-field listening – e. g., [37].

Before describing the decoding process which the auditory system performs when transforming the two one-dimensional signals at the eardrums back into a three-dimensional space representation, some notes should be made on the head-related coordinate system, Fig. 4.2, which is being used throughout this chapter. In this coordinate system [6], the inter-aural axis intersects the upper margins of the entrances to left and right ear canals. The origin of the coordinate system is positioned on the inter-aural axis, halfway between the entrances to the ear canals. The horizontal plane is defined by the

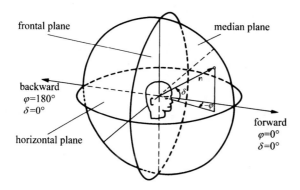

Fig. 4.2. Head-related coordinate system [6]

inter-aural axis and the lower margins of the eye sockets, while the frontal plane lies orthogonally on the horizontal plane intersecting the inter-aural axis. The median plane is orthogonal to the horizontal plane, as well as to the frontal plane and therefore, so to say, cuts the head in two symmetrical halves.[3] The position of a sound source is described using the three polar coordinates, azimuth, φ, elevation, δ, and distance, d. If δ is zero and d is positive, the sound source moves anti-clockwise through the horizontal plane with increasing φ. At $\varphi=0°$ and $\delta=0°$, the sound source is directly in front of the listener, intersecting the horizontal and median planes. If φ is zero and d is positive, the sound source moves in front of the listener with increasing δ, upwards along the median plane and downwards behind the listener. Taking the early model from [50], ITDs of the same magnitude form hyperbolas in the horizontal plane, and at greater distances the shell of a cone in three-dimensional space is apparent, the so-called "cones of confusion". Hence, there exist multiple positions with identical ITDs – yet, despite of this, the ITDs are still very reliable cues to determine the left–right lateralization of a sound source.

In a detection experiment with sinusoidal tones being masked by noise, it was revealed that the auditory system analyzes sounds in overlapping frequency bands [36]. The widths of these frequency bands always correspond to a constant distance of approximately 2 mm on the basilar membrane. The relationship of frequency and distance on the basilar membrane between the place of maximum deflection and the helicotrema is approximately logarithmic. Therefore, the frequency bands become broader with the frequency. Meanwhile it is known that inter-aural cues are analyzed in such frequency bands as well. This fact is especially important for ILD evaluation, because of their strong frequency dependence. By evaluating the ILDs and ITDs across

---

[3] It should be noted that the head is not perfectly symmetrical. For this reason slight inter-aural time and level differences are also measured for sound sources in the median plane

several frequency bands, an unequivocal sound-source position can be easily determined for most directions.

The median plane plays a special role in binaural psycho-acoustics, as the inter-aural cues are very small here and cannot reliably be used to resolve positions. In this case, other cues become important [2]. It was found that, for different elevation angles, the spectrum of the signals to the two ears is characteristically boosted or attenuated in different frequency bands, due to diffraction and scattering at the head and outer ears. It is assumed that the auditory system performs a spectral analysis to determine the position within the median plane. This fits to the finding that, for sinusoidal and other narrow-band sound sources, the positions of the auditory events are formed in positions for which the signals to the ears show local maxima for broad-band sound sources, so called "directional bands". These spectral cues are called "monaural cues", since one ear only is sufficient to resolve them.

In many realistic situations we have to deal with more than one sound source, e. g., as the desired source may be partly masked by concurrent sound sources. In general, there are two kinds of concurrent sources, namely, coherent and incoherent ones – although any degree in between is possible as well. Coherent sources are, e. g., generated by reflections of the desired primary source on obstacles, such as, walls, ceiling, furniture. Incoherent sound sources are emitted independently from the first sound source, such as traffic noise, people talking in the background. The effect of coherent, reflected sound sources has been investigated thoroughly – see [6, 67] for detailed reviews. Usually, the first reflections are observed in nature within a few milliseconds after the primary source arrives. If the delay between the primary source and a single reflection is less than 1 ms, only one auditory event is perceived, which is usually located in-between both sound sources, an effect which is known as "summing localization". However, for delays above 1 ms, the auditory event is located at the position of the primary source, independently from the position of the reflection. This phenomenon is called the "precedence effect". The precedence effect enables humans to localize the primary sound source in a reverberant environment. When the delay exceeds the "echo threshold" of about 4.5 to 80 ms, the effective value depending on the kind of signal, then a reflected source is perceived separately as a repeated auditory event, the "echo".

4.3 Computational Analysis of Binaural Cues

4.3.1 Simulation of the Peripheral Auditory System

Before binaural cues can be adequately analyzed, the signals have to be processed by a stage that simulates the auditory periphery, namely, outer, middle, and inner ear. The general structure of a localization model with all the

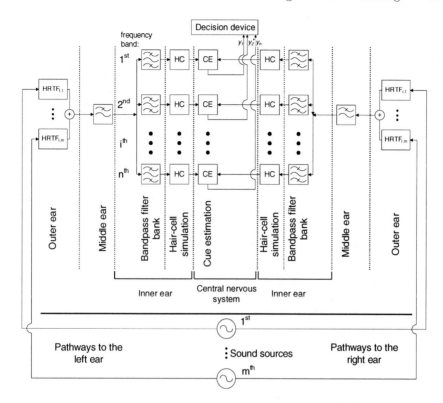

Fig. 4.3. Typical architecture of a model for human-sound-localization simulation

necessary stages is shown in Fig. 4.3. The outer ears are often simulated by filtering the signals with the "head-related transfer functions", HRTFs [45,78], corresponding to a specific sound-source position in space. Alternatively, signals that were recorded with a dummy head can be processed. Many binaural models were designed to simulate experiments based on stimulus presentation through headphones. In these cases, the influence of the outer ears can be simply ignored and the two headphone signals can be sent directly to the left and right channels of the next stage. The middle ear can be efficiently simulated by a sloppy band-pass filter [51], but most binaural models do not explicitly simulate this stage, because its existence is not required to show many common effects that are related to binaural hearing.

The analysis of sound in separate frequency bands is inevitable, unless narrow-band signals are used that are narrower than the auditory filter at the corresponding center frequency. Based on the concept of *Fletcher*'s auditory frequency bands, *Zwicker & Feldkeller* measured the effective width of the critical bands in psycho-acoustic masking experiments for 24 frequency bands from 50 to 13500 Hz. Below 500 Hz, the filters are approximately 100-

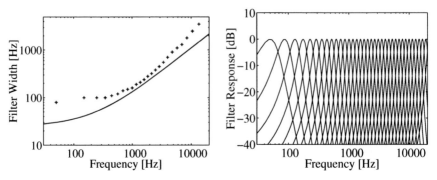

Fig. 4.4. In the **left** panel, the dependence of the filter band width on the center frequency is shown for a band-pass-filter bank after *Zwicker & Feldtkeller*, *+signs*, and the GAMMATONE-filter bank, *solid line*. In the **right** panel, the magnitude responses of a GAMMATONE-filter-bank implementation with 36 bands are depicted

Hz wide, above that frequency the bandwidth widens almost logarithmically with frequency – see Fig. 4.4, left panel. The idea of critical bands was later related to inner-ear filters [82]. Nowadays, often a GAMMATONE-filter-bank implementation is used [83]. The particular implementation of a GAMMA-TONE filter-bank as used throughout this chapter, is shown in Fig. 4.4, right panel. It consists of 36 bands per channel. The dependence of bandwidth on center frequency is slightly different from [105], due to differences in the psycho-acoustic measurement method – compare Fig. 4.4, right panel.

It should be noted that both band-pass-filter banks were derived using monaural signals. In [61] the question of whether the filter shapes have to be chosen differently for binaural tasks than for monaural ones is dealt with. The author concludes from his experiments that both monaural and bin-aural filters can be treated as equally wide. Furthermore, he points out that some previous investigations came to a different conclusion only because they measured different psycho-acoustic features for the monaural and binaural conditions.

The simulation of many binaural tasks requires the modelling of the hair-cell behavior. For numerous applications, it is sufficient to include a half-wave rectifier, simulating that the hair-cell behaviour basically codes only one of the two half-waves of the signal [84], further, a low-pass filter with a cut-off frequency of about 1000 Hz. The latter has to be implemented to take into account of the fact the hair cells do not resolve the fine structure of the signal at higher frequencies. Other binaural tasks, especially detection tasks, require a more detailed simulation of the hair-cell functioning, namely, modelling of the stochastic firing rate of the nerve fibers [30, 72] and the adaptive, time-variant hair-cell response [19–21, 26].

It is noteworthy that, even though all elements of the auditory periphery, namely, outer ear, middle ear, and basilar membrane, have been discussed separately here, their functions cannot be strictly separated in the model algorithms. Generally spoken, in those algorithms that were derived with psycho-acoustic methods only the overall response of the auditory system to the ear signals, as measured through the perception of the subject, is available as a basis for modelling. For instance, adaptation stages, e. g., [19, 20, 26], do not only simulate the non-linear behavior of the hair cells, but include as well non-linear effects of the central nervous system, of the basilar membrane, and in certain cases even of the middle ear.

Physiological methods allow measurements almost anywhere in the auditory pathway, e. g., electro-physiological recording in single hair cells. Uncertainties remain because only the response of a single cell or an ensemble of few cells can be measured, and from these results assumptions about the overall function of the whole cell complex are derived. Inner hair-cell models, e. g., [72], were established on the basis of single-cell-recording data and are able to simulate the response for different types of cells to the ear signals. Despite the models being originally designed to simulate the behaviour of single cells, they have frequently been implemented successfully into psycho-acoustic models where they simulate a complete hair-cell population – e. g., [49, 91].

After the simulation of the hair cells, the binaural parameters are extracted. For the analysis of inter-aural cues, both inputs for the left and right ears are necessary, as shown in Fig. 4.3, whereas the analysis of monaural cues only requires the input for one ear. Thus, the monaural cues can be separately determined for the left and the right channels. The next sections are dedicated to computational methods to analyze binaural cues.

4.3.2 Inter-Aural Time Differences

The first physiology-related algorithm to estimate ITDs was proposed back in 1948 by *L. A. Jeffress* [54]. The model consists of two delay lines that are connected by several coincidence detectors as shown in Fig. 4.5. A signal arriving at the left ear, $y_l(m)$, with m being the index for time, has to pass the first delay line, $l(m, n)$, from left to right. The variable n is the index for the coincidence detectors at different internal delays. A signal arriving at the right ear, $y_r(m)$, travels on the other delay line, $r(m, n)$, in the opposite direction. The discrete implementation of the delay lines can be described as follows,

$$l(m + 1, n + 1) = l(m, n) \ \ldots \ 1 \leq n < N \wedge l(m, 1) = y_{l(m)}, \quad (4.1)$$
$$r(m + 1, n - 1) = r(m, n) \ \ldots \ 1 < n \leq N \wedge r(m, N) = y_{r(m)}, \quad (4.2)$$

with N being the number of implemented coincidence cells. The time, t, and the internal delay, τ, can be easily estimated from the indices, m and n, and the sampling frequency, f_s, as follows: $t = (m - 1)/f_s$ and $\tau = (n - (N + 1)/2)/f_s$. A coincidence detector, $c(m, n)$, is activated when it receives simultaneous inputs from both delay lines at the positions that it is connected

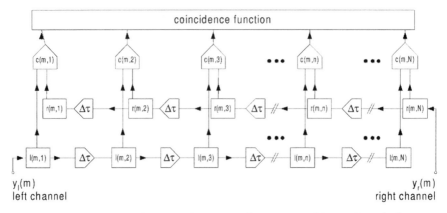

Fig. 4.5. Coincidence mechanism as first proposed by *Jeffress* [54]

to. Each of the coincidence detectors is adjusted to a different ITD, due to the limited velocity of propagation of the signals on the delay line. For example, a sound source located in the left hemisphere will arrive at the left ear first, and therefore the signal will travel a greater distance on the delay line than the signal for the right ear before both of them activate the coincidence detector for the corresponding ITD.

Actually, *Jeffress* himself has never specified explicitly how two spikes would coincide. However, it was pointed out later [93] that the coincidence model can be considered to be an estimator for the inter-aural cross-correlation function, ICC. To this end it was assumed that many parallel coincidence detector cells exist which are tuned to the same ITD. Then, the probability that two spikes from two opposite channels will activate a specific coincidence cell is given by the product of the number of spikes in those left and right channels, the inter-aural delay of which matches the internal delay of the coincidence cell. This product also appears in the running cross-correlation function, which is defined for a discrete system as follows,[4]

$$\Psi_{y_{l,r}}(m,n) = \frac{1}{\Delta m} \sum_{m'=m}^{m+\Delta m} c(m',n) = \frac{1}{\Delta m} \sum_{m'=m}^{m+\Delta m} l(m',n)r(m',n), \quad (4.3)$$

with $c(m,n) = l(m,n)r(m,n)$ and the assumption that the amplitudes in the left and right channels are proportional to the number of spikes. In (4.3) a rectangular window of the length Δm was chosen within which the cross-correlation function for each time interval is calculated. Often other window shapes, e. g., *Hanning* window, triangular window, exponential window, are used. The duration of the time window, $\Delta t = \Delta m / f_s$, can be determined in

---

[4] Note that $l(m',n)$ and $r(m',n)$ have to be determined recursively from (4.1) and (4.2). The classical continuous form of the running cross-correlation function is
$$\Psi_{y_{l,r}}(t,\tau) = \int_{t'=t}^{t+\Delta t} y_l(t'-\tau/2) \cdot y_r(t'+\tau/2) \, dt'$$

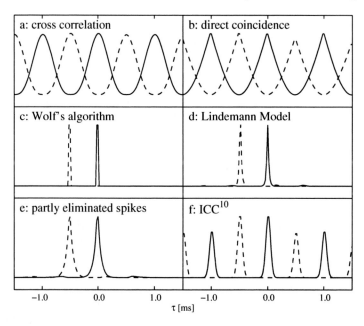

Fig. 4.6. Examples for the outputs of different coincidence detectors for a 1-kHz sinusoidal signal. (**a**) cross-correlation algorithm, (**b**) direct coincidence algorithm, spikes always interact when passing each other, (**c**) *Wolf*'s algorithm, (**d**) *Lindemann*'s algorithm, (**e**) coincidence algorithm with partly eliminated spikes after coincidence, (**f**) cross-correlation algorithm taken to the power of 10 – 0-ms ITD, *solid lines*, 0.5-ms ITD, *dashed lines*

psycho-acoustical experiments measuring binaural sluggishness. Values come out to be on the order of tenths to hundreds of milliseconds depending on the listener and measurement method [39–42,62]. *Sayers & Cherry* [90] used the inter-aural cross-correlation, ICC, to determine the ITDs, and in 1978 a computational lateralization model based on ICC and the simulation of the auditory periphery was introduced independently in [4,93]. Besides the common cross-correlation function, there are alternative ways to implement the coincidence detectors. Figure 4.6 shows the results of different implementations, namely, the inter-aural cross-correlation functions of two 1-kHz sinusoidal tones with ITDs of 0 ms and 0.5 ms are depicted. In contrast to the models reported in [4,93], another study [99] assumed that two spikes from opposite channels would always coincide when they pass by each other on the delay lines. In this case the output of the coincidence function is not the product of the amplitudes in the left and right channel for each delay time, but rather the minimum of those two amplitudes. The signal amplitude, then, correlates with the number of spikes within the time interval of Δt as follows,

$$c_\mathrm{d}(m, n) = \min\left[l(m, n),\ r(m, n)\right]. \tag{4.4}$$

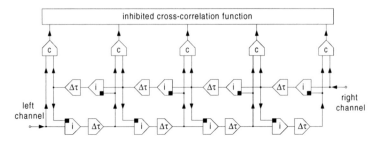

Fig. 4.7. Structure of the *Lindemann* algorithm

The output characteristics of this algorithm, Fig. 4.6 b, is quite similar to the output characteristics of the cross-correlation algorithm, with the exception that the peaks are slightly narrower at the top. The author [99] further assumed that two spikes would be cancelled out after they coincide. This approach, however, is very sensitive to inter-aural level differences and, thus, the signals in the left and right channels have to be compressed in amplitude beforehand. For this reason, a hair-cell model taken from [30] was employed to transform the signal-level code into a rate code. The output of this algorithm is shown in Fig. 4.6 c. In contrast to the cross-correlation algorithm, the peaks are very narrow and the side peaks have vanished.

A similar effect was achieved a few years earlier by introducing contra-lateral inhibition elements into the model [65,66]. The implementation of the inhibition elements is achieved by modifying the computation of the delay lines from (4.1), (4.2) to

$$l(m+1, n+1) = l(m,n)[1 - c_s \cdot r(m,n)] \ \ldots \ 0 \le l(m,n) < 1, \quad (4.5)$$
$$r(m+1, n-1) = r(m,n)[1 - c_s \cdot l(m,n)] \ \ldots \ 0 \le r(m,n) < 1, \quad (4.6)$$

with c_s being the static inhibition constant, $0 \le c_s < 1$. Now the signals in both delay lines inhibit each other before they meet and reduce the amplitude of the signal in the opposite channel at the corresponding delay unit. As can be seen in Fig. 4.6 d, the side peaks that are found in the plain cross-correlation algorithm, Fig. 4.6 a, are eliminated in this way. *Wolf*'s algorithm becomes more similar to *Lindemann*'s, if only a certain percentage of the spikes is cancelled – Fig. 4.6 e. In this case, it is not even necessary to use a probabilistic hair-cell model. If only a smaller amount of the spikes is eliminated, there is, qualitatively spoken, only a little difference in whether the spikes are inhibited/cancelled before or after they coincide. It should be noted that the outcome of these inhibitory algorithms is dependent on the ILDs.

In the simulation of binaural hearing, it is sometimes advantageous to reduce the peak widths of the cross-correlation curves when determining the position of the peak. Besides employing an inhibition stage, a peak reduction can be achieved by taking the signal to a power greater than one. Figure 4.6 f shows this procedure for the power of 10. However, by this approach the side peaks are hardly reduced.

When using the cross-correlation algorithm, not only the position of the cross-correlation peak, but also its normalized height – the so-called inter-aural cross-correlation coefficient, IACC, or inter-aural coherence, IC, can be used to gain information about the auditory spatial properties of the environment, e. g., an enclosed space. It can be determined by taking the maximum of the normalized cross-correlation function,

$$\Psi_{y_{l,r}}(t, \tau) = \frac{\int\limits_{t'=t}^{t+\Delta t} y_l(t - \tau/2) \cdot y_r(t + \tau/2)\, dt}{\sqrt{\int\limits_{t'=t}^{t+\Delta t} y_l(t)\, dt \cdot \int\limits_{t'=t}^{t+\Delta t} y_r(t)\, dt}}. \tag{4.7}$$

The cross-correlation coefficient correlates strongly with "spaciousness" or "auditory source width", i. e. psycho-acoustic measures for the spatial extent of auditory events – and important indicators for room acousticians [5, 8, 79]. Spaciousness decreases with an increasing correlation coefficient and, therefore, with the height of the cross-correlation peak.

In contrast to *Jeffress'* predictions, recent findings in mammals showed that most cells in the medial superior olive, MSO, are tuned to inter-aural phase differences of 45° [13,69]. It has therefore been suggested [70] that the ITDs are not determined by the activity peak in the corresponding cells of a topographical cell array, but rather by the ratio between two cell populations tuned to inter-aural phase shifts of −45° and 45°. So far, it has not yet been proven that such an alternative sound-localization model is a better description of how ITDs are analyzed in the auditory system of mammals. For birds, namely the barn owl, mechanism similar to the *Jeffress* model have been found in electro-physiological experiments. However, mammals and the barn owl show great differences in processing ITDs. For example, barn owls are able to resolve the fine structure of the signals up to 9 kHz when encoding ITDs [60].

4.3.3 Inter-Aural Level Differences

Inter-aural level differences, α, are usually computed directly, from the ratio of the signal powers, $P_{l,r}$, in both the left and the right channels, after the simulation of the auditory periphery. They are calculated independently for each frequency band, k, as a function of time, m,

$$\alpha_k(m) = 10 \log_{10} \left[\sum_{m=m'}^{m'+\Delta m} P_{k,l}(m') \right] - 10 \log_{10} \left[\sum_{m=m'}^{m'+\Delta m} P_{k,r}(m') \right]. \tag{4.8}$$

The equation describes an excitation-inhibition, EI, process, because the input of the left channel has a positive sign "excitation", while the sign for the

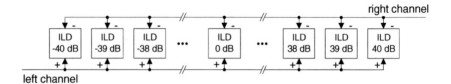

Fig. 4.8. Model structure using EI cells

input of the right channel is negative, "inhibition". The terms excitation and inhibition refer to the response of a neuronal cell. An excitatory cell input increases the firing rate of the cell, whereas an inhibitory cell input reduces it. For this reason, the cross-correlation algorithm is an excitation-excitation, EE, process, as the inputs of both the left and the right channels have a positive sign. Physiological-related algorithms exist also for the ILD analysis. For example, in [88] a cell population of EI cells is simulated. In this model, each cell is tuned to a different ILD as shown in Fig. 4.8. One possible way to simulate the activity of the EI cells, $E(k, m, \alpha)$, at a given ILD, α, is as follows,

$$E_k(m, \alpha) = \exp\left[\left(10^{\alpha/\mathrm{ILD_{max}}} \sqrt{P_{k,l}(m)} - 10^{-\alpha/\mathrm{ILD_{max}}} \sqrt{P_{k,r}(m)}\right)^2\right], \quad (4.9)$$

with $P_{k,l}(m)$, $P_{k,r}(m)$ being the power in the left and right channels and k referring to the kth frequency band. $\mathrm{ILD_{max}}$ is the maximal ILD magnitude that the cells are tuned to. Another model solution was proposed in [14] where the aim was to simulate binaural detection experiments. The EI cells were tuned to both ILDs and ITDs, resulting in a two-dimensional grid of cells with all the possible combinations of different ITDs and ILDs. More algorithms, which use combined analysis of ITDs and ILDs will be introduced in Sect. 4.4 – see, e. g., [65, 93].

4.3.4 Monaural Cues

A model proposed in 1969/1970 [3] analyzes monaural cues in the median plane as follows. The powers in all frequency bands are compared to each other and a logical decision is made on whether the signal was presented from the front, from above or from behind. It has been shown that models based on spectral-cue evaluation can be improved by observing the first and second derivative of the spectrum in frequency [102]. To this end these differences were measured by observing the powers in a various frequency bands. An internal representation becomes obsolete, given that the signals have a rather smooth frequency spectrum without strong frequency dips. The latter are often caused by room reflections, i. e. the so-called comb-filter effect. As psycho-acoustical investigations showed, the perceptual effects of these dips can be partly eliminated by the auditory system, an effect which is called binaural decolouration – see [18, 89, 103].

In contrast to models which do not require a reference spectrum, it is often assumed that listeners use an internal representation of the sound to which the ear signals are compared to in order to estimate the monaural cues. A database with the internal representations of a high number of common sounds has not been implemented in monaural model algorithms so far. Some models exist, however, that use an internal representation of a single reference sound, e.g., for click trains [53] and for broad-band noise [47]. As to the analysis of inter-aural cues, available model algorithms usually achieve results which are similar to the performance of the auditory system itself. Yet, for monaural cues, the localization performance of human subjects is generally much better than the performance of the present models.

4.4 Decision Process

Inter-aural cues cannot be determined directly by means of psycho-acoustical methods. Instead, perceptual parameters like the perceived lateralization are measured, as these correlate well with inter-aural cues. Binaural models usually include a decision stage to predict perceptual parameters from analyzing inter-aural cues. In a localization model, the perceptual parameter of interest would be the position of the auditory event. When modelling binaural hearing, simulation of human performance and not the best technically-possible performance is aimed at. For this reason, the performance of the algorithms is sometimes degraded, for example by the implementation of internal noise, in order to explicitly consider the limitations of the human auditory system. In localization tasks, the position of the auditory event often deviates in a characteristic way from the presented direction. Even though this effect could be often avoided in computational algorithms, simulating those deviations is sometimes of great interest, because they can help reveal how the auditory system works.

4.4.1 Lateralization Models

Many models have been established to predict the perceived left/right lateralization of a sound which is presented to a listener through headphones with an ITD or an ILD, or both. Usually, those sounds are perceived inside the head on the inter-aural axis with a distance from the center of the head. This distance, the so-called lateralization, is usually measured on an interval or ratio scale. The simplest implementation of a decision device is to correlate the perceived lateralization with the estimated value of a single cue, e.g., the position of the cross-correlation peak in one frequency band. This is possible, because the laterality, the perceived lateral position, is often found to be nearly proportional to the value of the analyzed cue.

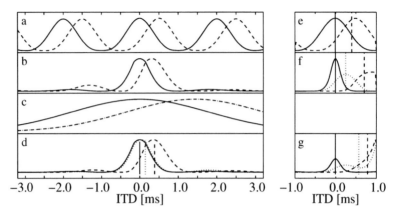

Fig. 4.9. Results for the position-variable model [93], **left** panels, and the contra-lateral-inhibition model [65], **right** panels, for a 500-Hz sinusoidal signal with different combinations of ITD and ILD, namely, 0 ms/0 dB, *solid line*, 0 ms/15 dB, *dotted line*, and 0.5 ms/15 dB, *dashed line*

The decision device has to be more complex if different cues are to be analyzed or if one cue under consideration is observed in different frequency bands. A very early model that integrates information from different cues, ITDs and ILDs, is the position-variable model of Stern and Colburn [93]. The authors were able to predict the perceived lateralization for a 500-Hz sinusoidal tone for all combinations of ITDs and ILDs. The left panels of Fig. 4.9 show the results of the position-variable model for three different combinations of ITDs and ILDs. The ITDs are measured using the cross-correlation algorithm of Fig. 4.9 a. Afterwards, the cross-correlation curve is multiplied by a delay-weighting function in order to enhance the output for small ITDs, see Fig. 4.9 b. This is done to take into consideration that the number of neural fibers which are tuned to small ITDs is higher than those tuned to large ITDs. The delay-weighting function [22] is shown in Fig. 4.10, left panel, dotted curve. The influence of the ILDs is represented in a 2nd function of *Gaussian* shape and constant width of 1778 μs, as depicted in Fig. 4.9 c. The peak position of this second weighting is varied with the ILD of the signal. For this purpose, the signal's ILD, α, is calculated according to (4.8) and transferred into a corresponding ITD, τ, using a function which can be derived from psycho-acoustical data as follows,

$$\tau = 0.1\,\alpha - 3.5 \cdot 10^{-5}\alpha^3 \ [\text{ms}]. \tag{4.10}$$

Finally, this function is multiplied with the weighted cross-correlation function from Fig. 4.9 b, and the centroid of the resulting function correlates with the perceived lateralization, Fig. 4.9 d.

A second approach to combine the influences of ILDs and ITDs in order to predict the perceived lateralization was accomplished in [65]. The basic algorithm from this study was already introduced in Sect. 4.3.2. Figure 4.9 e shows the cross-correlation algorithm without static inhibition. The results change and, most importantly, become dependent on the ILDs when the inhibitory stages are included, see Fig. 4.9 f. Since the displacement, which was caused by the occurring ILD, was too small, additional monaural processors where introduced to the model, Fig. 4.9 g. The author did not estimate the simple cross-correlation product, $c(n, m) = l(n, m) \cdot r(n, m)$ with static inhibition anymore, but rather

$$c(n, m) = [(1 - w(n)) \cdot l(n, m) + w(n) \cdot r(n, m)] \dots$$
$$\cdot \ [(1 - w(N + 1 - n)) \cdot r(n, m) + w(N + 1 - n) \cdot l(n, m)], \ (4.11)$$

with $w(n) = 0.035 \cdot e^{-n/6}$. The value $w(n)$, the ratio of monaural processing, increases with the ITD. Using monaural processors, the signal in one channel can lead to activation in the delay line of the other channel and, therefore, the coincidence cells are also activated if the signal is presented monotically only. In [65] it was also noted that this modification increases the displacement of the centroid in similar way as is described in [93], where the range of the analyzed ITDs had to be extended to larger values, e. g., ±3.0 ms). The initial idea to use monaural processors was due to the observation that the auditory event might fall apart into a time image and an intensity image, when the ITD and ILD deviate too much from the existing natural combinations that are found in the HRTFs. Experienced listeners can indeed distinguish between the time and the intensity image, and so does the *Lindemann* model. In the model, the time image is determined by the local peak of cross-correlation curve, the intensity image by the contribution of the monaural processors.

4.4.2 Weighting Functions

In order to predict the position of auditory events for a large number of sound source positions accurately, it is vital to combine the information gained for all analyzed cues in an appropriate manner. Each cue is usually analyzed in different frequency bands and in various time frames. A common approach is to weight the information differently throughout time and frequency. A third type of weighting, the method of internal-delay weighting [22] was already introduced in the previous section. In [96] and [91] internal-delay weighting functions from different psycho-acoustic experiments have been estimated, similar to the solution reported in [22], see Fig. 4.10, left panel. In [96] the internal-delay weighting curve was determined as a function of the center frequency. In Fig. 4.10 the weighting curve for 500-Hz center frequency is shown.

For signals whose bandwidth exceeds the bandwidth of the corresponding filter, it is often necessary to introduce a frequency-weighting function as well. Figure 4.10, right panel, shows three frequency weighting functions that were derived from different psycho-acoustical measurements and proposed in different studies, namely, [1, 85, 94].

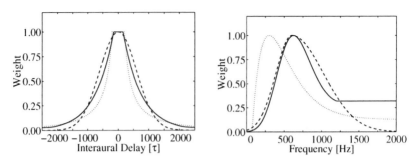

Fig. 4.10. Delay weighting, **left** panel, taken from [22], *dotted line*, from [91], *solid line*, and from [96], *dashed line*. Frequency weighting, **right** panel, from [85], *dotted line*, from [94], *solid line*, and from [1], *dashed line*

For the remaining dimension of cross-correlation models, i. e. time, there exists no simple weighting function, i. e. for weight vs. time, because the time weighting strongly depends on the characteristics of the sound and the environment, e. g., on room acoustics. If stationary signals are analyzed, the model output is often averaged over the whole signal duration, without applying a weighting function, to generate an ITD-frequency correlogram, e. g., [94]. For non-stationary signals, the current location of the auditory event is frequently determined by implementing a so-called sluggishness filter, such as to reduce the temporal resolution of the model. In [65, 99] a weighting function has been applied to the cross-correlation algorithm which is multiplied by the cross-correlation functions of the corresponding time slots as shown in the following term,

$$\Psi(m,n) = \sum_{m'=1}^{m} c(m',n) \cdot e^{-(m-m')/T_{\text{int}}}, \qquad (4.12)$$

with the integration time, T_{int}, chosen in [65] to be 5 ms. In a similar approach [95] the ITD-time correlogram is smoothed along the time axis, using an exponentially-shaped low-pass filter with a cutoff frequency of 5 Hz. To consider the non-linear, time-invariant response of the auditory system, algorithms as described in [26,72] or [19,20] can be included into the preprocessing stage of the model.

A weighting approach that combines frequency- and internal-delay weighting was proposed in [94]. The authors, there, concluded that those peaks in the ITD-frequency correlogram, which have a similar position across several frequency bands, k, so-called "straightness", should be enhanced. Straightness weighting is achieved by multiplying the outputs, $c(k, m, n)$, of the ITD-time correlograms or several adjacent frequency bands after they have been smoothed with a sluggishness filter as follows,

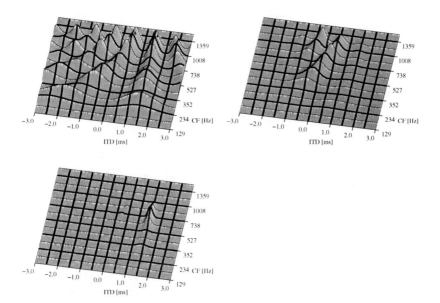

Fig. 4.11. Output of *Stern et al.*'s model [94] to a bandpass noise, 700-Hz center frequency, 1000-Hz bandwidth, 1.5-ms ITD. The **top-left** panel shows the output without centrality and straightness weighting, the **top-right** panel with centrality weighting only, and the **bottom-left** panel with centrality and straightness weighting

$$c_{\text{sw}}(k, m, n) = \prod_{k-\Delta k_l}^{k+\Delta k_h} c_{\text{s}}(k, m, n). \tag{4.13}$$

Usually, Δk_l and Δk_h are set to a value of 1 or 2. For very low-frequency bands, the value Δk_l has to be locally adapted to fit the relation $k - \Delta k_l \geq 1$. Further, the relation $k + \Delta k_h \leq k_{\text{max}}$ has to be considered for high-frequency bands. Note that the outputs for the adjacent frequency bands are not orthogonal since the auditory filters overlap in frequency – compare Fig. 4.4. Figure 4.11 illustrates the influence of the straightness-weighting functions. The top-left panel shows the ITD-frequency correlogram for a bandpass noise, 700-Hz center frequency, 1000-Hz bandwidth, 1.5-ms ITD, before weighting. In the top-right panel, the same correlogram is shown after centrality weighting was included using the weighting function from [22] – Fig. 4.10, left panel, dotted line. It can be clearly seen that the values for small magnitudes of the internal delay are now enhanced. The bottom-left panel shows the results for the same correlogram after centrality and straightness weighting according to Δk_l=2, Δk_h=2 have been applied. Note that the side peaks at negative ITDs vanished, because their positions vary with the period of the center frequency of the frequency band.

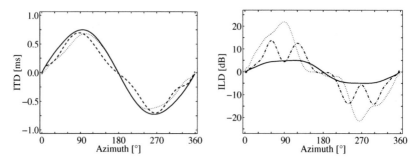

Fig. 4.12. Inter-aural time differences, **left** panel, and inter-aural level differences, **right** panel, for different frequency bands, i.e. band 8, f_c=434 Hz, *solid line*, band 16, f_c=1559 Hz, *dashed line*, band 24, f_c=4605 Hz, *dotted line*

4.4.3 Localization Models

In contrast to lateralization models, localization models are optimized for signals that are presented in a free sound field rather than through headphones. In free field, the signals are filtered by the outer ears and the auditory events are, thus, usually perceived as externalized in three-dimensional space. Frequently the free-field environment is simulated by means of a headphone-based virtual environment by convolving the presented signals with HRTFs. In this way internal auditory events along the inter-aural axis, as observed in many early headphone-based experiments, can be avoided. The first localization models were restricted to the frontal half of the horizontal plane, and the azimuth angle of the analyzed signal was determined using the inter-aural cues. A relatively simple way of doing this is to apply an axis transformation, e.g., [37]. The frequency dependent relationship of ITDs and/or ILDs and the azimuth angle can be determined from a catalog of HRTFs which contains the HRTFs for several adjacent angles in the horizontal plane. Spatial maps of these relationships can be set up, using one map per analyzed frequency band.[5]

In Fig. 4.12, one-dimensional examples of such spatial maps are depicted. In the left panel, the relationship between the ITDs and the azimuth in the horizontal plane is shown for three different frequency bands. In the right panel, the relationship between the ILDs and the azimuth is illustrated. Given the usage of such maps, the output of the cross-correlation algorithm on the basis of ITDs can be re-mapped on the basis of the azimuth in the horizontal plane as shown in Fig. 4.13. Unfortunately, ambiguities often occur. For example, as seen in Fig. 4.13, the ITD-based analysis cannot reveal if the

---

[5] It should be noted at this point that, although neurons that are spatially tuned were found in the *inferior colliculus* of the guinea pig [48] and in the primary field of the auditory area of the cortex of the cat [52,73], a topographical organization of those type of neurons could not be shown yet

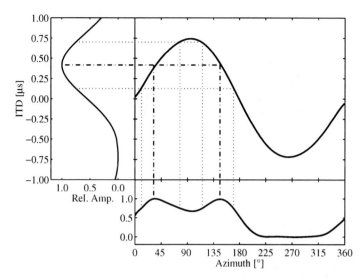

Fig. 4.13. Re-mapping of the cross-correlation function from ITD to azimuth angle, shown for the frequency band 8 centered at 434 Hz. The signal was presented at 30° azimuth, 0° elevation

sound was presented from the front or the rear hemisphere. If the full 3-D space is considered in the analysis rather than the horizontal plane, these areas of ambiguities are determined by the previously mentioned cones of confusion. To resolve these ambiguities, ILD cues have to be processed as well, and for reliable localization, especially within the median plane, the monaural cues are also of importance. In general, only the combination of ITDs and ILDs across several frequency bands allows an unequivocal estimate of the sound-source position. For this reason, humans face difficulties in localizing narrow-band signals.

The process the auditory system undergoes in combining the single cues to a single – or multiple – auditory event is not trivial. This holds in particular as many psycho-acoustical details are still unknown, e. g., how the single cues have to be weighted in general. It also remains unclear whether the ITD and ILD cues are combined before or after they are spatially mapped. An early approach to simulate human localization in the horizontal plane has been described in [37]. There, the previously mentioned *Lindemann* model [65] has been extended to optimally process the natural combinations of ITDs and ILDs as they occur in typical free-field listening situations. For this purpose, he introduced weighting, i.e. compensation factors, into the delay line. Now, each coincidence cell is tuned to a natural combination of ITD and ILD as found in the HRTFs for different angles of incidence and frequencies, see Fig. 4.14. In this approach, the ILD of the signal is just eliminated when it meets in both delay lines at the "corresponding" coincidence detector.

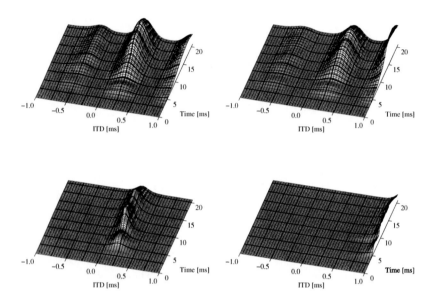

Fig. 4.14. Output of the algorithms of the *Lindemann–Gaik* model [37, 65] for a broad-band-noise signal presented at 30° azimuth, in the **top** panels for the frequency band 14 centered at 1172 Hz, in the **bottom** panels for the frequency band 22 centered at 3539 Hz. In the **left** panels, the natural combinations of ILDs and ITDs in the HRTFs are considered in the inhibition process, the **right** panels show results for the *Lindemann* algorithm without *Gaik*'s extension

This adjustment is necessary, because the contra-lateral inhibition shifts the cross-correlation peak for signals with existing ILDs sideward. To be able to compensate the ILDs for various directions, the compensation factors have to be implemented between every two coincidence detectors. Finally, the ITDs simply have to be re-mapped on the basis of azimuth, to estimate the azimuthal angle of the sound source.

As was demonstrated in [6,8], this kind of model can be used to generate the so-called binaural activity maps of room impulse responses, see Fig. 4.15. Binaural-activity maps show characteristic features of a room, e. g., arrival time and lateral positions of reflections, and are a useful tool in assessing the perceptual acoustical quality of a room. For the estimation of binaural-activity plots, the dynamic inhibition and a binaural sluggishness filter have to be avoided in order to make the reflections clearly visible. The maximum peak in each time frame was scaled to Decibels in the figure, in order to optimize the visualization of the late reflections. From the two binaural-activity maps shown, the characteristic properties of different types of rooms become clearly apparent – e. g., the larger initial-time-delay gap in the case of the concert hall.

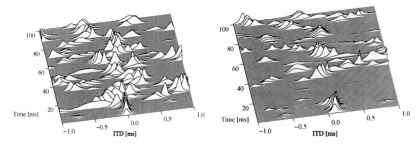

Fig. 4.15. Binaural-activity maps of a small seminar room, **left** panel, and a concert hall, **right** panel. The maps, derived by using the model from [37,65], are shown for the frequency band centered at 867 Hz

An alternative approach to estimate the position of a sound source is to train a neural network to estimate the auditory event from the inter-aural cues rather than to combine the cues analytically, e. g., [47,53]. When applying such a method, the neuronal network has to be trained on test material. The advantage of this procedure is that often very good results are achieved for stimuli that are very similar to the test material. The disadvantages are, however, the long time necessary to train the neural network and that the involved processing cannot easily be described analytically.

Those types of localization models that analyze both ITDs and ILDs either process both cues in a combined algorithm, e. g., [93], or evaluate both cues separately and combine the results afterwards, in order to estimate the position of the sound source, e. g., [47,53,76,77]. In [53] it is demonstrated that, for filtered clicks, both the ITDs and ILDs contribute very reliable cues in the left–right dimension, while in the front–back and the up–down dimensions, ILDs are more reliable than ITDs. The findings are based on model simulations using a model that includes a neural network. The network was trained with a back-propagation algorithm on 144 different sound-source positions in the whole sphere, the positions being simulated using HRTFs. The authors could feed the neural network either with ITD cues, ILD cues or both. Monaural cues could also be processed.

4.5 Detection Algorithms

Many psycho-acoustical investigations are dedicated to understanding the human ability to detect a target sound in noise. Often the minimum sound-pressure level is determined above which a signal that is masked by noise with different inter-aural attributes can be detected. For this purpose, several models have been developed, particularly the geometric vector model [55–57], the "equalization-cancellation", EC, model [23,31–34] and models based on inter-aural cross correlation, ICC, [29,71,80,90]. It had been demonstrated [23] that

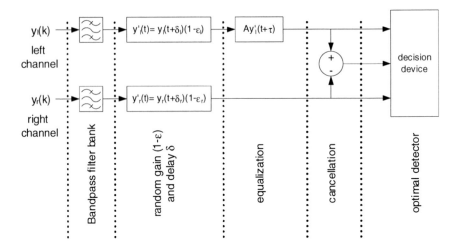

Fig. 4.16. Structure of the EC model

the models based on ICC algorithms are actually very similar to the EC models. Yet, recently, it has been discussed that under special circumstances both algorithms may lead to different results [14, 17].

The structure of the EC model is depicted in Fig. 4.16. The model, despite having only two adjustable parameters, is able to predict the threshold level in a large number of detection experiments. In the initial step, the signal, consisting of the target signal and the masker, is sent through an auditory band-pass filter bank. Afterwards, the signal is distorted independently in the left and right channels by a random gain factor, ϵ, and a random time jitter, δ. The parameters ϵ_l, ϵ_r, δ_l, and δ_r are statistically independent zero-mean *Gaussian* variables. The parameters ϵ_l and ϵ_r have the same variance, i.e. σ_ϵ^2. The variance of both δ_l and δ_r is equal to σ_δ^2. The values of σ_ϵ^2 and σ_δ^2 are adjusted such as to optimize the EC model in predicting binaural masking-level differences, BMLDs.[6]

In the next step, both channels are "equalized", i.e. the amplitude of the masking noise in both channels is adjusted to the same value by a gain factor, A, and the phase difference between both channels is eliminated optimally through a time delay, τ. In the "cancellation" process, one channel is subtracted from the other one in order to eliminate the masking noise and, subsequently, increase the signal-to-noise ratio, SNR. The performance of the process is limited by the random gain factor, ϵ, and time jitter, δ. Similar to the findings in psycho-acoustical experiments, the more the binaural parameters of target and masker differ from each other, the more gain in SNR

---

[6] Binaural masking-level differences, BMLDs, describe the changes in the masked-threshold levels when listening with both ears as compared to listening with only one ear

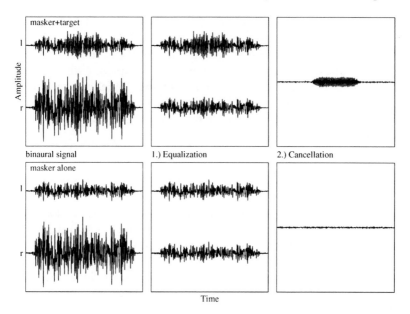

Fig. 4.17. Demonstration of the EC model for a masker–target pair, **top** panels, and a masker without target, **bottom** panels

increases. The decision device, which finally decides whether the signal is detected or not, is often implemented as an optimal detector according to signal-detection theory [43, 44]. Durlach [32] proposed that the auditory system would automatically select from the monaural or binaural detector the one with the better performance. Figure 4.17 demonstrates the performance of the EC model for a sinusoidal tone which is embedded into a narrow band noise. In the top panels, a target–masker pair is shown. The target, a 500-Hz sinusoidal tone is presented diotically, the narrow-band-noise masker with an ILD of 10 dB, 1-Bark-band wide, 500-Hz center frequency. In the bottom panels, the same stages of processing are shown for the masker without a target. The left panels show the headphone signals, and, obviously, it is not clearly visible which of the two panels contain a target. The middle panels show the signal after equalization and the right panels after the cancellation. Now, it is apparent that the signal in the upper panel contains the target.

Frequently, not the detection-threshold level per se, but detection thresholds of changes in the binaural parameters, for example, changes in the ITD, are of interest [22]. The results of experiments regarding this question show that the sensitivity for changes of the ITD are smaller for small inter-aural delays than they are for large ones. These findings lead to the introduction of the previously-mentioned delay weighting [22]. In order to improve the prediction of detection experiments a novel algorithm with a two-dimensional array of EI cells for every frequency band has recently been introduced [14–16]. The new model is roughly based on the idea of the EC model. Each EI cells is tuned to a different characteristic ITD–ILD pair. The authors were able to include several types of weighting algorithms into their model, namely, delay weighting, frequency weighting and a non-linear temporal weighting using an adaptive signal compression stage [26]. A similar model was independently proposed in [101], where there the same adaptive signals compression was combined with a binaural detection model, yet, in this case, with the classical EC algorithm.

4.6 Localization in Multiple-Sound-Source Scenarios

To take into account that listeners have to analyze concurrent sound sources in various everyday-listening situations, a number of binaural localization models exist that are specialized to localize a test sound in the presence of distracting sound sources. In this context the previously mentioned model [65] is able to simulate the precedence effect. The model in its original form could predict the outcome for experiments with click pairs, but was recently modified to simulate the precedence effect for stationary narrow-band signals as well [12]. In Fig. 4.18 the response of the model to a narrow-band noise burst in presence of one reflection in comparison with the simple running cross-correlation algorithm is depicted. The inhibitory stages of the model keep the cross-correlation function on the track of the primary signal. In addition to the static inhibition, the author has also introduced a dynamic inhibition [66] which he defined as follows,

$$\phi(m,n) = c(m-1,n) + \phi(m-1,n) \cdot e^{-T_\mathrm{d}/T_\mathrm{inh}}[1 - c(m-1,n)], \quad (4.14)$$

with the time delay between two taps of the delay lines, T_d, and the fade-off-time constant of the nonlinear low pass, T_inh. When using the dynamic inhibition algorithm, (4.5) and (4.6) have to be modified to

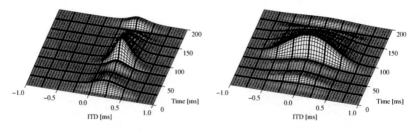

Fig. 4.18. Output of the algorithm from [65], **left** panel, compared to the plain ICC algorithm without inhibition, **right** panel, for a narrow-band noise, 500-Hz center frequency, 100-Hz bandwidth, 0.3-ms ITD, in the presence of a single reflection, −0.3-ms ITD, 1.5-ms inter-stimulus interval

$$l(m + 1, n + 1) = l(m, n)[1 - c_s \cdot r(m, n)](1 - \phi(m, n)), \qquad (4.15)$$
$$r(m + 1, n - 1) = r(m, n)[1 - c_s \cdot l(m, n)](1 - \phi(m, n)). \qquad (4.16)$$

In [99] a different approach to simulate the localization dominance of the precedence effect has been taken. There, the ITDs in the onset of envelope of signal rather than the ITDs in the fine structure of the stationary signal were determined. The ITDs of the onset slopes are measured by comparing those points in time at which the slope in the left and right channels exceed a threshold. The resulting from this approach proved to be more robust in a reverberant environment than a common cross-correlation algorithm. However, recent experiments indicated that even in a reverberant environment, e. g., a small seminar room, human listeners often determine the sound-source position from the stationary cues rather than from analyzing the onset of the signal, [10, 11]. Further is has been shown recently that, for short test impulses such as clicks, localization dominance can be simulated using a simple cross-correlation algorithm without inhibition stages when a hair-cell model is included in the pre-processing stage [49]. To this end an adaptive hair-cell model [72] was employed. Parts of the precedence effect are, thus, understood as a results of sluggish processing in the auditory periphery. Further models to simulate the adaptive response of the auditory system to click trains, "build-up of the precedence effect", can be found in [28, 104].

In the meanwhile, models have been put forward to simulate the localization of a sound source in the presence of further, incoherent sound sources, e. g., [11, 76, 77]. For certain conditions these specialized models can reach human localization ability, while the localization results of most other computational models are strongly degraded by concurrent sound sources. In [76, 77] improvement in robustness have been achieved by weighting the individual cues by use of a *Bayes* classifier. This classifier works with information about the statistical parameters of the expected noise. Another promising approach, [35], achieves robust localization in complex listening scenarios by focusing the analysis of the binaural cues on time instants and frequencies with high inter-aural coherences. In a current model of this author the influence of the concurrent sound source on the overall cross-correlation function is estimated from that part which precedes the test signal and is, afterwards, eliminated [11].

In another class of models, termed cocktail-party processors, the information on the location of the sound sources is used to segregate them from each other, e. g., [9, 64]. These algorithms can be used to improve the performance of speech recognition systems [86]. For further improvement of binaural models, it has been proposed to implement an expert system, which completes the common signal-driven, bottom-up, approach [7, 8]. The expert system should include explicit knowledge on the auditory scene and on the signals – and their history. This knowledge is to be used to set up hypotheses and to decide if they prove to be true or not. As an example, front-back differentiation can be named. The expert system could actively test whether the hypothesis that the sound source is presented in the front is true or false, by employing "auditory scene analysis", ASA, cues. It could further analyze the monaural spectrum of the sound source in order to estimate the influence of the room in which the sound source is presented, determine the inter-aural cues and even evaluate cues from other modalities, for example visual cues. The expert system would evaluate the reliability of the cues and weight them according to the outcome of the evaluation. In future, once the computational power increases furthermore and more knowledge on the auditory system was gained, one can expect that binaural models will become more complex and include the simulation of several binaural phenomena rather than the simulation of only a few specific effects.

Acknowledgement

This chapter was compiled out at the Institute of Communication Acoustics, Ruhr-University Bochum. I would like to thank *J. Blauert* and my colleagues at the institute for their valuable support. Part of this work was financed by the Deutsche Forschungsgemeinschaft, DFG. Further, I would also like to acknowledge that the MATLAB toolboxes of (i) *M. A. Akeroyd*, (ii) *R. M. Stern* and (iii) *M. Slaney* were used for parts of the simulation, namely,

(i)   http://www.biols.susx.ac.uk/home/Michael_Akeroyd/download2.html
(ii)   http://www-2.cs.cmu.edu/afs/cs.cmu.edu/user/rms/www/BinauralWeb/
(iii) http://rvl4.ecn.purdue.edu/~malcolm/interval/1998-010/

List of Abbreviations and Mathematical Symbols

A	gain factor
α	inter-aural level difference . . . also denoted ILD
c	coincidence detector
c_s	static inhibition constant
c_d	direct coincidence detector
δ	elevation, random time jitter of internal noise
e	activity of EI-cell
ϵ	random gain factor of internal noise
f_c	center frequency
f_s	sampling frequency
IACC	inter-aural cross-correlation coefficient
ICC	inter-aural cross correlation
ISI	inter-stimulus interval
ILD	inter-aural level difference
ITD	inter-aural time difference
k	index for frequency band
l	left delay line

m	discrete time
n	index for internal delay
N	number of coincidence detectors
P	power
φ	azimuth
Ψ	cross-correlation function
r	right delay line
σ	variance
t	time
T	time constant
τ	internal delay
w	ratio of monaural processing
$y_{l,r}$	signal at left/right eardrum

References

1. Akeroyd M A, Summerfield A Q (1999) A fully temporal account of the perception of dichotic pitches, Br J Audiol 33:106–107
2. Blauert J (1969) Untersuchungen zum Richtungshören in der Medianebene bei fixiertem Kopf. Doct diss. Rhein-Westf Techn Hochsch Aachen, Aachen
3. Blauert J (1969/70) Sound localization in the median plane, Acustica 22:205–213
4. Blauert J, Cobben W (1978) Some consideration of binaural cross correlation analysis. Acustica 39:96–104
5. Blauert J., Lindemann W (1986) Auditory spaciousness: some further psychoacoustic analysis. J Acoust Soc Amer 80:533–542
6. Blauert J (1997) Spatial hearing – the psychophysics of human sound localization. MIT press, Cambridge MA
7. Blauert J (1999) Binaural auditory models: architectural considerations. 18th Proc 18th DANAVOX Symp 189–206. Scanticon, Kolding
8. Blauert J (2005) Analysis and synthesis of auditory scenes. Chap 1 this vol
9. Bodden M (1992) Binaurale Signalverarbeitung: Modellierung der Richtungserkennung und des Cocktail-Party-Effektes. Doct diss. Ruhr-Univ Bochum, Bochum. VDI Publ, Düsseldorf
10. Braasch J, Hartung K (2002) Localization in the presence of a distracter and reverberation in the frontal horizontal plane. I. Psychoacoustical data. ACUSTICA/acta acustica 88:942–955
11. Braasch J (2002) Localization in the presence of a distracter and reverberation in the frontal horizontal plane. II. Model algorithms. ACUSTICA/acta acustica 88:956–969
12. Braasch J, Blauert J (2003) The precedence effect for noise bursts of different bandwidths. II. Comparison of model algorithms. Acoust Sci & Techn 24:293–303
13. Brand A, Behrend O, Marquardt T, McAlpine D., Grothe B (2002) Precise inhibition is essential for microsecond inter-aural time difference coding. Nature 417:543–547

14. Breebaart J, van de Par S, Kohlrausch A (2001a) Binaural processing model based on contra-lateral inhibition. I. Model setup. J Acoust Soc Amer 110:1074–1088
15. Breebaart J, van de Par S, Kohlrausch A. (2001b) Binaural processing model based on contra-lateral inhibition II. Dependance on spectral parameters. J Acoust Soc Amer 110:1089–1104
16. Breebaart J, van de Par S, Kohlrausch A (2001c) Binaural processing model based on contra-lateral inhibition. III. Dependance on temporal parameters. J Acoust Soc Amer 110:1105–1117
17. Breebaart J, van de Par S, Kohlrausch A (2002) On the difference between cross-correlation and EC-based binaural models. In: Calvo-Manzano A, Pèrez-López A, Santiago J S (eds) Proc Forum Acusticum 2002, Sevilla. CD-ROM
18. Brüggen M (2001) Coloration and binaural decoloration in natural environments, Acustica/acta acustica 87:400–406
19. Buchholz J M, Mourjopoulos J (2002a) A computational auditory masking model based on signal dependent compression. I. Model description and performance analysis, ACUSTICA/acta acustica 90:873–886
20. Buchholz J M, Mourjopoulos J (2002b) A computational auditory masking model based on signal dependent compression. II. Mathematical concepts and model simulations, ACUSTICA/acta acustica 90:887–900
21. Colburn H S (1973) Theory of binaural interaction based on auditory-nerve data, I. General strategy and preliminary results on inter-aural discrimination. J Acoust Soc Amer 54:1458–1470
22. Colburn H S (1977) Theory of binaural interaction based on auditory-nerve data, II. Detection of tones in noise. J Acoust Soc Amer 61:525–533
23. Colburn H S, Durlach N I (1978) Models of Binaural Interaction. In: Carterette E C, Friedman M P (eds) Handbook of Perception, Vol IV, Hearing 467–518. Academic Press, New York
24. Colburn H S, Latimer J S (1978) Theory of binaural interaction based on auditory-nerve data, III. Joint dependence on inter-aural time and amplitude differences in discrimination and detection. J Acoust Soc Amer 64:95–106
25. Colburn H S (1995) Computational Models of Binaural Processing. In: Hawkins H L, McMullen T A, Popper A N , Fay R R (eds) Auditory Computation 332–400. Springer, New York,
26. Dau T, Püschel D, Kohlrausch A (1996) A quantitative model of the effective signal processing in the auditory system. I. Model structure. J Acoust Soc Amer 99:3615–3622
27. David E E, Guttman N, von Bergeijk W A (1959) Binaural interaction of high-frequency complex stimuli. J Acoust Soc Amer 31:774–782
28. Djelani T (2001) Psychoakustische Untersuchungen und Modellierungsansätze zur Aufbauphase des auditiven Präzedenzeffektes. Doct diss. Ruhr-Univ Bochum, Bochum
29. Dolan T R, Robinson D E (1967) Explanation of masking-level differences that result from inter-aural intensive disparities of noise. J Acoust Soc Amer 42:977–981
30. Duifhuis H (1972) Perceptual analysis of sound, Doct diss. Techn Hogesch Eindhoven, Eindhoven
31. Durlach N I (1960) Note on the Equalization and Cancellation theory of binaural masking level differences. J Acoust Soc Amer 32:1075–1076

32. Durlach N I (1963) Equalization and Cancellation theory of binaural masking-level differences. J Acoust Soc Amer 35:1206–1218
33. Durlach N I (1966) On the application of the EC model to interaural jnds. J Acoust Soc Amer 40:1392–1397
34. Durlach N I, Colburn H S (1972) Binaural signal detection: equalization cancellation theory. In: Tobias J V (ed) Foundation of modern auditory theory 365–466. Academic Press, New York
35. Faller C, Merimaa J (2004) Source localization in complex listening situations: Selection of binaural cues based on interaural coherence. J Acoust Soc Amer 116:3075–3089
36. Fletcher N H (1940) Auditory patterns, Rev Mod Phys 12:47–65
37. Gaik W (1993) Combined evaluation of inter-aural time and intensity differences: Psychoacoustic results and computer modeling. J Acoust Soc Amer 94:98–110
38. Goldberg J M, Brown P B (1969) Response of binaural neurons of dog superior olivary complex to dichotic tonal stimuli: Some physiological mechanism of sound localization. J Neurophysiol 32:613–636
39. Grantham D W, Wightman F L (1978) Detectability of varying inter-aural temporal differences. J Acoust Soc Amer 63:511–523
40. Grantham D W, Wightman F L (1979) Detectability of a pulsed tone in the presence of a masker with time-varying inter-aural correlation. J Acoust Soc Amer 65:1509–1517
41. Grantham D W (1982) Detectability of time-varying inter-aural correlation in narrow-band noise stimuli. J Acoust Soc Amer 72:1178–1184
42. Grantham D W (1982) Discrimination of dynamic inter-aural intensity differences. J Acoust Soc Amer 76:71–76
43. Green D M, Swets J A (1966) Signal detection analysis of the equalization and cancellation model. J Acoust Soc Amer 40:833–838
44. Green D M, Swets J A (1988) Signal detection theory and psychophysics, Peninsula Publishing, Los Altos
45. Hammershøi D, Møller H (2005) Binaural technique: basic methods for recording, synthesis and reproduction. Chap 9 this vol
46. Harris G G (1960) Binaural interaction of impulsive stimuli and pure tones. J Acoust Soc Amer 32:685–692
47. Hartung K (1998) Modellalgorithmen zum Richtungshören, basierend auf Ergebnissen psychoakustischer und neurophysiologischer Experimente mit virtuellen Schallquellen, Doct diss. Ruhr-Univ Bochum, Bochum
48. Hartung K, Sterbing S J (2001) A computational model of sound localization based on neurophysical data. In: Greenberg S, Stanley M (eds) Computational models of auditory function. IOS Press, Amsterdam
49. Hartung K, Trahiotis C (2001) Peripheral auditory processing and investigations of the "precedence effect" which utilize successive transient stimuli. J Acoust Soc Amer 110:1505–1513
50. von Hornbostel E M, Wertheimer M (1920) Über die Wahrnehmung der Schallrichtung, in: Sitzungsber Akad Wiss, Berlin, 388–396
51. Hudde H (2005) A functional view on the human peripheral hearing organ. Chap 3 this vol
52. Imig T J, Irons W A, Samson F R (1990) Single-unit selectivity to azimuthal direction and sound pressure level of noise bursts in cat high frequency auditory cortex. J Neurophysiol 63:1448–1466

53. Janko J A, Anderson T R, Gilkey R H (1997) Using neural networks to eval-
    uate the viability of monaural and inter-aural cues for sound localization. In:
    Gilkey R H, Anderson T R (eds) Binaural and spatial hearing in real and
    virtual environments, Mahwah, New Jersey, 557–570
54. Jeffress L A (1948) A place theory of sound localization. J Comp Physiol
    Psychol 41:35–39
55. Jeffress L A, Blodgett H C, Sandel T T, Wood C L (1956) Masking of tonal
    signals. J Acoust Soc Amer 28:416–426
56. Jeffress L A, McFadden D (1971) Differences of inter-aural phase and level in
    detection and lateralization. J Acoust Soc Amer 49:1169–1179
57. Jeffress L A (1972) Binaural signal detection: Vector theory. In: Tobias J V
    (ed) Foundation of modern auditory theory 351–368. Academic Press, New
    York
58. Joris P J, Yin T (1995) Envelope coding in the lateral superior olive. I. Sen-
    sitivity to inter-aural time differences. J Neurophysiol 73:1043–1062
59. Joris P J (1996) Envelope coding in the lateral superior olive. II. Charac-
    teristic delays and comparison with responses in the medial superior olive. J
    Neurophysiol 76:2137–2156
60. Knudsen E I, Konishi M (1978) Center-surround organization of auditory
    receptive field in the owl. Science 202:778–780
61. Kohlrausch A (1986) Auditory filter shape derived from binaural masking
    experiments. J Acoust Soc Amer 84:573–583
62. Kollmeier B, Gilkey R H (1990) Binaural forward and backward masking:
    Evidence for sluggishness in binaural detection. J Acoust Soc Amer 87:1709–
    1719
63. Kuhn G F (1977) Model for the inter-aural time differences in the azimuthal
    plane. J Acoust Soc Amer 62:157–167
64. Lehn K (2000) Unscharfe zeitliche Clusteranalyse von monauralen und inter-
    auralen Merkmalen als Modell der auditiven Szenenanalyse. Doct diss. Ruhr-
    Univ Bochum, Bochum, VDI Verlag, Düsseldorf
65. Lindemann W (1986) Extension of a binaural cross-correlation model by
    contra-lateral inhibition. I. Simulation of lateralization of stationary signals.
    J Acoust Soc Amer 80:1608–1622
66. Lindemann W (1986) Extension of a binaural cross-correlation model by
    contra-lateral inhibition. II. The law of the first wave front. J Acoust Soc
    Amer 80:1623–1630
67. Litovsky R Y, Colburn H S, Yost W A, Guzman S J (1999) The precedence
    effect. J Acoust Soc Amer 106:1633–1654
68. Macpherson E A, Middlebrooks J C (2002) Listener weighting of cues for
    lateral angle: The duplex theory of sound localization revisited. J Acoust Soc
    Amer 111:2219–2236
69. McAlpine D, Jiang D, Palmer A R (2001) A neural code for low-frequency
    sound localization in mammals, Nat Neurosci 4:396–401
70. McAlpine D, Grothe B (2003) Sound localization and delay lines – do mam-
    mals fit the model?, Trends in Neuroscience 26:347–350
71. McFadden D (1968) Masking-level differences determined with and without
    inter-aural disparities in masker intensity. J Acoust Soc Amer 44:212–223
72. Meddis R, Hewitt M J, Shackleton T M (1990) Implementation details of a
    computational model of the inner hair-cell auditory-nerve synapse. J Acoust
    Soc Amer 87:1813–1816

73. Middlebrooks J C, Pettigrew J D (1981) Functional classes of neurons in primary auditory cortex of the cat distinguished by sensitivity to sound localization. J Neurosci 1:107–120

74. Middlebrooks J C (1992) Narrowband sound localization related to external ear acoustics. J Acoust Soc Amer 93, 2607–2624

75. Mills A W (1958) On the minimum audible angle. J Acoust Soc Amer 30:237–246

76. Nix J, Hohmann V (1998) Lokalisation im Störgeräusch auf der Basis der Statistik binauraler Parameter. In: Fortschr Akust, DAGA'98 474–475, Dtsch Ges Akust, Oldenburg,

77. Nix J, Hohmann V (2004) Sound source localization in real sound fields based on empirical statistics of interaural parameters, submitted to J Acoust Soc Amer

78. Novo P (2005) Auditory virtual environments. Chap 11 this vol

79. Okano T, Beranek L L, Hidaka T (1998) Relations among inter-aural cross-correlation coefficient (IACC$_E$), lateral fraction (LF$_E$), and apparent source width (ASW) in concert halls. J Acoust Soc Amer 104:255–265

80. Osman E (1971) A correlation model of binaural masking level differences. J Acoust Soc Amer 50:1494–1511

81. van de Par S, Kohlrausch A (1997) A new approach to comparing binaural masking level differences at low and high frequencies. J Acoust Soc Amer 101:1671–1680

82. Patterson R D (1974) Auditory filter shape. J Acoust Soc Amer 74:802–809

83. Patterson R D, Allerhand M H, Giguère C (1995) Time-domain modeling of periphal auditory processing: A modular architecture and software platform. J Acoust Soc Amer 98:1890–1894

84. Pickels J O (1988) An introduction to the physiology of hearing, 2nd edition. Academic Press, London

85. Raatgever J (1980) On the binaural processing of stimuli with different inter-aural phase relations, Doct diss. Delft Univ Techn, Delft

86. Rateitschek K (1998) Ein binauraler Signalverarbeitungsansatz zur robusten maschinellen Spracherkennung in lärmerfüllter Umgebung. Doct diss. Ruhr-Univ Bochum, Bochum. VDI Verlag, Düsseldorf

87. Lord Rayleigh (1907) On our perception of sound direction, Philos Mag 13:214–232

88. Reed M C, Blum J J (1990) A model for the computation and encoding of azimuthal information by the lateral superior olive. J Acoust Soc Amer 88:1442–1453

89. Salomons A (1995) Coloration and binaural decoloration of sound due to reflections, Doct diss. Delft Univ Techn, Delft

90. Sayers B M, Cherry E C (1957) Mechanism of binaural fusion in the hearing of speech. J Acoust Soc Amer 29:973–987

91. Shackleton T M, Meddis R, Hewitt M J (1992) Across frequency integration in a model of lateralization. J Acoust Soc Amer 91:2276–2279

92. Steinhauser A (1877) The theory of binaural audition, Phil Mag 7:181–197 & 261–274

93. Stern R M, Colburn H S (1978) Theory of binaural interaction based on auditory-nerve data. IV. A model for subjective lateral position. J Acoust Soc Amer 64:127–140

94. Stern R M, Zeiberg A S, Trahiotis C (1988) Lateralization of complex binaural stimuli: A weighted image model. J Acoust Soc Amer 84:156–165 (erratum: J Acoust Soc Amer 90:2202)

95. Stern R M, Trahiotis C (1995) Models of binaural interaction. In: Moore, B C J (ed) Handbook of Perception and Cognition, Volume 6: Hearing, Academic Press, New York, 347–386

96. Stern R M, Shear G D (1996) Lateralization and detection of low-frequency binaural stimuli: Effects of distribution of internal delay. J Acoust Soc Amer 100:2278–2288

97. Thompson S P (1877) On binaural audition, Phil Mag 4:274–276

98. Thompson S P (1882) On the function of the two ears in the perception of space, Phil Mag 13:406–416

99. Wolf S (1991) Untersuchungen zur Lokalisation von Schallquellen in geschlossenen Räumen, Doct diss. Ruhr-Univ Bochum, Bochum

100. Woodworth R S, Schlosberg H (1962) Experimental Psychology, 349–361. Holt Rinehard Winston, New York

101. Zerbs C (2000) Modelling the effective binaural signal processing in the auditory systems. Doct diss. Univ Oldenburg, Oldenburg

102. Zakarauskas P, Cynader M S (1993) A computational theory of spectral cue localization. J Acoust Soc Amer 94:1323–1331

103. Zurek P M (1979) Measurements on binaural echo suppression. J Acoust Soc Amer 66:1750–1757

104. Zurek P M (1987) The precedence effect. In: Yost W A, Gourevitch G (eds) Directional hearing 85–115. Springer, New York

105. Zwicker E, Feldtkeller R (1967) Das Ohr als Nachrichtenempfänger, S Hirzel Verlag, Stuttgart

5 Audio–Visual Interaction in the Context of Multi-Media Applications

Armin Kohlrausch[1,2] and Steven van de Par[1]

[1] Philips Research Laboratories, Eindhoven,
[2] Department of Technology Management, Technical University Eindhoven, Eindhoven

Summary. In our natural environment, we simultaneously receive information through various sensory modalities. The properties of these stimuli are coupled by physical laws, so that, e.g., auditory and visual stimuli caused by the same event have a specific temporal, spatial and contextual relation when reaching the observer. In speech, for example, visible lip movements and audible utterances occur in close synchrony, which contributes to the improvement of speech intelligibility under adverse acoustic conditions. Research into multi-sensory perception is currently being performed in a number of different experimental and application contexts. This chapter provides an overview of the typical research areas dealing with audio–visual interaction[3] and integration, bridging the range from cognitive psychology to applied research for multi-media applications. A major part of this chapter deals with a variety of research questions related to the temporal relation between audio and video. Other issues of interest are basic spatio-temporal interaction, spatio-temporal effects in audio–visual stimuli – including the ventriloquist effect, cross-modal effects in attention, audio–visual interaction in speech perception and interaction effects with respect to the perceived quality of audio–visual scenes.

5.1 Introduction

In the daily perception of our environment we experience a strong relationship between information obtained through each of the sensory modalities. For instance, when we see somebody speaking, we usually have little difficulty of linking the auditory speech signal to this specific speaker, even if there are numerous other voices present simultaneously. The ability to integrate information from the separate sensory modalities is obviously an important property of our perceptual system. It is based on the fact that sensory information within the various individual modalities is usually not independent. Dependencies in multi-sensory stimuli typically result from the underlying – physical – processes that generate the stimuli. For example, in order to produce certain audible speech utterances, a human speaker normally needs to make visible lip movements.

---

[3] The terms "audio–visual" and "auditory-visual" are used without distinction in this chapter. The first one is more often used in technology, the second one in sensory research

Integration of multi-sensory information implies that, at certain levels of processing, the individual senses are not represented independently, but that their activity is merged to yield information that could not be obtained from a representation of the individual senses, e. g., the judgment of synchrony between an auditory and a visual event. Dependencies of perception in one modality on stimuli in another modality have been demonstrated in numerous studies on inter-sensory biases. An example of such inter-sensory bias is the influence of eye orientation on the perceived direction of a sound source [38]. When visually fixating a point away from the straight ahead direction, i. e. eccentric gaze orientation, the perceived location of an auditory source is shifted by several degrees in the direction opposite to the eye orientation. Another example for the interaction of stimuli from two different sensory modalities is the ventriloquist effect [26,81]. This term describes the situation where the perceived location of an auditory object is influenced by the presence of a related visual object. When there is a slight mismatch in the location of the visual and the auditory stimulus, the perceived locus of the auditory stimulus is shifted towards the visual stimulus.

Basic measurements of integration, like sensitivity to asynchrony between auditory and visual stimuli and measurements of interaction like inter-sensory bias and multi-sensory illusions, as well as aspects of the attentional capacity in various sensory modalities have been addressed by cognitive psychologists already for a long time. Another area with a considerable research effort on audio–visual interaction is speech production and perception. Being inspired by questions about how we integrate visual and auditory cues when listening to speech [40–42], this area becomes increasingly interesting for applications in telecommunications, e. g., video-conferencing, in user interfaces and in animations, where synthetic speech is being accompanied by animations of "talking heads". With an increased interest in multi-media applications, the question of perceived overall quality of reproduced AV stimuli has received considerable interest in application laboratories. Related to this quality aspect are investigations in which the perceived quality of auditory scenes, e. g., for sound-design applications and room-acoustic-quality evaluations, is measured with and without accompanying visual information.

Audio–visual, AV, interaction effects are presently investigated in a number of scientific contexts. This diversity of research has the advantage that the same question, such as the consequence of asynchrony between sound and image, is addressed from a number of different viewpoints. The disadvantage is that the material is scattered across many different disciplines and journals and that published articles often reveal a lack of knowledge of related work, if this has been performed and, eventually, been published in a different context. It is therefore one of the aims of this chapter to provide an overview of this research field by including a large number of references from a variety of sources, giving the interested reader a basis for further study.

The material presented in this chapter is roughly organized along an axis from more fundamental- to more applied-research questions. Section ,5.2 discusses basic spatio-temporal observations of audio–visual interaction, including illusions. Section 5.3 introduces cross-modal effects in attention and Sect. 5.4 presents some results related to audio–visual speech perception. The area of quality judgments is covered in Sect. 5.5. Section 5.6, finally, gives an in-depth overview of the perceptual consequences of temporal asynchrony between the audio and the video component of AV scenes.

5.2 Basic Observations of Spatio-Temporal Interaction in Audio-Visual Stimuli

We start this review with basic observations about the mutual influence of the auditory and the visual modalities. We will use a more strict interpretation of the term interaction than the more general one in the title of this chapter. We interpret interaction between two sensory modalities as the alteration of a certain percept in one sensory modality due to the presence of a stimulus in another sensory modality – following [87]. Such interaction effects are particularly strong if the stimulus in one modality allows an ambiguous interpretation, but they are also found in non-ambiguous stimuli.

5.2.1 Interaction Between Stimuli Allowing an Ambiguous Interpretation

An example of a stimulus allowing for such an ambiguous interpretation has been described in [63]. Two indistinguishable visual objects are moving along intersecting trajectories – see Fig. 5.1. The moment at which both objects overlap can be perceived in different ways. Both objects can continue along their original trajectory, a case denoted as "streaming", or they can reverse their direction, a case denoted as bouncing [63]. In this visual-only condition, observers have a strong preference to interpret this condition as "streaming". This interpretation can, however, be influenced by introducing a number of pause frames at the moment of intersection, during which the relative position of the two objects does not change. Already with a pause duration of one frame, the number of "bouncing" responses increased from about 12%, at standard condition without a pause, to about 82%.

This phenomenon has also been investigated in a multi-sensory context. To study it, visual stimuli were combined with a short click, which could occur either synchronously with the visual impact or 150 ms before or after the visual impact [62]. The presence of the sound significantly increased the percentage of "bouncing" responses. This increase was largest for a synchronous representation, and smallest for a presentation of the sound after the visual impact. There was an asymmetry with respect to the effect of the audio with

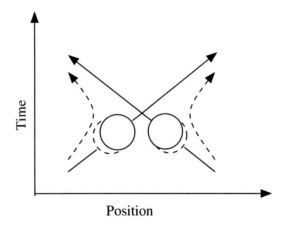

Position

Fig. 5.1. Schematic representation of a visual scene allowing an ambiguous interpretation. The indistinguishable visual objects, disks, move along intersecting trajectories. At the moment of intersection, the scene can be interpreted in two ways. The two objects continue along their original trajectories, indicated by the *continuous lines*, or they bounce back, as schematically indicated by the *dotted lines*

a sound coming before the visual overlap being more effective than a sound coming after the visual overlap. This interaction has been further investigated for a wider range of audio–visual delays by varying the relative delays from about -1000 ms, i. e. audio leading, to +1000 ms, audio lagging [83]. In this way a temporal tuning curve for the percentage of "bouncing" responses as a function of the delay was derived. Increased "bouncing" responses were observed for audio delays in the range of -250 ms to +100 ms, again with a stronger effect for audio being advanced with respect to the moment of visual impact – see [62]. The authors of the latter study also tried to understand the underlying mechanism of this inter-modal effect. They showed that increased bouncing responses and temporal tuning curves could not only be induced by adding a sound stimulus, but also by other transient stimuli like somatosensory vibrations and short visual flashes [67, 68]. On the other hand, the effect of the auditory click on the interpretation of the audio–visual scene could be reduced if it was embedded in a regular series of identical clicks, thus making it part of an auditory stream [84]. These observations led the authors to an explanation that the likeliness of perceiving the visual motion as bouncing or as streaming was dependent on the amount of attentional resources available to the observer at the moment of visual impact. They stated that a reduction in the attentional capacity for the observation and interpretation of the visual trajectories leads to an increased percentage of "bouncing" responses. A transient stimulus presented in addition to the visual trajectories would reduce the level of attention available to follow the visual trajectories, and in such a case, observers were more likely to give the response "bouncing". This role of attention is supported by the observation that the presentation of the ambiguous visual stimulus in the visual periphery together with a difficult central task in the fovea also leads to an increased percentage of bouncing responses [85].

5.2.2 Illusory Interaction Effects

An audio–visual interaction effect based on a non-ambiguous visual stimulus has been described recently [64, 66]. The visual stimulus consisted of one shortly-flashed white disk, while the concurrently presented acoustic stimulus consisted of one or more short tonal signals with a temporal-onset separation of 64 ms. Subjects were asked to report how many visual flashes they perceived. The basic observation was that the number of reported visual flashes increased with the number of presented acoustic beeps. This number was about one flash for one beep, increased to just above two flashes for two beeps and reached an asymptote of about 2.4 flashes for three and four beeps. Control experiments included conditions, in which one to four visual flashes, with the same temporal onset separation as used for the beeps, were presented without any beeps or together with only one beep. In both cases the number of observed flashes increased with the number of presented flashes up to an average of about 3.5 observed flashes for four presented flashes. One conclusion from this study is that the observers give the same response whether they observe one flash together with two beeps or two visual flashes combined with zero or one beep. Additional measurements of visually-evoked potentials for the same stimuli [65] revealed that the activity in the visual cortex to a single flash could be modulated by the simultaneous presentation of acoustics beeps. Furthermore the brain activity corresponding to the illusory flash was very similar to the activity for a physically present flash, indicating an interaction of auditory and visual stimuli at a neural level usually considered to be modality specific. [4]

Another example of a perceptual illusion of an audio–visual scene is the *McGurk* effect [42]. This effect is a demonstration of how we deal with conflicting information in two sensory modalities. In this condition, two consonant-vowel utterances are presented simultaneously to the observer, one visually and one acoustically. The two utterances of the type [ba], [da], [ga] contain the same vowels, but differ in the consonants. Subjects are instructed to report what utterance they hear. When [ga] is presented visually and [ba] auditorily [42], only 2% of the adult subjects reported to hear [ba], while 98% reported to hear [da] – which, in fact, was presented neither acoustically nor visually. Younger subject tended to have a higher percentage of [ba] responses.

The authors of the above mentioned study pointed out that the auditory [ba] stimulus contains features for [ba] as well as for [da]. The visual

---

[4] A demonstration of such an illusory stimulus is available at the following Internet address: http://neuro.caltech.edu/~lshams/

[ga] stimulus, on the other hand, contains features for [ga] and also for [da]. Thus, features in both modalities only match when [da] is assumed to have occurred, which may be the reason why this utterance is perceived. A similar, though somewhat less salient, effect was observed when the combination of [pa], auditory stimulus, and [ka], visual stimulus, was presented. In this case, subjects tended to perceive [ta]. For more information on how this perceptual result can be modelled the reader is referred to [40, 41]. [5]

5.2.3 Temporal-Rate Disparities

Already since several decades, the mutual influence in perceived rate between visual "flicker" and auditory "flutter" is known. In this context, subjects were asked to match the rates of regularly-modulated stimuli within and between the auditory and visual modalities [19]. A first observation was that the precision of rate matching was much lower for across-modality than for within-modality matches. The more interesting observation from this study relates to the fact that a change in the physical rate of auditory flutter did influence the perceived rate of visual flicker, while the reverse, visual flicker influencing auditory flutter, was not observed. Unfortunately, this so-called driving effect could not yet be established quantitatively.

In a more recent investigation however [86], experiments have been performed in which subjects have been asked to judge the modulation rate of a video, an audio, i. e. mono-sensory, or an audio-video, bi-sensory, stimulus by means of direct magnitude estimation. Rates of 2, 4, 6, 8, and 10 Hz were used. In the first experiment, it appeared that for a fixed presented modulation rate, the judged modulation rate was higher for the video stimulus than for the audio stimulus. By interpolation from the data of that study – presented in Fig. 5.1 – one can conclude that a modulation rate of the audio stimulus of about 7 Hz was rated equally fast as a modulation rate of 4 Hz for the video stimulus. In a bi-sensory condition with equal rates for the audio and the video component, the judged rate for the video component decreased as compared to the video-only condition. For example, 6 Hz modulation in a bi-sensory condition was rated equally high as 4 Hz in a video-only condition. In contrast, the judged audio rate was hardly influenced by the presence of the video stimulus. In a second experiment, the audio and video modulation rates were different. In a similar fashion there was a considerable effect of the audio stimulus on the visually perceived rate, while the opposite situation showed none or only a rather small effect. These results are in line with arguments of a modality-appropriate processing, in which the auditory system is seen as better suited for temporal processes than the visual system. Such a notion is supported by the maximum rate of temporal variations that the different sensory systems are able to deal with. Visual-flicker fusion occurs at

---

[5] A demonstration of the *McGurk* effect can be found at the following WEB address: http://www.media.uio.no/personer/arntm/McGurk_english.html

rates above about 50 to 100 Hz [36, 90]. The auditory system is able to detect amplitude modulations of high-frequency tonal carriers for modulation rates of up to 600 Hz [32]. This implies that temporal acuity is higher for the auditory system than for the visual system and that in an AV context the auditory system may be dominant in determining rate perception.

5.2.4 Spatial Disparities

A valuable means to study multi-sensory interaction is by creating a multi-sensory stimulus which provides conflicting information in two sensory modalities. In such cases it is often observed that, despite the conflicting information, the percept does not segregate into isolated percepts in each modality, but that instead, by altering the conflicting perceptual features in each modality, the conflict is at least partially resolved and the stimulus is perceived as a unity.

A classical example of such an effect is known as the *ventriloquist effect* [26, 81]. In a respective study [54], conflicting spatial information was presented via the visual and auditory modality. This was achieved by presenting the visual stimulus through a prism which created an azimuthal shift. In a pointing task subjects had to indicate the perceived direction of the auditory and visual stimuli when they were presented in isolation as well as when they were presented simultaneously. It appeared that for the condition in which visual and auditory stimuli were presented simultaneously, the perceived auditory location was substantially altered towards the location where the visual stimulus alone would be perceived. The perceived visual location, on the other hand, was hardly altered by the presence of the auditory stimulus. This spatial-interaction effect decreases if a delay between the auditory and visual temporal patterns is introduced [55, 82].

The perceptual consequences of a mismatch between the locations of visual and auditory stimuli have also been studied for a discrepancy in the distances. For example, it has been the task of subjects to decide whether an auditory rattling stimulus was within a reachable distance [10]. Besides in a sound-alone condition, subjects also had to perform the task while simultaneously looking at a coherent visual stimulus. This visual stimulus did not produce any sound and was placed either 8 cm in front of the auditory stimulus or 8 cm behind it. When the visual stimulus was closer, subjects were more likely to judge that the auditory stimulus was within reach than in the audio-alone condition.

Auditory perception of larger distances beyond several meters is known to not be very precise and the perceived distance typically underestimates the physical distance [9, 50]. Due to this inaccuracy in auditory perception of distance, it might be expected that also for larger distances visual information influences perceived auditory distance. Such a dependence on the presence or absence of visual information might be of relevance when judging room acoustic properties, and it also becomes relevant for the reproduction of audio and video in three dimensions – such as in stereoscopic television, 3-D TV.

The available literature about the visual influence on auditory distance perception does not allow a clear conclusion. In one study along these lines observers were asked to judge the apparent auditory distance of sound sources created by a room-acoustics processor such that the stimuli contained distance cues based on the properties of a room [49]. Visual images of the same room were presented via a 3-D-projection screen. The perceived auditory distances, indicated by the subjects on a numerical scale, increased with increasing perceived depth of the 3-D visual stimulus. This effect became, however, only significant after a certain training period of the subjects, indicating the need for a certain familiarization with the specific stimulus. In contrast, in another study [89], no visual capture at all could be found for the judgment of auditory distance . The author emphasizes that in his experiments, which were performed in a real room with a reverberation time of 0.3 s and distances between observers and sound sources of 1 to 5 m, the acoustic stimuli were rich in distance cues, and that observers did indeed produce relatively accurate distance judgments in the absence of visual information. In the visual condition, observers could only see the closest of the five loudspeakers such that a strong visual capturing effect should change the distance judgments towards that of the visible loudspeaker. Yet, such a shift was not observed and the only interaction effect was an increased accuracy in the judged distance.

In a related study [11], the perceptual effects of combining 3-D audio with 2-D video was investigated. This study was motivated by the possibilities of reproducing real three-dimensional acoustic fields using the technique of "wave-field synthesis", WFS, . The question was whether this technique can be used in a video-conferencing application where the video information is reproduced on a two-dimensional screen. In such a visual reproduction scheme, several of the visual distance cues like binocular parallax are lost and thus, the visual perception of space by the observer does typically not agree with the original perspective. Combined with a true auditory-depth percept, using the WFS technique, a possible mismatch between A and V can occur and the study tried to establish its importance. The experiments revealed that lateral displacements of the observer compared to the original viewpoint transform differences in distance between image and sound into differences in direction. Such differences are both noticeable and annoying. By making use of the less accurate auditory-distance percept, the authors propose as solution to generate the 3-D audio with a reduced depth. In their set-up, this would reduce the amount of angular disparity created by distance discrepancies and, thus, the annoyance caused by the mismatch in perceived direction.

Two recent studies have investigated interaction effects with respect to sensitivity in changing distances and depth perception. The first of these [91] measured sensitivity to changes in size of a visual object and to changes in the level of an acoustic stimulus. Both these perceptual parameters are related to the perception of distance, where an approaching object is signalled by an increase in size and an increase in level, respectively. The authors investigated whether observers were particularly sensitive to "ecologically valid" changes in AV stimuli. This term was used to characterize changes where both the audio and the image part change according to either an approaching or a retracting object. "Invalid" changes are combinations where, for instance, the audio level increases and, at the same time, the image size decreases. Their data of just-noticeable differences indeed show a behavior in line with this expectation. When audio and video are combined the gain in performance, as compared to uni-modal presentation, was much larger for a valid than for an invalid combination.

The second study [31] dealt with motion after-effects which can be created by prolonged viewing of a moving visual pattern. Their visual pattern consisted of squares presented to the left and the right eye, which, when viewed stereoscopically, fused to one square which appeared to move towards the observers. This stimulus, when viewed for a few minutes, did create an after-effect in the auditory domain, namely, after visual adaptation, a sound with a constant amplitude was perceived as decreasing in loudness. The auditory after-effects were even stronger when the adapting stimulus consisted of an AV stimulus changing in compatible directions, i.e. an "ecologically valid" combination – as defined before. On the other hand, for an "invalid" combination of audio and video in the adapting stimulus, the after-effect disappeared almost completely. The authors also showed that an auditory adapting stimulus did not induce a visual motion after-effect.

Both studies thus provide evidence that the percept of motion, or change in distance, is closely linked between the auditory and visual modalities.

As mentioned before in Sect. 5.2.3, the apparent dominance of one sensory modality over another might be a result of differences in the suitability of the different modalities for perceptually coding a certain stimulus feature. According to the modality appropriateness hypothesis, the sensory modality which is more suitable, usually established by the accuracy of coding of this stimulus feature, is assumed to be the dominant sensory modality [86]. In a comparison of auditory and visual perception, [29] arrived at the notion of vision as the generator of the concept of space and the auditory system as a time-keeper. As for space, where little influence of audio on video is observed, this view is supported by the difference in spatial acuity which is about 1 min of arc for vision [27] and 1 degree for hearing [46]. It should, however, also be mentioned that this association of space and vision on the one side, and time and hearing on the other, is not shared by all scientists – or is considered to be too simplistic. Some of the arguments of this discussion can be found in [20, 35].

5.3 Cross-Modal Effects in Auditory-Visual Attention

One area with a long history of research efforts on AV interaction is attention. This chapter, therefore, cannot give an adequate overview over past and present research in this area. Instead, we will just mention a few recent studies in which the potential role of across-modal attention for specific application domains is emphasized. The topic becomes increasingly relevant for developments in user-system interaction, where system information is no longer provided via the visual modality alone, but also via auditory displays [8,21,56] and tactile feedback [30]. In such applications it is often implicitly assumed that, by using more than one sensory modality, the information transfer rate to the subject can be increased. It is obvious that certain features of the auditory system, like its nearly constant omnidirectional sensitivity, can be advantageous, e. g., when alerting signals come from a direction away from the direction of focussed visual attention. But the question remains of whether there are, besides advantages, also costs involved when subjects have to monitor more than one sensory modality.

We want to mention two types of experimental questions that are used in studies on AV attention. In the first type of tasks, subjects have to make a speeded response to an upcoming stimulus. It is tested whether an additional prior cue about the modality of the stimulus leads to a faster and/or more accurate response. For example, in [70] auditory and visual stimuli were presented in random order from one out of four positions which were located in a rectangle around the point of visual fixation. In a series of experiments, subjects had to react as quickly as possible by giving a colour or intensity discrimination or a spatial-discrimination response. At the beginning of each trial, a symbolic visual cue indicated the modality of the next stimulus and subjects were instructed to attend to this modality. This cue was valid in 80% of the trials. Responses were faster and more accurate for those trials, in which the modality cue was valid. In those trials, where cued modality and actual modality of the test stimulus differed, reaction times increased by about 8%. These results show that attentional resources for processing auditory and visual stimuli are not – completely – independent and that shifting attention from an expected modality needs some extra time.

In another study [17], the attentional capacity for processing stimuli were compared, namely, for those occurring in the same, and those occurring in different sensory modalities. Subjects were presented with streams of visual and/or auditory stimuli and they had to identify occasional targets embedded in these streams. When subjects had to monitor two streams in the same modality for the occurrence of deviant stimuli, e. g., a high and a low voice, performance decreased relative to the control situation where they only had to monitor one stream. The decrease was strongest when the two target stimuli occurred within a few hundred milliseconds of each other. Thus, for stimuli within a single modality, there was a clear effect of dividing attention. In the mixed-modality condition, subjects had to monitor a visual and an auditory stream. Here, the performance in the divided-attention condition was only slightly lower than in the control situation. In particular, there was no influence of the relative timing of the stimuli in the dual-task condition, indicating that identification of a stimulus in one modality did not impair the ability to identify a concurrent stimulus in another modality.

A further investigation compared the tasks of detection and identification in an AV stimulus condition [7]. Subjects were presented with 900-ms AV stimuli with a constant intensity. The target was a 20-ms change, either up or down, in intensity, occurring in the middle of the long stimulus. In the detection task, subject had to indicate whether or not they observed a change. In the identification task, in addition, they had to identify the direction of the change, either increase or decrease. Subjects were instructed to apportion their attention to a certain percentage to one modality and the remaining attention to the other modality. In the detection task, there was no significant loss in performance compared to the control situation with only one modality, independently of how attention was distributed across the two modalities. There was, however, a clear trade-off in performance in the identification task. When attention had to be attributed equally to vision and audition, performance dropped in both modalities relative to the control condition. Thus, the more difficult task of *identifying* a stimulus in one modality could not be performed without a loss in identification performance in the other modality.

Another aspect of across-modal attention is the role of space. Can subjects direct their attention in different modalities towards different places, or does attention operate on one common representation of space for all modalities? Experiments have been reported in which subjects had to judge the elevation, i. e. up or down, of acoustic or optical targets which could occur at the right or left side [74]. Subjects were cued visually to attend covertly, i. e. without moving their eyes away from the fixation point in the middle, to either the left or right side. Since the cue was given in the middle, shifts in attention should only occur voluntarily, namely, endogenously, and not as a consequence of a spatially located cueing stimulus, exogenously. When covert attention matched the side of the stimulus, response times were shorter than for trials without any cue. The longest response time was observed in trials when covert attention matched the side opposite to the stimulus. Similar results were obtained when auditory targets were presented in the same locations as the visual targets.

When cueing was presented which only was informative on one modality, e. g., the visual one, response times to stimuli in the other modalities were still biased in favour of the cued side. This is an example suggesting that covert attention in one sensory modality is not independent of covert attention in another sensory modality. Implications of this research on spatial attention for the design of multi-modal interfaces and the design of multi-modal warning signals have been discussed by the same authors [71–73].

For those readers interested in this area of research, we can recommend a number of recent review articles on this topic [13, 14, 16, 71].

5.4 Auditory and Visual Speech Perception

In the human processing of speech, the acoustic signal usually provides enough information for a high level of speech intelligibility. But under difficult acoustic conditions, like at a cocktail party, additional visual information, e. g., as resulting from seeing the movements of the speaker's mouth, can contribute considerably to the intelligibility of speech [41, 80]. For example, subjects had to recognize words which were presented with different levels of background noise [79]. In one condition the percentage of correct-word recognition was measured for auditory stimuli alone. In another condition the speaker's face was also visible. Improvements of the number of correct answers were small for speech-to-noise ratios of 0 dB where speech intelligibility was nearly perfect in both conditions. However, at lower speech-to-noise ratios improvements of 40 to 80% were measured, depending on the size of the vocabulary that was used.

The study reported above was later extended to conditions where the audio signal was delayed with respect to the visual speech signal [43]. Such a condition is relevant for people wearing a digital hearing aid or a cochlear implant. These devices need time to process incoming audio signals, which are thus no longer synchronous with the visual image as for normally-hearing persons. From their experiments, the authors conclude that for delays of up to 80 ms no detrimental effect of audio delay on the word-recognition rate is observed. Other investigators [51] found the same limit of 80 ms and showed in addition that even at delays of 300 ms, the combined AV recognition rate is still higher than the rate for either of the two modalities alone.

Further, the just-noticeable AV delay in speech-like stimuli has been determined [43]. The visual stimulus was a *Lissajou* pattern resembling a pair of lips, the acoustical stimulus was a gated 120-Hz triangular wave. Thresholds were determined with a three-interval forced-choice procedure with adaptive variation of the AV delay. The average detection thresholds of the 10 subjects were -78.5 ms, audio first, and +137.8 ms, video first.

The improvement in speech intelligibility due to the presence of visual speech in addition to auditory speech has been shown to be lost when audio and video are not synchronous [43]. The *McGurk* effect – see Sect. 5.2.2 – is another example where auditory and visual information is integrated and, interestingly, this effect also seems to be vulnerable with regard to asynchrony. However, the effect could still observed for delays in the auditory signal by as much as 180 ms [48].

In addition to synchrony, spatial alignment of the visual and audible information seems to be a factor determining speech intelligibility, as revealed as follows. Two concurrent speech signals were presented through one single loudspeaker [15]. In addition, the visual signal, corresponding to one of the audible speech signals was presented on a monitor. Interestingly, the correct perception of the speech improved from 58% to 75% when the monitor was displaced laterally with respect to the loudspeaker instead of being located just above it. The author attributed this effect to an attentional benefit that was gained due to the already mentioned ventriloquist effect [27]. When the source of the audio signal, i. e. the loudspeaker, was spatially separated from the source of the video, the monitor, the part of the acoustic speech signal that corresponded to the visual speech signal was subject to the ventriloquist effect. The physical difference in location is reduced by a shift in the perceived location of the acoustic signal towards the visual signal. In this way, the two concurrent speech signals are perceived at different locations. This allows for an attentional benefit which resulted in improved speech intelligibility.

5.5 Perceived Quality of Auditory-Visual Stimuli

In this section, we will address the practically relevant problem of perceived quality of AV reproduction systems. For such systems it is not sufficient to know the perceived qualities of the audio and the video signals separately, but one also has to understand how these two influence each other and how they contribute to the overall AV quality. When we became interested in this problem some years ago, we were aware of informal claims of interaction between the two modalities, like an improvement in perceived image quality by an increased sound-reproduction level. It turned out, however, to be very difficult to find formal scientific studies that had addressed this issue. In the following, we will summarize the results of some studies which have been published in recent years. Additional discussions and references can be found in [88].

Within the framework of high-definition television, the question was investigated of how much spatial disparity between the sound and the video image was acceptable [34]. Pictures were displayed on a 182-cm television projector, which covered a viewing angle of about 30°, horizontally. Sound was reproduced by one out of ten loudspeakers which were placed in front and on the left side of the subject including the rear direction. The program consisted of a person reading a magazine. Subjects had to indicate on a 5-point impairment scale how much they were annoyed by the mismatch in the spatial location between picture and sound. It turned out that acoustic-research engineers were more critical about spatial disparity than non-experts. Taking as a threshold the transition between response categories 2, "perceptible, but not annoying", and 3, "slightly annoying", the former had their threshold at an angular discrepancy of 11° while the latter tolerated a discrepancy of up to 20°. These results were taken as evidence that, for HDTV reproduction, it would be sufficient to reproduce the frontal sound via stereo speakers with an opening angle of 60° in a standard-stereo set-up plus a center channel placed directly underneath the screen.

The intention of an experiment within the framework of the European Community's MOSAIC project was to compare judgements of picture quality with and without accompanying sounds. The picture material consisted of 30 sequences of 50 s, which were taken from a digital video-tape recording of a humoristic tale – in D1 format. Video sequences were presented with three different qualities, namely, uncompressed D1-reference, bit-rate reduced to 4 Mbit/s, and bit-rate reduced 2 Mbit/s. The sound-track was a mix of the original live sound with an appropriate Hi-Fi music background. Subjects had to rate the technical overall scene quality on a 10 point scale in a single stimulus paradigm. Two different groups of observers participated, one judging the video only material and the other one judging the video with accompanying sound. The results showed that every sequence received a higher quality score when it was accompanied by sound. This was true despite a great variety of video-quality judgements for the individual items, due to different bit-rates and content. The author concluded that "subjects' ability to detect picture impairments is lowered when a good-quality sound programme is added". From additional experiments, with impairments of both the audio and the video path, it was concluded that "an audio impairment is more detrimental to quality assessment than a video impairment". Another investigation studied the subjective attributes that contribute to the AV experience in a home theater [4, 88]. They emphasize the role of the AV system in communicating the intentions of the author/producer of a program to the viewer. The first paper [88] derives four perceptual dimensions for which the relation between sound and picture is more important than the effect of one of these modalities alone. These dimensions are "action", "motion", "mood", and "space". In the experiments [4], the contributions of sound and picture to the "space" dimension was assessed by varying screen size and various aspects of the audio reproduction, e. g., adding a sub-woofer, different distance between stereo speakers, and two different surround-sound systems. The subjects viewed five different film scenes of between 30 s and nearly 2 min duration and responded to five questions related to their experience of space. Subjects also had to indicate whether sound or picture contributed stronger to the perception of space. An increasing screen size improved the impressiveness of the reproduction and increased the – relative – contribution of picture on the overall impression of space. But even for the widest screen size of 94 cm, the contribution of the two modalities was judged to be equal, which emphasizes the influence of a good surround-sound reproduction for the overall perception of space. Besides giving a lot of experimental findings, this paper [4] is also a valuable source for technical details and the discussion of problems in evaluating AV quality.

In a study performed in a cooperation between the Institute for Perception Research, IPO, and Philips Research Laboratories, both at Eindhoven, the contribution of changes in presented video and audio quality on the perceived overall quality was measured [33, 45]. As in some other studies cited in this section, one intention of this study was to find out whether an effect of both audio and video quality on the perceived overall quality could reliably be measured. Three scenes of approximately 20 s each were used. In one scene, which showed a live performance of a Dutch artist, the video and audio signal had a functional relationship. For the two other scenes a musical fragment was chosen which seemed appropriate to accompany the video in terms of rhythm and atmosphere. The three video qualities were realized by choosing different MPEG2 bit-rates. Besides an unprocessed original, bit-rates of 3 Mbit/s and 2.3 Mbit/s were used. The three audio qualities were reached by varying the bandwidth characteristic and the perceived width of the stereo image. For the audio quality expected to be the best, a left, a right and a center loudspeaker were used in the price range typical for audio Hi-Fi sets. The angle between the right and the left loudspeakers was 60°. For the two other audio quality levels, the loudspeakers accompanying the HDTV television set were used. In one scheme – expected to give the lowest quality – normal-stereo reproduction was used. For the third audio quality, a source widening scheme known under the name "incredible sound" was used [1]. The difference between the two latter schemes was thus the apparent width of the stereo image. These qualities were combined to give nine different AV qualities. Subjects were asked to rate the overall quality of the stimuli on a continuous scale from 1 to 5. Increasing either the video or the audio quality always increased the overall perceived quality of the stimulus. Interestingly, the change in overall quality by going from the highest to the lowest video quality was about the same as by going from the highest to the lowest audio quality. Thus, at least for the stimuli used in this test, one can conclude that a bit-rate reduction to 2.3 Mbit/s in the video signal reduces the overall quality by about the same amount as it is increased by replacing the "low-quality" TV speakers by three Hi-Fi speakers.

Mutual interaction between video and audio qualities was also the topic of a study performed by KPN research as a contribution to the "study group 12" of the International Telecommunication Union, ITU, [5,6]. Here two questions were addressed, namely, how do audio and video quality interact and how are they combined to a single audio–visual quality. This study extended earlier work done, by, e. g., Bellcore, Swisscom and British Telecom. Stimuli were presented in five different modes. These modes where: a video only mode, in which the video quality had to be judged, VV, and, similarly, an audio only mode, AA. Further, an audio–visual mode for which three different quality judgements were requested, namely, video quality only, AVV, audio quality only, AVA, and, finally, over-all audio–visual quality, AVAV. The material consisted of four commercials of 25 s duration. Four different video and four different audio qualities were generated by spatial filtering of the luminance signal, video impairment, and by various amounts of bandwidth limitation for the audio signals. Judgements were given on a 9-point absolute-category-rating scale.

The data showed a rather strong influence of video quality on the judged audio quality. When the video quality was highest, the judged audio quality was about half a mean opinion score higher, $+0.5$ MOS, than in the audio only mode. On the other hand, when the lowest video quality was chosen, the MOS score for the audio quality was lower by about - 0.7 MOS as compared to the audio-only mode. Thus, dependent on the levels of video quality, the MOS scores for the perceived audio quality varied by 1.2 MOS. The influence of audio on video, by comparing AVV and VV, was much smaller and covered an MOS range of only 0.2. A comparison of the contributions of audio and video to the overall audio–visual quality showed a much stronger influence of the video quality. The correlation between video and audio–visual quality turned out to be 0.9, while the overall correlation for audio was only 0.35. One possible explanation for findings, also mentioned by the authors, is that the four levels of video degradation led to a much larger variation in perceived video quality than the audio degradations did for the perceived audio quality. But even when this difference was accounted for, a much stronger contribution of video quality to the audio–visual quality remained present in the data.

The influence of video quality on audio perception was also studied in [23]. The authors combined eight different video clips, each lasting about 10 s and showing virtual-reality fly-throughs with two different narratives, related to the content of the clips. Video quality was degraded by either distorting edges, blurring or adding noise, and audio degradations were realized by either band limitation or addition of noise. The subjects performed two sessions, one in which they judged audio quality alone, and one in which overall AV quality was judged. Answers were given on a 5-point quality scale. The data show a similar trend as the ITU study does [6]. With a stronger degraded video signal, the perceived audio quality decreases. But there is one clear difference. It was found that, when no video is present, the perceived audio quality is always judged lower than with video – even when the video signal was highly degraded. In the ITU study, a high video quality increased the scores for audio quality relative to judgements on audio-only presentation, while a low video quality decreased the judgements on the audio quality.

A related study, also performed at the British Telecom Laboratories [58], used a similar set-up as the ITU study [6]. Four audio qualities were combined with nine video qualities to give 36 AV-quality combinations. Video degradation was realized in a similar way as in [23] while audio degradations were based on MPEG coding at rather low bit-rates and, also, by cascading a codec up to five times. The material consisted of 6-s AV speech clips showing the face and shoulders of a speaker. Subjects gave their responses on a 5-point quality scale. The data show a strong interaction effect between audio and video quality which was, in contrast to the results of the ITU study [5], symmetric. In other words, this study also observed a strong effect of presented audio quality on judged video quality. The MOS score for the original audio track was 4.8 when the sound was accompanied by the original video signal, and decreased to a value of 2.3 when the lowest video quality was shown. The influence of the presented audio quality on the judged video quality was of a similar size. A reason for these different outcomes from the BT and the ITU studies might be that in the former, the A and V components were highly correlated – same speech source for video and audio – while in the latter audio and video were not correlated.

In a thesis related to the perceived realism in "virtual environments" across-modal effects in perceived quality were investigated [78]. The test material included still pictures, presented by a 50-cm monitor together with music. Image quality was manipulated by choosing three different levels of pixel density, while audio quality manipulations were realized by limiting the audio bandwidth by choosing sampling frequencies of 44.1, 17 and 11 kHz. In agreement with other studies, the author of the study observed that, when the quality is high for both modalities, the presence of high-quality audio increased the perceived image quality compared to the image-only condition. In contrast to other findings, he reported that, when a low-quality auditory display is combined with a medium or high-quality visual display, the perceived audio quality decreased. Thus, in contrast to the previously mentioned studies which always observed a positively-pulling effect of high visual quality, this study found a repelling effect on the perceived audio quality.

In summary, all studies mentioned here show a clear mutual influence of the auditory and visual modality on the perceived quality. The question of which of the two modalities contributes more to the overall AV quality cannot be answered un-equivocally. Depending on the choice of stimuli, i.e. video or still pictures and/or causal relation between audio and video being present or absent, very different amounts and even different signs for the across-modal influence have been found. Given this potential for across-modal quality effects and the volume of the multi-media consumer market, it is surprising that – as revealed by a rather small number of well-documented experiments – so little attention is being paid to this relevant question.

5.6 Sensitivity to Temporal Asynchrony in AV Stimuli

5.6.1 Perceived Temporal Relations in Simple Stimuli

Already in the 19th century, questions of perceived AV asynchrony were addressed experimentally. One motivation for this research were accuracy differences that had occurred in astronomical observations [37, 61]. In an early study *Exner* measured the smallest timing difference which had to be introduced between two stimuli presented to two different sensory modalities, such that these stimuli were perceived as being asynchronous [18]. Among the many different combinations of two stimuli in this study, there were also combinations of acoustic and optic stimuli, namely, the sound of a bell and the image of an electric spark. After each presentation of an AV stimulus with a certain delay, the subjects had three response alternatives: spark first; bell sound and spark simultaneous; spark second. The response patterns expected in such a task are explained schematically in the right panel of Fig. 5.2. Here,

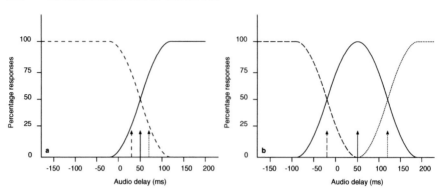

Fig. 5.2. Schematic response patterns in temporal-order judgment tasks. The pattern in the **left** panel is based on two response categories, i. e. audio first, *dashed curve*, and video first, *continuous*. The pattern in the **right** panel is based on three response categories, audio first, *dashed*, synchronous, *continuos*, and video first, *dotted* – see text for further details

the percentages of responses in the three response categories are plotted as a function of the audio delay in the AV stimulus. Negative delays are used for conditions where the audio signal precedes the video signal, and positive values for conditions where the video signal precedes the audio signal. *Exner* found that the smallest detectable delay depended on the order of the acoustic and the optic stimuli. When the acoustic stimulus was presented first, the value was -60 ms. When the optic stimulus was presented first, the value was 160 ms. Thus, the point of subjective equality, PSE, computed as the mean between the two detection thresholds, was reached for an audio delay of 50 ms.

In the context of temporal-order judgments in different sensory modalities, the perceived order for short audio–visual stimuli was studied in [22]. After each presentation of an AV pair, subjects had to indicate whether the audio or the video was presented first. The response pattern for such a task is depicted in the left panel of Fig. 5.2. The 50% point of the responses, PSE, see middle arrow in the figure, occurred at a small positive delay value of about +5 ms, visual before auditory. Taking the 25%- and 75%-response values as indicators for the transition between synchronous and asynchronous, the authors found values of about -25 and +25 ms. Thus, in contrast to the early data, the two situations, auditory first and visual first, lead to the same absolute values, and these are much smaller than the values obtained in [18]. Thus, it was discussed whether differences in the experimental procedures and the different stringency of the threshold criterion might be reasons for the much larger values found in the later experiments.

A further investigation [60] compared reaction times for auditory and visual stimuli with temporal-order judgments for short AV stimuli. Five different AV delays were presented using the method of constant stimuli, and subjects had two response categories, i.e., audio first or flash first. The analysis of this latter experiment was similar to the procedure used in [22], but gave quite different results. The point of subjective equality for the two subjects was reached for AV delays of -40 ms and -46 ms, i. e. the auditory stimulus had to precede the visual stimulus. This result contrasts with the observation that the reaction time to the auditory stimulus was about 45 ms shorter than the reaction time to the visual stimulus. Assuming that a shorter reaction time indicates some kind of faster internal processing, the point of subjective equality was expected to occur at positive AV delays, i. e. when the visual stimulus is presented first.

This discrepancy led to a repetition of this experiment with a very similar experimental paradigm [28]. The new study confirmed the finding reported in [60], namely, that the reaction time for the visual stimulus was about 50 to 60 ms longer than for the auditory stimulus. However, the temporal order judgment revealed quite different results. The point of subjective equality for AV stimuli occurred for positive delays in a 35-to-60 ms range of values for the three subjects, indicating that, for subjective equality, the visual stimulus had to be presented *before* the auditory one.

In an experiment focussing on the perceptual processing of stationary and moving visual stimuli [2], subjects were asked to perform temporal-order judgments between a short audio signal and either a moving or a 125-ms stationary visual stimulus. The response categories of the experiments were audio first or video first. From the data rendered, the point of equal-response probability, 50%, was calculated. Eight different AV delays were presented to the 16 subjects. The experiment did not reveal any difference between the values for moving and stationary visual stimuli. On average, the auditory stimulus had to be presented 15 ms *before* the visual stimulus to be judged as being simultaneous with the onset of the visual stimulus. This outcome goes in the same direction as the data given in [60] where, using the same response categories, a negative value was found for the point of subjective equality as well.

Another study [77] focused on measuring the point of subjective equality for a large number of observers and compared the values over a period of 1 day. The stimuli consisted of a red LED and a 250-Hz tone delivered via headphones. The response categories were synchronous and asynchronous. From 1000 presented trials, one value for the PSE was derived. For the 17 subjects, this value varied between -21 ms and +150 ms, with a mean of +51 ms. The authors emphasize two aspects of these data. First of all, the value of the PSE is observer-specific, and secondly, it is much more stable over the period of 24 hours than the reaction time. In an additional measurement, the authors also compared PSE values for AV stimuli presented in a near, 0.5 m, and a far, 3.5 m, condition. The far condition contains an additional audio delay of 11 ms caused by the travel time of sound. The results show that the PSE value, not corrected for the additional delay, differs between the two conditions, thus indicating that observers do not discount the extra physical delay. This result is also supported by the data from [59] presented in the next subsection.

In [52], thresholds for auditory-visual asynchrony detection were measured using an adaptive procedure. The visual stimulus was a disk displayed on a computer monitor that accelerated linearly downwards until it apparently bounced against a bar and moved upwards. In each trial, two intervals were presented. In one interval audio and video were synchronous, in the other they were asynchronous. In the synchronous interval, the auditory tonal stimulus started at the moment of visual incidence with a sharp onset followed by an exponential decay. In the asynchronous condition, the same auditory stimulus started either before or after the visual incidence. Detection thresholds in the audio leading condition had an average value of -29 ms, while larger values of 80 ms were obtained in the audio delayed condition.

Besides for rather simple audio–visual stimuli, temporal-order judgment have also been obtained using speech stimuli. For example, speech samples [69] constructed as consonant-vowel-consonant series, CVC, were employed to perform temporal order judgments. In a first experiment, subjects had to respond after each syllable whether the audio was first or the video. The 50% score values varied considerably across the 10 syllables, namely, from -8 ms to -229 ms, but all values were negative, indicating that perceived synchrony was reached when the audio signal *preceded* the video signal. In the second measurement, the same stimuli were used, but now the response alternatives were "synchronous" and "asynchronous". In a similar way as in the three-category procedure shown in the right panel of Fig. 5.2, such data show two transitions between synchronous and asynchronous as a function of the physical delay. The mean of the two 50% points was taken as synchronization point. Also these synchrony values varied considerably across syllables, i. e. -131 ms to +78 ms. Yet, the mean synchronization point in experiment 2 was shifted by +94 ms in the direction of greater audio delay, as compared to experiment 1. These results are remarkable in at least two respects. First of all, experiment 1 showed a huge negative value for the point of subjective equality, i. e. synchrony – in fact, these values are the most negative ones we have found so far in the literature. In addition, the large difference between the two experiments may indicate that these procedures do not measure the same effects, even if in many papers they are not discussed separately.

In summary, the majority of studies found an asymmetry in the perception of the temporal relation in AV stimuli. This asymmetry can be expressed by means of the point of subjective equality, which occurs in most studies for a positive delay of the auditory stimulus of about 30 to 50 ms relative to the visual stimulus. Only a few studies cited here, e. g., [2, 60, 69], found an opposite result, namely, that for subjective equality the auditory stimulus has to be presented before the visual stimulus. This discrepancy has been

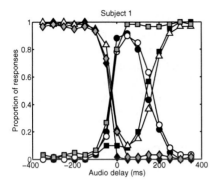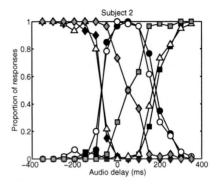

Fig. 5.3. The results of temporal-order judgments using three different response sets are shown for two subjects. The *black symbols* show results for allowing three different response categories, namely, audio first, *diamonds*, synchronous, *circles*, and video first, *squares*. The *white symbols* indicate data which were obtained with two response categories, i. e. synchronous, *circles* and asynchronous, *triangles*. Further, the *gray data* were obtained with a classical temporal-order paradigm with two response categories, audio first, *diamonds*, and video first, *squares*

discussed as being due to the different amounts of naturalness of the stimuli used in the experiment [3].

The authors of this chapter have recently proposed an alternative explanation based on experiments in which different experimental procedures were compared directly [53]. Using the same abstract AV stimuli and the same subjects, the three procedures described in Fig. 5.2 were used to measure the points of subjective equality. The panels in Fig. 5.3 show the response patterns of two subjects for the different categories as a function of the delay of the audio relative to the video. The black symbols show results for a measurement allowing three response categories. The synchrony curve in the middle is clearly off-set away from 0 ms towards positive audio delays. This result agrees with the data indicated by the white symbols, which were obtained in a measurement with two response categories, "synchronous" and "asynchronous". In contrast, the data obtained by the classical temporal-order paradigm, gray symbols, allowing response categories "audio first" and "video first", show a deviant behaviour. The results show large differences between the two subjects which might indicate different response strategies. With the first strategy, even when subjects perceive the stimuli as being synchronous, they try to divide their responses equally over the two categories "audio first" or "video first". Following such a strategy, the category transition occurs in the middle of the synchronous curves, as shown by the results of subject 2 in the right panel of Fig. 5.3. With the second response strategy, subjects only respond "audio first" when audio is perceived as leading, and they respond "video first" in all other cases, including those where the stimulus is perceived as synchronous. Such a strategy leads to a point of subjective equality at negative audio delays and has the advantage of the transition between the categories being sharper – as shown in the left panel of Fig. 5.3. Thus, in our opinion, the different values of PSE seen in the literature, and in particular the cases with a negative audio delay, are probably caused by the fact that in the temporal-order paradigm the response strategy of the subjects is more variable as compared to procedures which allow to give responses in the category "synchronous" [52, 53].

5.6.2 Asynchrony Detection and Quality Degradation for Video Reproduction

We have discussed the perceptual consequences of asynchrony between audio and video in much detail, because it is of particular relevance in multi-media applications. In this context, AV synchrony has mainly been studied with respect to two perceptual aspects, i.e. the detectability of asynchrony and the effect of AV asynchrony on perceived quality.

Thresholds for detecting asynchrony for two types of natural video stimuli have been measured in [12], with a speech utterance and a hammer hitting a peg. The videos were presented such that, starting with physical synchrony between audio and video, the AV delay was continuously decreased or increased, respectively, with a rate of 51 ms/s, until subjects detected asynchrony. For both stimuli, the thresholds were larger for the video-first condition. In addition, thresholds were generally larger for the speech signal as compared to the impact event. The actual threshold values are given in Tab. 5.1.

In a related context the influence of AV delay on perceived impairments of TV scenes was measured [57]. A video of a talk show with several speakers was utilized. Subjects were presented 50-s sections of the show with a fixed AV delay and they had to judge upon the decrease in perceived quality on a 5-point impairment scale. A decrease of half a point on this scale was defined as the threshold of detectability. These values were about -40 ms for the audio advanced and 120 ms for the audio delayed. In the same study, the acceptability thresholds were defined as those delays, for which the quality

Table 5.1. Asynchrony detection thresholds for two different video scenes and two conditions, audio delayed and video delayed. The 3rd row, labelled PSE, indicates the values of the points of subjective equality, i.e. the points in time midways between the thresholds given in the 1st row and the 2nd row

	Voice	Hammer
Audio delayed	+258 ms	+188 ms
Video delayed	-131 ms	-75 ms
PSE	+63.5 ms	+56.5 ms

decreased by 1.5 points. These values were -90 ms and 180 ms, respectively. The results from this study are reflected in the most recent recommendation of the ITU, Rec. ITU-R BT.1359, for the relative timing of sound and vision for broadcasting.

In an unpublished thesis [59] the perceived quality for a number of video scenes as function of the AV delay was assessed. On average, the quality for a delay of +80 ms was judged to be slightly higher than that for a synchronous condition. Interestingly, this judgment was the same for both lay-persons who had no experience with film-editing and subjects who had several years of experience – from the film university Potsdam-Babelsberg. The author also obtained quality judgments for two scenes showing impact events, but filmed from either a short, 5 m, or a large, 33 m, distance. In the distant scene, the recording already contains an audio delay of about 100 ms due to the limited speed of sound. The quality judgments differed between these two scenes. For the close scene, an extra 80-ms audio delay led to a higher perceived quality than synchronous recording while, for the distant scene, the synchronous condition received a higher quality score – see also below.

Besides for TV, timing has also been studied in other application contexts. For example, synchrony sensitivity has been investigated in order to develop a multi-modal perceptual model for multi-media and tele-presence products and services [24, 25]. The stimuli consisted of two different impact events, one with a short and one with a long visual cue, and a 4.5-s speech signal. Subjects were asked to indicate whether they observed a synchronization error in the scene. The results showed the same asymmetry between audio lead and audio lag as mentioned in the other studies. The thresholds, defined as the delay for which 50% of the subjects reported a synchronization error, was about -100 ms for audio lead and 175 ms for audio lag. In the audio-lead condition, the threshold for speech was somewhat lower than for the impact events, while the opposite trend was found in the audio-lag condition.

In the context of the requirements for multi-media presentations, very similar questions were addressed in [76]. In this study the detectability of lip-synchronization errors for three different views of a speaker was measured, i.e. a head view, a shoulder view and a body view – indicating increasing distance to and decreasing size of the face of the speaker. Subjects had to indicate whether a given scene contained a synchronization error. This procedure corresponds to the method of constant stimuli with two response alternatives, synchronous and asynchronous, see Fig. 5.2. As explained in Sect. 5.6.1, for such a procedure thresholds can be defined as the delays leading to 50% asynchronous responses. These values were smallest for the head and largest for the body conditions. For the head condition, which comes closest to the speech stimuli used in other studies, the thresholds were about -90 ms, audio lead, and +110 ms for audio delayed. Thus also these data show a slight asymmetry towards more tolerance for the audio-delayed condition. This study also contains several additional data sets for other stimuli like synchronization between a visual pointer and an acoustic description of a scene, but these experiments go beyond the scope of this overview.

The sensitivity to detect the delay of the audio signal for applications in auditory displays was the topic of [47]. According to the analysis presented there, the computational requirements for sound-synthesis, modelling of environmental effects like reverberation, *Doppler* shifts, and the generation of 3-D sounds are at least 66 ms – based on 1998 technology. This led the authors to their experiments in which thresholds were determined for different AV stimuli, including speech, single impact events, and repeated impact events. The average thresholds were lowest for repetitive impact events, 172 ms, and highest for speech, 203 ms. For the various single impacts events, thresholds ranged from 177 to 192 ms. The lowest threshold value of any observer in this test was 100 ms, which indicates the available time budget for computation of the audio signal, given that no computation time is needed for the visual stimulus.

In summary, all these studies using natural video material as stimulus show a very consistent pattern. Audio–visual delays are much easier detected for an *advance* of the audio than for an audio *delay*. The point of subjective equality, defined as the midpoint between the detection of a video advance and an audio advance, lies in the range 30 to 50 ms.

5.6.3 An Explanation for the Asymmetric Sensitivity to AV Delays?

This overview has shown that the great majority of studies found a clear asymmetry for detecting AV asynchrony for audio lead on the one hand and audio lag on the other, independent of the specific stimulus condition and task of the subjects. The point of subjective equality occurred for a physical delay in the AV stimulus of about 30 to 50 ms. The discussion of the reason for this observation focusses in nearly all studies on the physical delay in the audio signal caused by the rather low speed of sound of 340 m/s. If the perceptual system was adjusted to perceive AV objects from a certain distance, the reference point for AV synchrony should not be the head of the observer, but the position of the object. Since the perception experiments reveal a synchrony offset of 30 to 50 ms, the distance of this reference point should be about 10 to 15 m.

Data presented in [39] allow to answer the question of whether such an adaptation is acquired during early exposure to audio–visual stimuli or whether is must be inherited. In the experiments reported, synchrony had to be judged for a click and the bouncing of a visual disk. Adult subjects and also very young children from 2 to 8 months of age took part. For both groups, the temporal synchrony window was asymmetric with a larger extension for the condition video first. The major difference between the children and the adult observers was the much larger synchrony range in the children's responses. Thus, if our perceptual sensitivity indeed reflects an adaptation to the physical laws, his data argue against an acquisition in early life. After all, at the age of a few months, young children probably have not had the chance to perceive and to adapt to – meaningful – audio–visual stimuli from a distance of 10 to 15 m.

Some neurophysiological observations support such a view. Auditory, visual and somato-sensory information converges on cells in the *superior colliculus*, SC. These cells respond optimally to AV stimuli when the input from auditory and visual sensory receptors occurs simultaneously. The delay for transduction from the receptors to the cells in the SC is, however, quite different for the two modalities. They are in the range of 10 to 30 ms for auditory stimuli and 55–125 ms for visual stimuli [44]. In order for AV stimuli to occur simultaneously at the level of the SC, auditory signals should reach the peripheral receptors 20 to 115 ms later than the visual stimulus, which corresponds to the range of latencies for 90% of the cell population. Thus, the neurophysiological transduction delay could compensate for a certain delay caused by the physical distance of an object. In order for such a scheme to work at variable distances, the cells should show a rather wide temporal-integration window and this is indeed observed in neurophysiological experiments [75].

Such a neurophysiological basis would also explain the finding reported in [59,77] for synchrony perception of distant AV objects. If subjects were able to recognize the distance in a recorded scene and compensate for the delay caused by this distance, the perceptual consequences of an extra-inserted audio delay should be the same for the close and the distant scene. However, in [77] the subjects did not discount the extra delay of 11 ms. And giving an additional 80-ms delay to a scene which already contained a "natural" delay of 100 ms, as in [59], leads to a condition where the overall delay is beyond the – physiologically-based? – delay tolerance and this is reflected in a decrease in the perceived quality for this scene.

5.7 Concluding Remarks

Considering the numerous studies described in this paper, it is clear that there are many situations where auditory and visual perception cannot be regarded as independent processes.

Integration of auditory and visual information is an important process which, for instance, contributes to speech intelligibility. In the development of technical applications knowledge about the ability of our perceptual system to integrate AV information can be used to arrive at proper system specifications. For instance, there is a certain range of AV delays for which the audio and video signals are perceived to be in synchrony and, also, for the exact placement of audio and video sources, there is a certain tolerance since our perceptual system can compensate for spatial disparities.

The observation that there are various ways in which auditory and visual perception can interact indicates that in multi-media applications attention has to be paid to such interaction. From AV-quality studies it can be learned that, for the quality assessment of applications where both audio and video are presented, it is very important to measure not only quality for audio and video separately, but to consider interaction as well. Also the ability to attend to a certain stimulus in one sensory modality can be impaired by stimuli in another sensory modality. However, there is strong indication of that there can be an attentional benefit in presenting information through both the auditory and visual modalities instead of through only one of these.

With future introduction of even more multi-media applications there is certainly a growing need to create and apply knowledge about the interplay between auditory and visual perception. We hope that this overview contributes to the insight on how multi-sensory aspects of perception may influence new multi-media applications.

Acknowledgment

The authors want to thank *P. Novo* and *I. Vogels* for their comments on an earlier version of the manuscript.

References

1. Aarts R (1995) On the design and psychophysical assessment of loudspeaker systems. Doct diss, Delft Univ Techn., Delft
2. Aschersleben G. Müsseler J (1999) Dissociations in the timing of stationary and moving stimuli. J Exper Psych 25:1709–1720
3. Aschersleben G (1999) Cognitive contributions to the perceptions of spatial and temporal events. 421–424. Elsevier Science, Amsterdam
4. Bech S, Hansen V, Woszczyk W (1995) Interaction between audio–visual factors in a home theater system: Experimental results. 99th Conv/ Audio Engr Soc, New York. Preprint 4096:K-7
5. Beerends J G, de Caluwe F E (1997) Relations between audio, video and audiovisual quality. Contr COM 12-19 to ITU-T Study Group 12. Int Telecom Union, ITU, Geneve
6. Beerends J G, de Caluwe F E (1999) The influence of video quality on perceived audio quality and vice versa. J Audio Engr Soc 47:355–362
7. Bonnel A M, Hafter E R (1998) Divided attention between simultaneous auditory and visual signals. Percept Psychophys 60:179–190
8. Bronkhorst A, Veltman J A, van Breda L (1996) Application of a three-dimensional auditory display in a flight task. Human Factors 38:23–33
9. Bronkhorst A, Houtgast T (1999) Auditory distance perceptions in rooms. Nature, 397:517–520
10. Brown J M, Anderson K L, Fowler C A, Carello C (1998) Visual influences on auditory distance perception. J Acoust Soc Amer 104:1798

11. Bruijn W de, Boone M (2002) Subjective experiments on the effects of combining spatialized audio and 2-D video projection in audio–visual systems. 112th Conv Audio Engr Soc, Munich. Preprint 5582

12. Dixon N F, Spitz L (1980) The detection of audiovisual desynchrony. Perception 9:719–721

13. Driver J Spence C (1998) Cross-modal links in spatial attention. Phil Trans R. Soc Lond. B 353:1319–1331

14. Driver J Spence C (1998) Attention and the crossmodal construction of space. Trends Cogn Sc, 2:254–262

15. Driver J (1996) Enhancement of selective listening by illusory mislocation of speech sounds due to lip-reading. Nature 381:66–68

16. Driver J, Spence, C (1998) Crossmodal attention. Current Opinions Neurobiol 8:245–253

17. Duncan J, Martens S, Ward R (1997) Restricted attentional capacity within but not between sensory modalities. Nature 387:808–810

18. Exner S (1875) Experimentelle Untersuchung der einfachsten psychischen Processe. III. Der persönlichen Gleichung zweiter Theil. Arch gesammte Physiol des Menschen und der Tiere. 11:403–432

19. Gebhard J W, Mowbray G H (1959) On discriminating the rate of visual flicker and auditory flutter. Amer J Psychol 72:521–528

20. Handel S (1988) Space is to time as vision is to audition: seductive but misleading. J Experim Psychol: Human Percept Perform 14:315–317

21. Hermes D J (1998) Auditory material perception. IPO Ann Progr Rep 33:95–102

22. Hirsh I J, Sherrick C E (1961) Perceived order in different sense modalities. J Experim Psychol 62:423–432

23. Hollier M P, Voelcker R (1997) Objective performance assessment: Video quality as an influence on audio perception. 103 [rd] Conv Audio Engr Soc, New York. Preprint 4590:L-10

24. Hollier M P, Voelcker R (1997) Towards a multi-modal perceptual model. BT Technol J, 14:162–171

25. Hollier M P, Rimell A N (1998) An experimental investigation into multi-modal synchronisation sensitivity for perceptual model development. 105st Conv Audio Engr Soc, San Francisco. Preprint 4790

26. Howard I P, Templeton, W B (1966) Human spatial orientation. Wiley, New York

27. Howard I P (1982) Human visual orientation. Wiley, New York

28. Jaśkowski P, Jaroszyk F, Hojan-Jezierska D (1990) Temporal-order judgment and reaction time for stimuli of different modalities. Psychol Res, 52:35–38

29. Julesz B, Hirsh I J (1972) Visual and auditory perception – an essay of comparison. In: David jr E E, Denes P B (eds) Human communication: A unified view. 283–340, Mc Graw Hill, New York,.

30. Keyson D V (1997) Touch in user interface navigation. Doct diss, Eindhoven Univ. of Techn, Eindhoven

31. Kitagawa N, Ichibara S (2002) Hearing visual motion in depth. Nature 416:172–174

32. Kohlrausch A, Fassel R, Dau T (2000) The influence of carrier level and frequency on modulation and beat-detection thresholds for sinusoidal carriers. J Acoust Soc Amer 108:723–734

33. Kohlrausch A, Messelaar A, Druyvesteyn E (1996) Experiments on the audio-video quality of TV scenes. In: Hauske G, Heute U, Vary P (eds) Worksh Quality Assessment in Speech, Audio and Image Communication 105–106

34. Komiyama S (1989) Subjective evaluation of angular displacement between picture and sound directions for HDTV sound systems. J Audio Engr Soc, 37:210–214

35. Kubovy M (1988) Should we resist the seductiveness of the space/time vision/audition analogy? J Experim Psychol: Human Percept Perform, 14:318–320

36. Landis C (1954) Determinants of the critical flicker-fusion threshold. Physiol Rev 34:259–286

37. Levitin D J, MacLean K, Matthews M, Chu L, Jensen E (2000) The perception of cross-modal simultaneity, or "The Greenwich observatory problem" revisited. In: Dubois, D M (ed) Computing anticipatory systems: CASYS'99 3rd Int Conf 323–329

38. Lewald J (1998) The effect of gaze eccentricity on perceived sound direction and its relation to visual localization. Hearing Res 115:206–216

39. Lewkowicz D J (1996) Perception of auditory-visual temporal synchrony in human infants. J Experim Psych: Human Percept Perform 22:1094–1106

40. Massaro D W (1998) Perceiving talking faces. From speech perception to a behavioral principle. MIT Press, Cambridge MA

41. Massaro D W (1987) Speech perception by ear and eye. Lawrence Erlbaum, Hillsdale NJ

42. McGurk H. MacDonald J (1976) Hearing lips and seeing voices. Nature, 264:746–748

43. McGrath M, Summerfield Q (1985) Intermodal timing relations and audio-visual speech recognition by normal-hearing adults. J Acoust Soc Amer, 77:678–685

44. Meredith M A, Nemitz J W, Stein B E (1987) Determinants of multisensory integration in superior colliculus neurons. I. Temporal factors. J Neurosc 7:3215–3229

45. Messelaar A P J M (1995) A preliminary test for measuring the audio-video quality of television scenes. Rep 1064, Inst Perception Res, IPO, Eindhoven

46. Mills A. W (1958) On the minimum audible angle. J Acoust Soc Amer 30:237–246

47. Miner N. Caudell T (1998) Computational requirements and synchronization issues for virtual acoustic displays. Presence, 7:396–409

48. Munhall K G, Gribble P, Sacco L, Ward M (1996) Temporal constraints on the *McGurk* effect. Perc Psychophys 58:351–361,

49. Nathanail C. Lavandier C. Polack J D. Warusfel O (1997) Influence of sensory interaction between vision and audition of the perceived characterisation of room acoustics. Proc. Int Comp Music Conf, ICMC 97, 414–417

50. Nielsen S (1991) Distance perception in hearing. Doct diss, Aalborg Univ, Aalborg

51. Pandey P C, Kunov H. Abel S M (1986) Disruptive effects of auditory signal delay on speech perception with lipreading. J Audit Res 26:27–41

52. van de Par S, Kohlrausch A, Juola J F (2005) Synchrony judgments and temporal discrimination thresholds for audio–visual stimulus pairs. Perc Psychophys :submitted

53. van de Par S, Kohlrausch A, Juola J F (2002) Some methodological aspects for measuring asynchrony detection in audio–visual stimuli. In: Calvo-Manzano A, Perez-Lopez A., Santiago J S (eds) Proc Forum Acusticum Sevilla 2002. CD–Rom

54. Pick H L, Warren D H, Hay J C (1969) Sensory conflict in judgements of spatial direction. Perc Psychophys 6:351–361

55. Radeau M, Bertelson P (1977) Adaptation to auditory-visual discordance and ventriloquism in semirealistic situations. Perc Psychophys 22:137–146

56. Rauterberg M (1998) About the importance of auditory alarms during the operation of a plant simulator. Interacting with Computers 10:31–44

57. Rihs S (1995) The influence of audio on perceived picture quality and subjective audio-video delay tolerance. In: Hamberg R, de Ridder, H (eds) Proc MOSAIC worksh: Advancanced Methodes for Evaluation of Television Picture Quality. Chap 13, 133–137. Inst Percept Res, IPO, Eindhoven,

58. Rimell A N, Hollier M P, Voelcker R M (1998) The influence of cross-modal interaction on audio–visual speech quality perception. 105rd Conv Audio Engr Soc, San Francisco. Preprint 4791:H-6

59. Rudloff I (1997.) Untersuchungen zur wahrgenommenen Synchronität von Bild und Ton bei Film und Fernsehen. Master thesis, Ruhr-Universität Bochum, Bochum

60. Rutschmann J, Link R (1964) Perception of temporal order of stimuli differing in sense mode and simple reaction time. Percept Motor Skills 18:345–352

61. Schaffer S (1988) Astronomers mark time: Discipline and the personal equation. Science in Context 2:115–145,

62. Sekuler R, Sekuler A, Lau R (1997) Sound alters visual motion perception. Nature 385:308,

63. Sekuler B, Sekuler R (1999) Collisions between moving visual targets: what controls alternative ways of seeing an ambiguous display. Perception 28:415–432,

64. Shams L, Kamitani Y, Shimojo S (2000) What you see is what you hear. Nature 408:788

65. Shams L, Kamitani Y, Thompson S, Shimojo S (2001) Sound alters visual evoked potential in humans. Neuro Rep 12:3849–3852

66. Shams L. Kamitani Y. Shimojo S (2002.) Visual illusion induced by sound. Cogn Brain Res 14:147–152

67. Shimojo S, Scheier C, Nijhawan R, Shams L, Kamitani Y, Watanabe K (2001) Beyond perceptual modality: Auditory effects on visual perception. Acoust Sci & Tech, 22:61–67

68. Shimojo S, Shams L (2001) Sensory modalities are not seprate modalities: plasticity and interaction. Current Opinion Neurobiol 11:505–509

69. Smeele P M T (1994) Perceiving speech: Integrating auditory and visual speech. Doct diss, Techn Univ Delft, Delft

70. Spence C, Driver J (1997) On measuring selective attention to an expected sensory modality. Perc Psychophys 59:389–403

71. Spence C, Driver J (1997) Cross-modal links in attention between audition, vision, and touch: Implications for interface design. Int J Cogn Ergonom 1:351–373

72. Spence C, Driver J (1999) Multiple resources and multimodal interface design. In: Harris D (ed), Engr Psychology and Cognitive Ergonomics. Vol 3. Ashgate Publishing, Hampshire, 305–312

73. Spence C, Driver J (1999) A new approach to the design of multimodal warning signals. In: Harris D (ed) Engineering Psychology and Cognitive Ergonomics. Vol. 4. Ashgate Publishing, Hampshire, 455–461

74. Spence C, Driver J (1996) Audiovisual links in endogenous covert spatial attention. J Expt Psychol Human Percept 22:1005–1030

75. Stein B E, Meredith M. A (1993) The merging of the senses. MIT Press, Cambridge MA

76. Steinmetz R (1996) Human perception of jitter and media synchronization. IEEE J Select Areas Comm 14:61–72

77. Stone J V, Hunkin N M, Porrill J, Wood R, Keeler V, Beanland M, Port M, Porter N R (2001) When is now? Perception of simultaneity. Proc. R. Soc Lond B-268:31–38

78. Storms R L (1998) Auditory-visual cross-modal perception phenomena. Doct diss, Naval Postgraduate School, Monterey CA

79. Sumby W H, Pollack I (1954) Visual contribution to speech intelligibility. J Acoust Soc Amer, 26:212–215

80. Summerfield A. Q (1979) Use of visual information in phonetic perception. Phonetica 36:314–331

81. Thurlow W R G, Jack C E (1973) Certain determinants of the "ventriloquist effect". Percept Motor Skills 36:1171–1184

82. Warren D H, Welch R B, McCarthy T J (1981) The role of visual-auditory "compellingness" in the ventriloquist effect: Implications for transitivity among the spatial senses. Percept Psychophys, 30:557–564

83. Watanabe K (2000) Crossmodal interaction in humans. Doct diss, California Inst Techn, Pasadena

84. Watanabe K, Shimojo S (2001) When sounds affects vision: Effects of auditory grouping on visual motion perception. Psychol Sc 12:109–116

85. Watanabe K, Shimojo S (1998) Attentional modulation in perception of visual motion events. Perception 27:1041–1054

86. Welch R B, DuttonHurt L D, Warren D H (1986) Contributions of audition and vision to temporal rate perception. Percept Psychophys 39:294–300

87. Welch R B, Warren D H (1989) Intersensory interaction. In: Boff K R, Kaufman L, Thomas J P (eds) Handbook of Perception and Human Performance. Chap 25, 1–36. Kluwer Academic, Dordrecht,

88. Woszczyk W, Bech S, Hansen V (1995) Interactions between audio–visual factors in a home theater system: Definition of subjective attributes. 99th Conv Audio Engr Soc, New York. Preprint 4133:K-6

89. Zahorik P (2001) Estimating sound source distance with and without vision. Optometry and vision science, 78:270–275

90. van der Zee E, van der Meulen A W (1982) The influence of field repetition frequency on the visibility of flicker on displays. IPO Ann Progr Rep 17:76–83

91. Zetzsche C, Röhrbein F, Hofbauer M, Schill K (2002) audio–visual sensory interactions and the statistical covariance of the natural environment. In: Calvo-Manzano A. Perez-Lopez A, Santiago J S (eds) Proc Forum Acusticum Sevilla 2002. CD–Rom

6 Psycho-Acoustics and Sound Quality

Hugo Fastl

Technical-Acoustics Group, Department of Human-Machine-Communication,
Technical University of Munich, Munich

Summary. In this chapter psycho-physical methods which are useful for both psycho-acoustics and sound-quality engineering will be discussed, namely, the methods of random access, the semantic differential, category scaling and magnitude estimation. Models of basic psycho-acoustic quantities like loudness, sharpness and roughness as well as composite metrics like psycho-acoustic annoyance will be introduced, and their application to sound-quality design will be explained. For some studies on sound quality the results of auditory evaluations will be compared to predictions from algorithmic models. Further, influences of the image of brand names as well as of the meaning of sound on sound-quality evaluation will be reported. Finally, the effects of visual cues on sound-quality ratings will be mentioned.

6.1 Introduction

Psycho-acoustics as a scientific field has a tradition of more than 2500 years. For example, already around 500 B.C. the Greek philosopher *Pythagoras* – with his monochord – had studied musical consonance and dissonance. These early experiments had all the ingredients of a psycho-acoustical experiment. Psychoacoustics needs a sound stimulus which can be described in the physical domain. For *Pythagoras* this physical quantity was the length of the string stretched out along his mono-chord and supported by a bridge. By varying the position of the bridge and while plucking both ends of the string he judged with his hearing system whether the resulting musical interval was consonant or dissonant, i. e. he judged on attributes of his auditory percept. In this way, he found out that for simple ratios of the string divisions – like 1:2, i. e. octave, 2:3, fifth, and 3:4, quart – consonant musical intervals were perceived.

In modern psycho-acoustics the procedures applied are very much the same as those that have already been used by *Pythagoras*. At first, acoustic, i. e. physical, stimuli are produced, nowadays usually with the help of sophisticated digital signal-processing algorithms. After D/A conversion the resulting signals are presented to subjects via headphones or loudspeakers. The subjects, then, are asked to judge upon attributes of what they hear, such as the pitch, the loudness or the tone colour of the perceived sounds.

The same principles are often applied in sound-quality engineering, however, in reversed sequence as follows. During extensive psycho-acoustic studies an optimum sound for a specific product is "tailored" – a so-called *target sound*. Consequently, it is the task of engineers to modify the physics of sound generation, e. g., in an industrial product, in such a way as to arrive at a sound which comes as close as feasible to the target sound.

In this chapter psycho-physical methods which are useful for both psycho-acoustic research and sound-quality engineering will be discussed. Algorithmic models for the estimation of basic psycho-acoustic quantities like loudness or sharpness as well as compounded metrics will be introduced, and their application to sound-quality design will be elaborated on. Further, some practical examples will be reported, e. g., concerning comparison of results from listening tests with predictions as rendered by algorithmic models, the influence of brand names, and the effect of visual stimuli on sound-quality judgements.

6.2 Methods

For sound-quality evaluation psycho-physical methods are in use which have already proven successful in psycho-acoustics. From a variety of possible methods, four more important ones have been selected for discussion in this chapter, namely, *ranking methods* – they indicate whether a product sounds better than the product of a competitor, the method of *the semantic differential* – it provides hints on what sounds are suitable to convey an intended message, e. g., as a warning signal – and *category scaling and magnitude estimation* – which can give an indication of how much the sound quality differs among products, which is often of particular relevance for cost/benefit evaluations.

6.2.1 The Ranking Procedure "Random Access"

A ranking procedure called "random access", which has proven very successful for the investigation of sound-quality [12], is illustrated in Fig. 6.1. In the example displayed, six sounds, denoted *A* through *F*, have to be ranked with

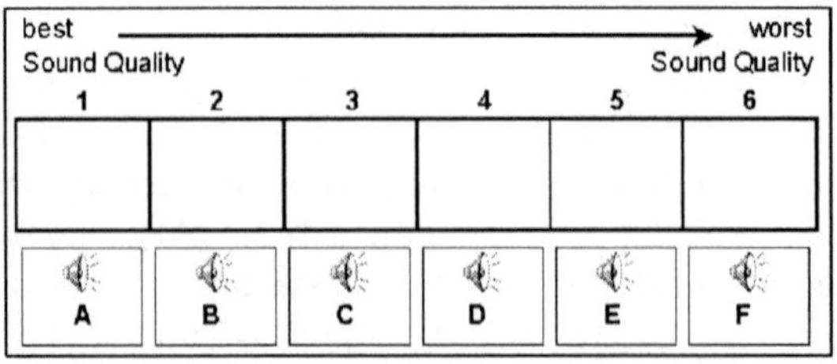

Fig. 6.1. Example for ranking of the sound quality by the method random access [15]

respect to their sound quality. When clicking on the loudspeaker icon, the respective sound, e. g., an idling motor, is heard. The task of the subject is to shift the icons, *A* through *F*, into one of the empty fields, denoted 1 through 6, in such a way that the sounds are finally ordered with respect to their sound quality. The subjects are free to listen to each individual sound as often as they like and to correct the sequence again and again, until they feel that a final status has been reached. This large freedom of the subjects, who have "random access" to the sounds to be ranked, is one of the reasons for this procedure to be preferred nowadays for ranking of sound quality.

6.2.2 The Semantic Differential

The method of "the semantic differential" is used to test what sounds are suitable for an intended purpose. In Fig. 6.2 an example of adjective scales is given, which has been used in an international study on the suitability of signals as warning signals [22]. It goes without saying that warning signals should have high loadings on adjectives like dangerous, frightening and unpleasant.

adjective scales	
loud	soft
deep	shrill
frightening	not frightening
pleasant	unpleasant
dangerous	safe
hard	soft
calm	exciting
bright	dark
weak	powerful
busy	tranquil
conspicuous	inconspicuous
slow	fast
distinct	vague
weak	strong
tense	relaxed
pleasing	unpleasing

Fig. 6.2. Semantic differential from an international study on warning signals [22]

6.2.3 Category Scaling

"Category scaling" is a preferred method for the assessment of the loudness of the sounds of products. Five-step scales as well as seven-step scales are usually employed, e. g., [10]. Figure 6.3 gives examples of five-step scales as well as seven-step scales as used for loudness assessment. In comparison to the five-step scale, the seven-step scale has in addition the steps "slightly

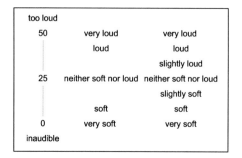

Fig. 6.3. Category scaling with seven categories, *right*, five categories, *middle*, and 50 subcategories, *left*

soft" and "slightly loud". For this reason the whole range from "very soft" up to "very loud" shows a finer grading in seven-step scales than it does in five-step ones.

A variant of category scaling which is frequently used in audiology as well as in noise-immission assessment, originates from a five-step scale. However, each step is subdivided into ten subcategories each, such leading to a 50-point scale [17]. The relation between the 50-point scale and the five-category scale is indicated in the left part of Fig. 6.3. Since the numerical representation of categories may induce a ceiling effect, the categories "inaudible" at the low end and "too loud" at the high end are sometimes added to the 50-point scale. "Inaudible", then, corresponds to zero loudness, while "too loud" may be related to any numbers higher than 50 – see [17]

6.2.4 Magnitude Estimation

Advantages of the method of "magnitude estimation" are that no ceiling effects shows up and that, theoretically, it has an infinite resolution, e. g., [39]. Magnitude estimation – with anchor sounds – is a frequently used method for sound-quality evaluation. Its procedure is illustrated by means of Fig. 6.4. Pairs of sounds are presented. The first sound, A, is called "anchor sound" and the second one, B, "test sound". Throughout an experiment the anchor sound is kept constant and the test sound is varied. A numerical value, for instance, 100, is then assigned to a predefined psycho-acoustic quantity of the anchor, A – e. g., to its loudness. The task of the subject, consequently, is to assign a numerical value also to the test sound, B. This value should represent the ratio of the magnitudes of the psycho-physical quantity under observation in the test sound with respect to that in the anchor sound. If, for example, the loudness of a test sound is perceived 20 percent softer than that of the anchor, the subject should give the response 80. Through magnitude estimates a ratio of the magnitudes of psycho-physical quantities is obtained directly, which is often of advantage for cost/benefit analyses.xxx Intra-individual as well as inter-individual differences of magnitude estimates usually come out within

Fig. 6.4. Illustration of sounds presented for magnitude estimation

10 percent variation. However, sometimes the choice of the anchor sound may influence the results of magnitude estimation significantly. Therefore, it is recommended to use at least two anchor sounds, one with a large magnitude of the psycho-physical quantity in question and the other one with a small magnitude.

Certainly, all psycho-physical methods as mentioned so far have their specific advantages and disadvantages. Random access and the semantic differential are used when a more "qualitative" description is aimed at. If a more "quantitative" assessment of sound quality is the goal, methods like category scaling and magnitude estimation are recommended. They provide data on the level of interval and ratio scales which can easily be processed further with parametric statistics. While traditional category scaling is confined to five or seven response categories, magnitude estimation – in principle – has an infinite resolution and can also provide absolute zeros. However, in magnitude scaling, effects of the frame of reference as well as influences of the choice of the anchor sound(s) have to be taken into account.

6.3 Modelling of Psycho-Acoustic Quantities

In sound-quality engineering basic psycho-acoustic quantities like loudness, sharpness, roughness, and fluctuation strength play an important role. Since the evaluation of those quantities in psycho-acoustic experiments can be quite time consuming, models have been proposed which simulate the formation of psycho-acoustic quantities. These models can be used to provide estimates to predict the magnitudes of these quantities from given input data on the physical, i. e. acoustic, level.

6.3.1 Loudness

As a rule, the "loudness" of a product sound strongly affects the sound quality of the product. Therefore a model of loudness which had been proposed by *Zwicker* already back in 1960 [35] has been improved [37, 38], and been extended in recent years – thereby including its applicability to persons with hearing deficits [3].

The basic features of *Zwicker*'s loudness model, see [39], are illustrated in Fig. 6.5. Essentially, there are three steps that form the kernel of the *Zwicker* model. In a first step, the physical frequency scale is transformed into the

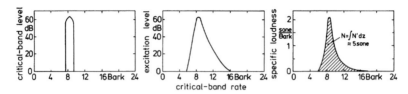

Fig. 6.5. Illustration of spectral effects as simulated in *Zwicker*'s loudness model [39]

psycho-acoustic *Bark* scale. The name *Bark* was chosen for the unit of this scale to honour the late Professor *Barkhausen* of Dresden for his merits with respect to the basics of loudness measurements. Figure 6.5 shows, in the left panel, a 1/3-octave-band noise centered at 1 kHz, displayed along the *Bark*scale. In the middle panel of Fig. 6.5 masking effects are accounted. These masking effects reflect a spectral broadening of the excitation within the cochlea, mainly due to inner ear mechanics. In particular, higher frequencies are masked by lower frequencies, an effect which is nowadays exploited in many practical applications, such as the GSM coding in mobile telephones or *mp3* coding in consumer electronics. The right panel in Fig. 6.5 shows a specific loudness/critical-band-rate pattern which is commonly denoted a loudness pattern or "*Zwicker* diagram". Simply speaking, the transition from the masking pattern in the middle of Fig. 6.5 to the loudness pattern at the right is obtained by taking the square root of sound pressure or the fourth root of sound intensity, respectively. Most important for practical applications is the fact that the area as covered by the loudness pattern, hatched in the figure, is directly proportional to the perceived loudness. This means that with this area being reduced by, say, 30%, it can be predicted that the associated loudness will also be reduced by 30%. This direct proportionality to perceived loudness is unique to this loudness-estimation procedure and cannot be obtained by alternative spectral-analysis systems, such as *Fourier* transforms, 1/3-octave-band analysis, wavelets, gamma-tone filters, etc.

Zwicker's loudness model has been standardized both in international [18] as well as in national [6] standards. In comparison to ISO 532 B of 1975, the latest revision of DIN 45 631, as of 1991, includes improvements with respect to the evaluation of sounds with strong low-frequency components.

Figure 6.5 illustrates the spectral processing of loudness, while temporal effects of loudness, see [39], are depicted in Fig. 6.6. The top panel shows the temporal envelope of tone impulses with a 100-ms duration, solid in the figure, or a 10-ms duration, dashed. The middle panel shows the temporal processing of loudness in each of the 24 channels of a loudness analyzer. One can clearly see that the decay of specific loudness is steeper after a short sound is switched off, in comparison to the decay after a longer sound. The lower panel in Fig. 6.6 shows the time dependency of total loudness, being summed up across all 24 channels. When being presented at the same sound-pressure

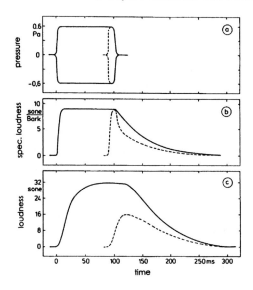

Fig. 6.6. Illustration of temporal effects in loudness processing [39]

level, sounds of 100-ms duration give rise to twice the loudness, i. e. 32 sone, as compared to sounds of 10-ms duration, namely, 16 sone. In contrast to spectral processing of loudness, see Fig. 6.5, for temporal processing, Fig. 6.6, it is not the total area under the loudness function, but the peak value of the loudness function which is of relevance.

Figure 6.7 illustrates an actual implementation [3] of a *Zwicker*-type loudness model. Essentials of spectral processing, as illustrated in Fig. 6.5,

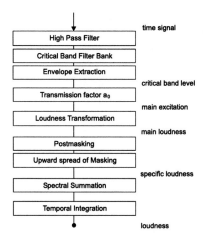

Fig. 6.7. Block diagram of the dynamic loudness model, DLM, as proposed in [3]

can be found in the critical-band-filter bank, the upward spread of masking and in the spectral summation applied. The temporal processing which has been illustrated in Fig. 6.6 is also represented in Fig. 6.7, namely, by the blocks marked envelope extraction, post-masking and temporal integration. Most important is the block denoted *loudness transformation*. As was discussed already in connection with Fig. 6.5 in simplified form, this block represents the fact that loudness is proportional to the square-root of sound pressure or the fourth root of sound intensity. A distinguishing feature of the new implementation – called dynamic loudness model, DLM – is that, by modification of the loudness-transformation block, loudness perception of both normal-hearing and hearing-impaired persons can be simulated [3]. This novel feature is of particular relevance for many practical aspects of sound-quality engineering, as in the ageing populations of industrialized countries a large part of prospective customers of a product will show mild to moderate hearing losses. Further, even a growing percentage of the younger generation has developed evident hearing deficits these days, frequently due to extremely loud leisure activities.

Figure 6.8 provides more details of loudness transformation in normal-hearing as well as hearing-impaired persons. The dashed curve shows the relation between level and loudness for normal-hearing persons. The dash-dotted curve would give the same relation for a person with a 50-dB hearing loss, provided that processing of loudness in the hearing system were linear. However, as illustrated by the solid curve in Fig. 6.8, according to a phenomenon which is known as "recruitment", the following can be observed: Loudness perception of hearing-impaired people "catches up" at high levels. This means that for impaired persons the gradient of loudness is very steep

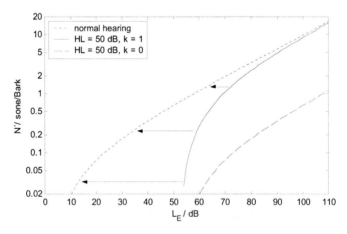

Fig. 6.8. Relation between level and loudness for normal hearing persons, *dashed*, and hearing impaired people, *solid*. The *dash-dotted* curve ignores the recruitment phenomenon [3]

just above the threshold of audibility. The consequence is that a small increase in level of a few Decibels can lead to a drastic increase of loudness, thus causing serious annoyance to the hearing-impaired person. Modern hearing aids try to compensate this effect by digital signal processing. Yet, even the most-advanced hearing instruments can (still) not restore normal hearing completely.

6.3.2 Sharpness

Besides loudness a further psycho-acoustic quantity, called "sharpness", plays a prominent role in sound quality. Sharpness, among other things, can be regarded as a measure of tone colour [1]. If the right amount of sharpness is added to a sound, e.g., the sound of an industrial product, this will give it a character of powerfulness. However, too much sharpness will render a sound aggressive. If the loudness pattern of a sound is available, its sharpness can be relatively easily estimated by calculation. The corresponding procedure is illustrated in Fig. 6.9. The left panel depicts the spectral distribution of a narrow-band noise, a broad-band noise and a high-pass noise. The right panel in Fig. 6.9 shows the loudness pattern as already known from Fig. 6.5. However, to account for the increased sharpness of high-frequency sounds, a weighting function, g, is to be applied. In order to derive sharpness from the resulting patterns, the first momentum is calculated. The respective values are indicated in the right panel of Fig. 6.9 by vertical arrows. It becomes clear from Fig. 6.9 that, when adding low frequencies to a high-pass noise, the center of gravity shifts downwards, thus leading to a smaller value of sharpness – compare the dotted and dashed arrows. This means for practical purposes in sound engineering that the sharpness and, hence, the aggressiveness of product sounds can be reduced by adding low-frequency components.

It should, however, be kept in mind that such an addition of low-frequency components also increases total loudness. Nevertheless, if the loudness of the original sound is not too high, the reduction in sharpness and, hence, aggressiveness can overcompensate the loudness increase in its effect on overall sound quality.

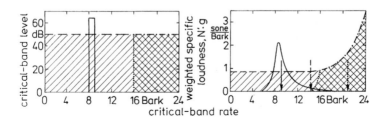

Fig. 6.9. Illustration of the model of sharpness [39].

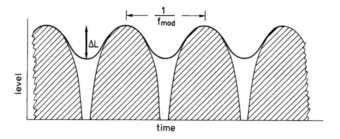

Fig. 6.10. Illustration of the input to the model of roughness [8].

6.3.3 Roughness

"Roughness", a further psycho-acoustic quantity, is used in sound-quality engineering, e. g., to stress the feature of "sportiness" in a car-engine sound. Roughness is governed by temporal variations of a sound and reaches a maximum for modulation frequencies around 70 Hz [28]. In essence, roughness can be described by the temporal-masking pattern of sounds [8]. This reasoning is illustrated in Fig. 6.10. The hatched areas show the temporal variation of a sound, modulated in amplitude by a degree of modulation of almost 100%, with the level being displayed as a function of time. Theoretically the troughs between the peaks reach a minimum near minus infinity on the Decibel scale. In practical applications, however, the minimum level is controlled by the dynamics of the hearing system, i. e. the modulation depth of the temporal-masking pattern, ΔL, reaches much smaller values due to the effects of post-masking. Post-masking is represented by the the decay of psycho-acoustic excitation in the hearing system.

This limited resolution of level is illustrated in Fig. 6.10 by the solid curve. The temporal distance of the peaks is inversely related to the modulation frequency. In principle, the roughness, R, of a sound can be described by the product of the modulation depth, ΔL, of the temporal masking pattern and the modulation frequency, f_{mod}.

$$R \approx \Delta L \cdot f_{\mathrm{mod}} \tag{6.1}$$

Since this product carries the unit $[R] = \mathrm{dB/s}$, the hearing sensation roughness is proportional to the speed of the variation of the temporal masking pattern.

6.3.4 Fluctuation Strength

The psycho-acoustic quantity "fluctuation strength" is similar to roughness. However, fluctuation strength reaches a maximum at modulation frequencies of about 4 Hz. The input to the model of fluctuation strength is the same

as the input to the model of roughness, as displayed in Fig. 6.10. In addition
to the modulation depth, ΔL, of the temporal-masking pattern, the relation
of the modulation frequency to a modulation frequency, f_{mod}, of 4 Hz is of
relevance. Therefore fluctuation strength, F, can basically be calculated as
follows:

$$F \approx \frac{\Delta L}{4\mathrm{Hz}/f_{\mathrm{mod}} + f_{\mathrm{mod}}/4\mathrm{Hz}} \tag{6.2}$$

Fluctuation strength plays a crucial role in the assessment of human
speech for the following reason. The envelope fluctuation of fluent speech
also shows a maximum around a modulation frequency of 4 Hz. This roughly
corresponds to the number of syllables pronounced per second. As one would
have expected from nature, the human speech organ indeed produces speech
sounds with dominant envelope fluctuations at a rate that the human hearing
system is most sensitive to.

6.3.5 Composed Metrics

A combination of psycho-acoustic quantities has proven successful for the
prediction of the perceived annoyance of sounds from noise emissions as well
as immissions [32]. The corresponding formula for this annoyance estimator,
PA, reads as follows.

$$PA \approx N_5 \cdot (1 + \sqrt{w_{\mathrm{S}}^2 + w_{\mathrm{FR}}^2}) \tag{6.3}$$

with
$N_5 \dots$ percentile loudness in sone

$$w_{\mathrm{S}} = (\tfrac{S}{\mathrm{acum}} - 1.75) \cdot 0.25 \lg(\tfrac{N_5}{\mathrm{sone}} + 10) \qquad \text{for } S > 1.75 \text{ acum}$$

describing the effects of sharpness, S, and

$$w_{\mathrm{FR}} = \frac{2.18}{(N_5/\mathrm{sone})^{0.4}} (0.4 \cdot \tfrac{F}{\mathrm{vacil}} + 0.6 \cdot \tfrac{R}{\mathrm{asper}})$$

describing the effects of fluctuation strength, F, and roughness, R.

The units acum, vacil and asper are related to sharpness, S, fluctuation
strength, F, and roughness, R, respectively. For details of their definition the
reader is referred to [39]. From the formula it becomes clear that loudness is
a dominant feature of annoyance. The percentile value, N_5, indicates that a
value near the maximum loudness is of importance for sound quality ratings.
However, sharpness as well as roughness and fluctuation strength may play
an important role as well. When thinking, for example, of a dentist's drill,
not only the loudness but, even more so, the sharpness is responsible for the

annoyance. Likewise, the tick-tack of a clock, in particular when heard during the night, is usually not annoying because of its loudness but because of the regularity and, hence, the fluctuation strength of the sound produced.

Although the model of psycho-acoustic annoyance proposed in [33] can account for various practical situations, it is certainly not designed to solve all questions of sound quality. Nevertheless, this model contains some relevant ingredients for sound-quality evaluation, namely, loudness, sharpness, roughness and fluctuation strength. However, the appropriate "recipe" for a mixture of psycho-acoustic quantities may vary for different families of product sounds and different applicational context.

Another composite model based on several psycho-acoustic quantities, as being put forward in [29], has proven successful to rate "sensory pleasantness" of sounds – in particular, the pleasantness of speech and music [30]. However, in this model, clearly audible tonal components receive a bonus, while, when dealing with noise-immission problems, tonal components are rather undesired and thus usually are assigned a penalty instead. Consequently, this model of sensory pleasantness is not recommended to estimate pleasantness related to noise immissions.

6.4 Sound Quality

Since this book contains a further chapter dealing with sound quality [19], we restrict ourselves here to a small selection of practical sound-quality-evaluation examples from our own laboratory.

Due to the electronics being implemented on modern road vehicles, it is now comparatively easy to control and adjust the engine's state of operation [25]. In our first example, displayed in Fig. 6.11, a *Diesel* motor was driven in either a "hard" or "normal" manner. Normal, here, means the adjustment as used in the current product. The advantage of the hard motor adjustment is that the engine is more fuel efficient. Obviously, the disadvantage is that it produces more noise. In psycho-acoustic experiments it was assessed whether the acoustic disadvantage of the fuel efficient hard motor adjustment can be reduced by absorptive measures. In one part of the study, frequencies from 1 kHz to 5 kHz where attenuated by 3 to 15 dB in 3 dB steps. In the other set of experiments, the whole spectrum was attenuated by 3 to 15 dB, again in 3 dB steps. The data displayed in Fig. 6.11 show the ranking of sound quality in medians, circles, as well as the inter-quartiles, bars. The crosses denote loudness predictions from acoustic measurements.

The results, as displayed in Fig. 6.11, show clearly that the motor with a "hard" motor adjustment obtains the poorest sound quality ranking, i.e. rank 12. However, the motor with the "normal" motor adjustment – as used in series vehicles – attains rank 4 in sound quality. Even better ranks, namely 1 to 3, result when the whole spectrum of the hard motor sound is attenuated by 9 to 15 dB.

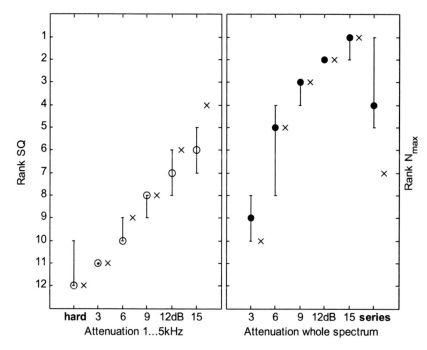

Fig. 6.11. Improvement of the sound quality of a *Diesel* motor with "hard" motor adjustment and simulated absorptive measures which cause different amounts of attenuation. *Circles* denote subjective sound quality estimates. *Crosses* mark loudness predicted from acoustic measurement [25]

The results depicted in Fig. 6.11 suggest that even when the sound of a hard motor adjustment is attenuated in the frequency range of 1 to 5 kHz by as much as 15 dB, the sound quality of a "normal" motor adjustment is still not achieved. Rather the complete spectrum of the motor sound for hard motor adjustment would have to be attenuated by as much as about 7.5 dB to just attain the sound quality of a present-day-series vehicle. Although it will not at all be easy, it seems quite worthwhile for engineers to take on this challenge of reducing the sound level, because of the higher fuel efficiency of the motor with hard motor adjustment.

The crosses displayed in Fig. 6.11 indicate the ranking of the physically-measured maximum loudness, N_{max}, produced by each sound. As a rule there is good agreement of the sound-quality ranking and the ranking of the maximum loudness. However, for an attenuation of 15 dB between 1 and 5 kHz sound quality attains only a rank of 6 while the ranking of maximum loudness attains rank 4. On the contrary, for the series motor, the ranking in loudness attains only rank 7, whereas the sound quality attains rank 4. This means that loudness alone cannot always predict sound-quality ratings. In spite of its larger loudness, rank 7, the sound quality of the motor of the series vehicle

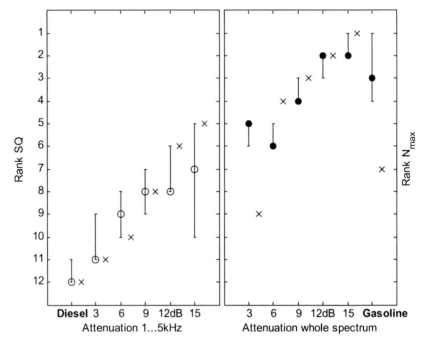

Fig. 6.12. Comparison in sound quality between a gasoline engine and a "normal" *Diesel* engine with additional absorptive measures applied [25]

is ranked higher in sound quality, rank 4. This is presumably due to its more familiar, "natural" sound. A similar argument holds true for the hard motor when attenuated by 15 dB in the frequency range from 1 to 5 kHz. Although the loudness, then, is significantly reduced, rank 4, the sound quality is still ranked lower, rank 6 – presumably because the resulting stimulus sounds quite "unnatural".

An even more striking example is given in Fig. 6.12. In this case, a series *Diesel* engine with the same attenuations as described above has been compared in sound quality to a gasoline engine [25]. Again circles indicate subjective rankings, crosses rankings estimated from acoustic measurement.

Again, the physical ratings, crosses in the figure, and the subjective evaluations are frequently in line - at least within the inter-quartile ranges. However, clear differences are obtained for the gasoline engine. While its loudness attains only rank 7, the sound quality attains rank 3. Despite its higher loudness, the sound of the gasoline engine is preferred. Presumably this is due to a particular sound attribute of *Diesel* engines which can be characterized by the term "*Diesel* nailing". The large discrepancy between sound-quality ranking and ranking of loudness for the sound with a 3-dB attenuation of the whole spectrum is not clear. Obviously it is not easy for the subjects to evaluate those sounds since - contrary to expectation - the sound which is only

3 dB attenuated gets a better ranking than the sound with 6 dB attenuation. Starting from a series *Diesel* engine, an attenuation of the whole spectrum of about 10 dB would be necessary to arrive at a sound quality similar to that of a gasoline engine.

6.5 The Meaning of Sounds

When evaluating sound quality, the meaning as assigned to sounds when listening to them may have an effect on judgements. In a global market it may thus be of some relevance to take possible cultural differences into account. Cross-cultural studies with listeners in Japan and Germany [21] showed that sometimes one and the same sound can be rated differently by subjects from different cultural backgrounds. For example, by German subjects, the sound of a bell was interpreted as the sound of a church-bell, leading to connotations such as "pleasant" or "safe". On the contrary, Japanese subjects were reminded by the bell sounds to sounds of a fire engine or a railroad crossing, leading to feelings as denoted by the terms "dangerous" or "unpleasant". In Fig. 6.13 [21] the corresponding data from a study with the method of semantic differential are displayed. Data for Japanese subjects are connected by solid lines, data of German ones by dotted lines.

The data for the German subjects suggest their feelings that the bell-sound is not frightening, but pleasant, safe, attractive, relaxed, and pleasing. Japanese subjects, however, feel that the bell sound is shrill, frightening, unpleasant, dangerous, exciting, busy, repulsive, distinct, strong, tense, and unpleasing.

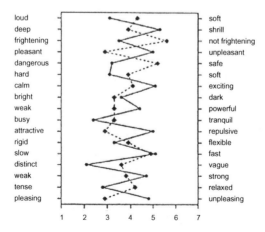

Fig. 6.13. Semantic-differential data for a bell sound. Data for Japanese subjects are connected by *solid lines*, for German ones by *dotted lines* – adopted from [21]

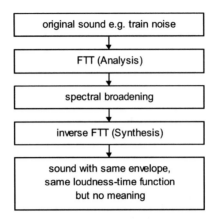

Fig. 6.14. Block diagram illustrating the procedure to remove information about the sound source from a stimulus [13]

In order to overcome undesired influences of meaning in psycho-acoustic experiments a procedure has been proposed [13] which largely removes the information about the sound source from a stimulus. The block-diagram displayed in Fig. 6.14 illustrates the correlated procedure. From an original noise, e. g., train noise, a spectral analysis is being performed by means of a *Fourier*-time transform, FTT. The FTT algorithm [31] is a spectral-analysis technique which, in contrast to, e. g., *Fourier* transforms, uses a sliding temporal window corresponding to a frequency-dependent bandwidth which mimics the the frequency resolution of the human hearing system. In the next steps, after spectral broadening and, hence, obscuring the spectral details, the sound is re-synthesized by means of an inverse FTT. In this way a sound with the same spectral and temporal envelope and, such, the same loudness/time function is created from which, however, information about the sound source has been removed.

The data displayed in Fig. 6.15 enable a comparison of the loudness-time functions of (a) original sounds, (b) the same sounds, but with the information about the sound source being removed.

The results as displayed in Fig. 6.15 clearly show that the goal of ending at identical loudness-time functions of original sounds and sounds without information about the sound source can well be achieved. With the procedure as outlined in Fig. 6.14 the information about the sound source can be removed from many signals which are important in our daily life [34]. However, some signals like for example FTT-processed speech sounds still have great similarity to the original sound. It is worth mentioning at this point, that algorithms to remove meaning from sound have also been proposed in speech technology to study prosody – "re-iterant speech", see, e. g., [24] for details.

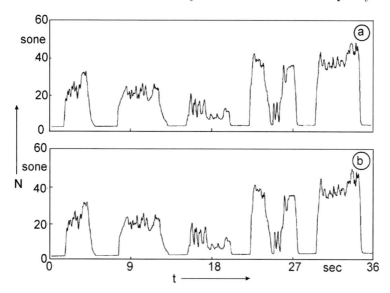

Fig. 6.15. Loudness-time function of (**a**) original sounds, and (**b**) the same sounds, but with the information about the sound source being removed by the procedure illustrated in Fig. 6.14 [13]

6.6 Image of Brand Names

When evaluating the sound quality of passenger cars, the image of the brand name of the cars has been shown to be of relevance. A well-known typical example along these lines is that the quality of a car is judged on the basis of the sound produced when closing a door. If the door sound is "tinny", this suggests that the whole vehicle is cheap and not at all solid. On the contrary, a full and saturated door-closing sound has a connotation of luxury. In a cooperation with colleagues from Osaka University Japan door-closing sounds have been studied with the method of the semantic differential. In addition to evaluating sound quality our subjects have been asked to guess the type of car and to guess the corresponding brand name of the car. Figure 6.16 shows data which are an excerpt of a larger study [23]. Data are given for a door sound which was allocated to a luxurious sedan, as well as for a door sound judged to stem from an economy sedan.

The data displayed in Fig. 6.16 suggest that the door sounds of luxury sedans are best described by adjectives such as deep, pleasant, heavy, dull, dark, powerful, calm, smooth, pleasing. On the contrary, adjectives related to the door sounds of economy sedans are best characterized as metallic, unpleasant, noisy, bright, shrill, rough, unpleasing.

Figure 6.17 gives an example of a histogram of the number of guesses associated to different brand names for luxurious sedans – based on door-

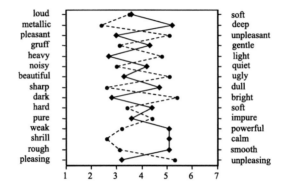

Fig. 6.16. Semantic differential for car door sounds judged to stem from a luxurious sedan, *solid*, or an economy sedan, *dashed* [23]

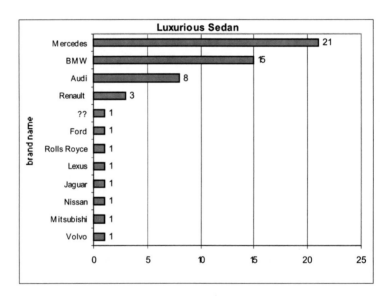

Fig. 6.17. Distribution of brand names associated with car door sounds of luxurious sedans [7]

closing sounds [7]. The twenty German subjects clearly related luxurious sedans to brand names like Mercedes, BMW or Audi.

The German Automobile Association, ADAC, regularly publishes a ranking of car manufactures. In our context it very interesting to compare this ranking of manufactures with the rating of car-door sounds as rendered from subjects in psycho-acoustic experiments [7]. The results are given in Table 6.1. When regarding these results, it is very interesting to note the strong relation of the ranking of car manufactures by the ADAC to the rating of the sounds

Table 6.1. Ranking of car manufactures by ADAC and rating of brand names by subjects for the different categories on the basis of sounds of closing car doors [7]

Ranking of car manufacturers	Rating 1-4 in each class by subjects
1. Mercedes	Luxurious 1
2. BMW	Sporty 1, Luxurious 2, Others 3
3. Audi	Luxurious 3
4. Volkswagen	Economy 1, Pick up 3, Others 4
5. Porsche	Sporty 2
6. Toyota	Economy 4
7. Peugeot	-
8. Smart	-
9. Renault	Luxurious 4
10. Ford	Economy 3, Pick up 4
11. Opel	Economy 2
12. Skoda	Others 2

of closing doors of cars. For example, the brand name Mercedes, being ranked first by the ADAC, gets the highest rating in the category of luxurious cars. BMW, which is ranked second by the ADAC, gets the best rating for sporty cars in the psycho-acoustic experiment, the second best for luxurious cars and the third best for others. Audi, number three in the ADAC ranking, gets the third rating of luxurious cars and so forth. Obviously, the brand name of a car strongly triggers the expectations about the sounds produced by a closing door.

6.7 Audio–Visual Interactions

Sound-quality ratings may depend not only on auditory stimuli but on input from other senses as well, for instance, from the visual system. In the following, two examples to support this view will be given.

The first deals with the influence of a visual image on the sound-quality rating of speech. In a concert hall, speech was radiated from the stage and recorded at different positions in the hall. In a first experiment subjects just listened to the recorded speech and rated its speech quality. In a further experiment, in addition to the acoustic presentation of the speech sounds, subjects where presented photos taken at the respective recording position, depicting the distance between the source and the receiving point. Fig. 6.18 shows a schematic plan of the ground floor, denoting the sound-source, S, and three positions, 1 through 3, of the receiver. The sounds as sent out by the speaker, S, where recorded on DAT tape at the positions 1, 2, or 3. In addition, photos where taken at these positions, showing the speaker on the stage and enabling the listeners to evaluate their distance from the receiving point.

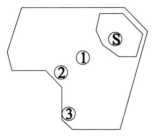

Fig. 6.18. Schematic plan of the ground floor of a concert hall with indications of the sound source, *S*, and three positions, 1 through 3, of the receiver [15]

Figure 6.19 gives the ratings of the speech quality for acoustic presentation alone, unfilled symbols in the figure, or with additional visual presentation, filled symbols. The data show that a visual image can influence the rated sound quality of speech. At position 1, which is relatively close to the source, the addition of the visual image causes the rating to degrade from fair to poor, medians taken. This may be due to the effect that the visually perceived distance to the speaker, which is relatively small, calls for a better speech quality which is not degraded by reverberation as in a concert hall. Since the concert hall was, of course, designed for classical music and, hence, has a reverberation time at mid-frequencies around 2 seconds, this reverberation is much too large for speech. For best intelligibility of speech a reverberation time below 1 second would be optimal [5]. At position 2, there is no influence of the visual image on the rating of speech quality. Obviously the subjects think that for such a larger distance from the speaker the quality is fair. Most interesting is the rating at position 3. Without visual information the speech quality is rated fair. However, with additional visual information the speech quality is even rated good. Obviously, given the large distance between the speaker, *S*, and the receiving point, 3, the subjects feel that for such an adverse situation the speech quality can be rated as being relatively good.

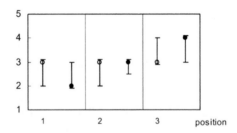

Fig. 6.19. Rating of speech quality in a concert hall at positions 1, 2, and 3 for acoustic presentation alone, *unfilled symbols*, and acoustic plus visual presentation, *filled symbols* [15]

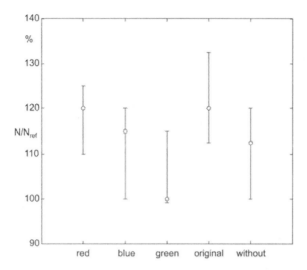

Fig. 6.20. Loudness rating of the sound from a passing train when photos of the train in different color are presented together with the sound [26]

Our second and last example deals with the influence of colour on the rating of the loudness of sound sources [26]. Sounds of a passing train have been presented either without visual input or with pictures of the same train, but painted in different colours. The related data are displayed in Fig. 6.20. Despite identical acoustic stimuli, the train sound is perceived as being softest when the train is painted in a light green. The loudness rating of this setting is taken as a reference in the further course of the experiment. According to the results displayed in Fig. 6.20, a train painted in red is perceived as being 20 percent louder than the same train in green.

The authentic painting of the train – a German high-speed, ICE, train – is white with a red stripe. In this case the relative loudness reached is also 120 percent of the reference. If the train sound is presented without visual input, it is perceived as somewhat softer than the presentation of sound plus original image – very similar to a train painted in light blue. In summary then, the colour of a product can indeed influence the loudness and, hence, the quality of its sound to some extent. Comparable cross-modal influences have been shown before for other kinds of visual cues [16, 20], and other auditory quality features – see, e. g., [27].

6.8 Outlook

The application of psycho-acoustic principles to sound engineering and sound-quality design has only recently become accepted as a useful analysis and design method. Although a solid psycho-acoustic basis had been available for quite a while, e. g., [36], applications of psycho-acoustics to noise and product-sound evaluation used to be rather sparse in the past – as noted, e. g., in [2]. Yet, since roughly a little more than a decade, application of knowledge from psycho-acoustics, e. g., [11], or even from musical acoustics, e. g., [15], has increased substantially in this context. Among other reasons, this situation is driven by economic necessities. In a global market with many competing products having almost the same functionality, the sound attached to a product can well become a decisive quality feature. Moreover, from the quality of the sound that a product produces, the user may extrapolate to the quality of the whole product – more or less consciously. It is thus to be expected that the application of psycho-acoustics in sound-quality evaluation and design will further increase.

Acknowledgment

The author wishes to thank all members of his group, *AG Technische Akustik*, for experimental support, stimulating discussions and editorial help. Much of the work reported here has been supported by the Deutsche Forschungsgemeinschaft, DFG.

References

1. von Bismarck G (1974) Sharpness as an attribute of the timbre of steady sounds. Acustica 30:159–172
2. Blauert J (1986) Cognitive and aesthetic aspects of noise engineering. Proc Internoise'86 vol I:5–14
3. Chalupper J, Fastl H (2002) Dynamic loudness model. ACUSTICA/acta acustica 88:378–386
4. Cederlöf R, Jonsson R E, Kajland A (1981) Annoyance reactions to noise and from motor vehicles – an experimental study. Acustica 13:270–279
5. Cremer L, Müller H A (1978) Die wissenschaftlichen Grundlagen der Raumakustik. Hirzel, Stuttgart
6. DIN 45631 (1991) Berechnung des Lautstärkepegels und der Lautheit aus dem Geräuschspektrum. Verfahren nach Zwicker E
7. Filippou T G, Fastl H, Kuwano S, Namba S, Nakamura S, Uchida H (2003) Door sound and image of cars. Fortschr. Akust. DAGA'03, Dt. Ges. f. Akustik, Oldenburg
8. Fastl H (1982) Beschreibung dynamischer Hörempfindungen anhand von Mithörschwellen-Mustern. Hochschul-Verlag, Freiburg
9. Fastl H (1987) A background noise for speech audiometry. J Audiol Acoust. 26:2–13
10. Fastl H, Kuwano S, Namba S (1996) Assessing the railway-bonus in laboratory studies. J Acoust Soc Japan (E) 17:139–148
11. Fastl H (1998) Psychoacoustics and sound quality metrics. In: Davies P, Ebbitt G (eds) Proc Sound-Quality Symp 3-10, Ypsilanti MI

12. Fastl H (2000) Sound quality of electric razors - Effects of Loudness. Proc Internoise 2000, Nizza, CD-ROM
13. Fastl H. (2001) Neutralizing the meaning of sound for sound-quality evaluations. In: Proc 17th Int Congr Acoust, ICA, Rome, CD ROM
14. Fastl H (2002a) Psychoacustics and sound quality. Fortschr Akust, DAGA'02, 765–766. Dt Ges f Akustik, Oldenburg
15. Fastl H (2002b) Sound design of machines from a musical perspective. In: Davies P, Ebbitt G (eds) Proc Sound-Quality Symp, Dearborn MI, CD-ROM
16. Fastl H (2004) Audio–visual interactions in loudnesss evaluation. 18th Int Congr Acoust, ICA, Kyoto, 1161-1166
17. Hellbrück J (1993) Hören. Hogrefe, Göttingen
18. ISO 532B (1975) Method for calculating loudness level
19. Jekosch U (2004) Assigning meaning to sounds: semotics in the context of product-sound design. Chap 8 this vol
20. Kaska J , Noack R, Mau U, Maas P (1985) Untersuchung der Wirkung von visuellen Umweltmerkmalen des Strassenbildes auf die Lästigkeit von Verkehrslärm an schweizer und deutschen Anwohnern. In: A Schick (ed) Rep 4th Oldenburg Symp Psychoacoust, Univ Oldenburg, Oldenburg
21. Kuwano S, Namba S, Fastl H, Schick A (1997) Evaluation of the impression of danger signals - comparison between Japanese and German subjects. In: Schick A, Klatte M (eds) Contributions to Psychological Acoustics. 115–128. BIS, Oldenburg
22. Kuwano S, Namba S, Schick A, Hoege H, Fastl H, Filippou T, Florentine M, Muesch H (2000) The timbre and annoyance of auditory warning signals in different countries. In: Proc. Internoise 2000, Nizza, CD ROM
23. Kuwano S, Fastl H, Namba S, Nakamura S, Uchida H (2002) Subjective evaluation of car door sound. In: Proc. Sound-Quality Sym., Dearborn MI
24. Mersdorf J (2000) Sprecherspezifische Parametrisierung von Sprachgrundfrequenzverläufen. Doct diss, Ruhr-Univ Bochum, Shaker, Aachen
25. Patsouras C, Fastl H, Patsouras D, Pfaffelhuber K (2002a) How far is the sound quality of a Diesel powered car away from that of a gasoline powered one? Proc Forum Acusticum Sevilla 2002, NOI-07-018-IP, CD-ROM
26. Patsouras Ch, Filippou T G, Fastl H (2002b) Influences of colour on the loudness judgement. Proc Forum Acusticum Sevilla 2002, PSY-05-002-IP, CD-ROM
27. Steven H (1981) Subjektive Beurteilung von Geräuschemissionen von Lastkraftwagen. In: Forsch Inst Geräusche Erschütterungen Aachen (ed) Report 105-05-104/02, Umweltbundesamt, Berlin
28. Terhardt E (1968) Über akustische Rauhigkeit und Schwankungsstärke. Acustica 20:215–224
29. Terhardt E (1984) Wohlklang und Lärm aus psychophysikalischer Sicht. In: Schick A, Walcher KP (eds) Ergebnisse des 3. Oldenburger Symposions zur Psychologischen Akustik. 403–409. P Lang AG, Bern,
30. Terhardt E (1985) Fourier transformation of time signals: Conceptual revision. Acustica 57:242–256
31. Terhardt E (1987) Audiokommunikation, Springer, Berlin-Heidelberg
32. Widmann U (1995) Subjektive Beurteilung der Lautheit und der Psychoakustischen Lästigkeit von PKW-Geräuschen. Fortschr Akust, DAGA'95, 875–878, Dt Ges Akust, Oldenburg,
33. Widmann U (1998) Aurally adequate avaluation of Sounds. In: Fastl H, Scheuren J (eds) Proc. Euro-Noise'98. 29–46, Dt Ges Akust, Oldenburg

34. Zeitler A, Fastl H, Ellermeier W (2004) Impact of source identifiability on perceived loudness. 18th Int Congr Acoust, ICA, Kyoto, 1491–1494
35. Zwicker E (1960) Ein Verfahren zur Berechnung der Lautstärke. Acustica 10:304–308
36. Zwicker E (1982) Psychoakustik. Hochschultext, Springer, Berlin
37. Zwicker E, Fastl H (1983) A portable loudness meter based on ISO 532 B. In: Proc. 11th Int Congr. Acoustics, ICA, vol 8:135–137, Paris
38. Zwicker E, Deuter K, Peisl W. (1985) Loudness meters based on ISO 532 B, with large dynamic range. Proc Internoise'85 vol II:1119–1122
39. Zwicker E, Fastl H (1999) Psychoacoustics. Facts and models. 2nd ed, Springer, Berlin–Heidelberg–New York

7 Quality of Transmitted Speech for Humans and Machines

Sebastian Möller

Institute of Communication Acoustics, Ruhr-University Bochum, Bochum

Summary. This chapter provides an overview of quality aspects which are important for telecommunication speech services. Two situations are addressed, the telephone communication between humans and the task-oriented interaction of a human user with a speech-technology device over the phone, e. g., a spoken-dialogue system. A taxonomy is developed for each situation, identifying the relevant aspects of the quality of service. The taxonomies are used for classifying quality features, as perceived by the users of the service, as well as parameters and/or signals which can be measured instrumentally during the interaction. For conversations between humans, relationships can be established between the parameters/signals and the perceptive quality features, thus allowing for prediction of the quality for specific application scenarios. Finally, future efforts necessary for establishing similar prediction models for task-oriented human-machine interaction over the phone are identified.

7.1 Introduction

Although the transmission of speech signals through telecommunication networks, namely, telephone networks, is a well-established procedure, and networks for this purpose have been commonly available for more than a century, the quality of the transmitted speech has become a new focus of attention in recent years. One major reason for this is the change that the traditional networks undergo with the advent of new coding and transmission techniques, e. g., mobile and packet-based, and new types of user interfaces. A further reason is the liberalization of the telecommunication market in many countries, which has resulted in interconnected networks and in a more economic use of transmission bandwidth.

The changes in the networks have had an impact on the quality of the transmitted speech signals and on the quality of the services which are available in such networks. Previously, telephone-speech quality was closely linked to a standard analogue or digital transmission channel of 300 to 3400 Hz bandwidth, terminated at both sides with conventionally shaped telephone handsets. Common transmission impairments were attenuation and circuit noise, as well as quantizing noise resulting from pulse-code modulation, PCM. The relatively low variability of the channel characteristics lead to stable user assumptions regarding the quality of transmitted speech being delivered over

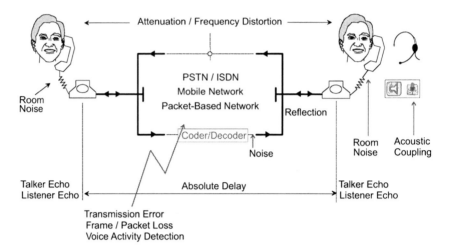

Fig. 7.1. Transmission-channel influences in a human-to-human telephone conversation. Indicated are the signal paths, *thick lines*, including 2-wire to 4-wire transitions and sources of different degradations

such networks. This situation completely changed when mobile networks and networks based on the internet protocol, IP, were set up on a large scale. New types of impairment result from low-bit-rate coding-techniques [16] and signal-processing algorithms causing considerable delays, which result in echoes and, consequently, communication difficulties. Common user interfaces show acoustic transmission characteristics which are very different from traditional telephone handsets, such as hands-free terminals in desktop or car environments or computer headsets.

In addition to the standard wire-line-telephone service, new types of services are now available to the human user. They offer speech communication in mobile situations or through the internet, or the interaction with some type of speech-technology device, e. g., for information retrieval, telephone banking or reservation tasks. When dialling a number, it is often not obvious for the users which type of service, e. g., wire-line, mobile, interaction with a human operator, or with a machine, they will be connected with. As a consequence, there is no stable reference available against which quality can be judged. In particular, the users' expectations and the level of quality which is offered by the service may diverge. Thus, a gap develops between the quality level which is expected from the side of the service operators, i. e. planning quality, and what is actually perceived by the users. In many cases, this leads to dissatisfied or even disappointed users.

In the following chapter, the quality of telecommunication services as based on speech transmission will be addressed in an integrated way for two scenarios. The first scenario is the human-to-human interaction, HHI,

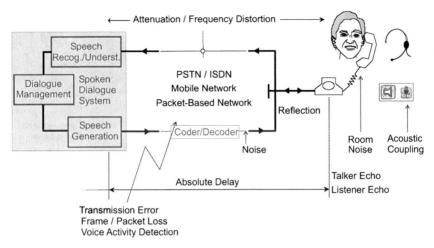

Fig. 7.2. Transmission-channel influences in human-machine interaction via telephone – compare to Fig. 7.1

over a telecommunication network, see Fig. 7.1. Different types of networks will be covered, as is discussed in Sect. 7.1.1. The second scenario is the interaction of a human user with a spoken-dialogue system, SDS, over the phone, thus a human-machine interaction, HMI. The HMI usually takes place to carry out a specific task, e.g., to obtain information from a database. This scenario is depicted in Fig. 7.2. As telecommunication networks will be confronted with both scenarios, it is important to consider the requirements imposed by both the human users and the speech-technology devices. The requirements will obviously differ, since the perceptive features influencing the users' judgment on quality are not identical to the technical specifications of a speech-technology device, e.g., an automatic speech recognizer, ASR. Nevertheless, the way the speech-technology device performs will significantly affect the users' perception of the quality of the HMI service. Therefore, the quality of speech transmission has a major influence on the quality of the service at large, see Sect. 7.1.2.

7.1.1 Speech Transmission in Telephone Networks

In actual situations, speech-transmission networks are composed of a chain of various transmission, switching and terminal equipment. To investigate the quality of such networks, it is important to take the whole transmission channel into account – from the mouth of the talker or the speech output generated by an SDS up to the ear of the listener or the input to an ASR device. The channel is influenced by the characteristics of each of its composing elements. A variety of equipment is available, and it would be impractical to

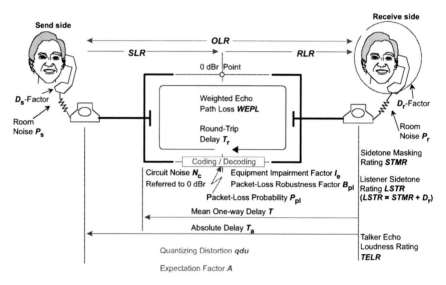

Fig. 7.3. Telephone reference connection of the E-model for transmission planning
– see [21] for details

study the perceptive effects of each individual piece of equipment anew for
all possible transmission-channel settings.

A better way, to be sure, is to set up a standard structure which is representative for the most relevant degradations of the transmission channel.
Such a schematic has been agreed upon by the Telecommunication Standardization Sector of the International Telecommunication Union, ITU-T,
for network planning purposes [21]. There, the relevant network parameters
are listed for a 2-wire/4-wire analogue or digital telephone connection as terminated at both sides with a telephone handset. The schematic is depicted in
Fig. 7.3. The main transmission path through the network, the electrical side-
tone path, i. e. the coupling of the talker's own voice through the telephone
handset, and the echo paths, both back to the talker and to the listener in
case of multiple reflections, are all taken into account.

All paths are described in terms of their contribution to speech or noise
loudness by means of frequency-weighted scalar values. This description is
advantageous because it reasonably simplifies the individual characteristics
of each piece of equipment. For the transmission paths, so-called loudness
ratings are calculated, describing the loudness reduction resulting from the
specific path [24]. Such loudness ratings are calculated for the main transmission path, namely, the send loudness rating, SLR, and the receive loudness
rating, RLR, for the side-tone path with respect to speech and ambient noise
transmission, $STMR$ and $LSTR$, and for the echo paths, $TELR$ for talker
echo, and $WEPL$ for listener echo. The paths are further characterized by
their mean overall delays, T_a, T, and T_r. Circuit and ambient room noise are

modelled by ideal noise sources, i. e. point sources of equivalent A-weighted or psophometrically weighted noise powers, N_c, P_s, and P_r. Non-linear speech codecs are not modelled in detail, but taken into account by a factor describing the additional degradation that they introduce – the so-called equipment impairment factor, I_e [23, 37]. Quantizing noise resulting from AD/DA conversion and PCM coding is quantified in terms of an equivalent number of cascaded log-PCM coding/decoding processes, qdu. Up to now, time-varying degradations due to lost frames or packets are still described in a very simplified way, i. e. via a packet-loss probability and a so-called robustness factor, taking into account the characteristics of the specific speech and channel codec, including error recovery strategies [21, 40].

This schematic is still under study for the reason that it does not yet cover all perceptively relevant characteristics of modern networks. In particular, adequate descriptions are still needed for terminal equipment other than standard handsets, e. g., short mobile handsets, hands-free terminals and headsets. In modern networks, non-stationary degradations are expected to have a major impact on the overall quality, such as frame/packet loss, speech clipping due to imperfect voice-activity detection, comfort-noise insertion, fading channels, and impulsive noise resulting from handovers in mobile networks. These degradations are still not adequately covered by the schematic of Fig. 7.3 as they are difficult to describe via instrumentally measurable parameters. The validity of quality-assessment and -prediction methods for the depicted networks, however, will suffer if not all perceptively relevant degradations can be adequately described.

7.1.2 Quality and Quality of Service

Even if all perceptively relevant elements of the speech transmission channel were well-known and adequately described, it would still be important to note that the quality of a service operating over this network is neither an inherent characteristic of the network, nor of the signals which are transmitted through it. Quality is always linked to the users, or – in more general terms – to perceiving human subjects and to the situational context in which quality is experienced. Thus, following the conception developed in [30], quality "happens" in a specific application scenario. Similarly to the notion of an auditory event – see, e. g., [4] – one can speak of a "quality event", taking place when the user perceives and judges the service. This idea is reflected in the definition of quality as given in [30] as follows.

> "*Quality* ... Result of judgement of perceived composition of a unit with respect to its desired composition. [...]
> *Perceived composition* ... Totality of features of a unit. Signal for the identity of the unit visible to the perceiver."

For the scenarios which are addressed in this chapter, the unit to be judged upon is the telecommunication service with which users interact through

HMI, or through which they communicate with a human communication partner in HHI. In this context, speech transmission quality forms just one aspect of a broader quality concept, namely, "quality of service", QoS, at large. Other constituents will be identified in Sect. 7.2 and in Sect. 7.3.

In principle, quality of service can be addressed from two different points of view. The first one is that of the service providers. They are, in fact, mainly interested in the effects of individual elements of the service and how they relate to the users' degree of satisfaction or acceptability. To this end, service providers make use of a definition of QoS given in [20].

> "*Quality of Service, QoS* ... The collective effect of service performance which determines the degree of satisfaction of a user of the service."

More precisely, QoS comprises four factors, namely, service support, service operability, "serveability", and service security. In this context, speech transmission quality forms just one aspect of serveability, among others.

The second point of view is that of the users. Users perceive characteristics, i.e. features, of the service, then compare these perceptions with some type of internal reference, and judge on them according to whether they fulfill these expectations and/or desires. Quality, thus, is the result of a perception-and-judgment process.

When investigating quality of service, it is important to take both points of view into account. Relationships have to be established between what the users expect and/or desire on the one hand, and their percepts and the network characteristics which are responsible for evoking these percepts, on the other hand.

In order to reflect these points of view, it is useful to distinguish between "quality elements" and "quality features" [30]. Whereas the former ones are system or service characteristics which are in the hands of the designer, and can thus be optimized by them for high quality, the latter ones are perceptual dimensions which form the overall picture in the mind of the user. In general, no stable relationship, which would be valid for all types of services, users and situations, can be established between the two. Nevertheless, it is possible and desirable to establish temporary relationships for individual situations, because this facilitates network and service design for optimum quality.

Perceived quality features – according to their definition – can only be assessed in auditory tests. Examples for issues observed include intelligibility, listening-effort, noisiness, speech-sound quality or sound quality in general. For HMI over the phone, additional features relate to the dialogue and task capabilities of the speech-technology device, e.g., perceived system understanding, naturalness of system behaviour, or contextual appropriateness of system answers. Quality elements can be divided into those relating directly to the system, and those relating to the performance of the system in a dialogic interaction with the user. System and dialogue measures are closely linked to the quality elements. We call all those system parameters and signals "system measures" which can be measured instrumentally, e. g., speech and noise signals at different points of the transmission channel, frequency responses of the transmission paths, delay times, noise levels, and loudness ratings. In the interaction with the user, "dialogue measures" can be logged, either instrumentally or with the help of experts and/or experts' transcriptions or annotations. Examples of dialogue measures include dialogue duration, the number of user and system turns, the number of barge-in attempts by user, the word error rate of the speech recognizer, or the contextual appropriateness of system responses.

In order to adequately describe the quality of service, an analysis of QoS aspects, taking both viewpoints into account, is necessary. This will be performed with the help of the two taxonomies which are described in Sect. 7.2 for human-human-interaction scenarios, HHI, and in Sect. 7.3 for human-machine-interaction scenarios, HMI, see Fig. 7.1 and Fig. 7.2, respectively. On the basis of this analysis, assessment methods and prediction models can be developed, see Sect. 7.4. Examples are presented which show the predictive power, but also the limitations of such models. It will be shown how experiences gained with HHI and models set up for HHI can serve as a basis for future quality prediction models in HMI which are still in the development stage.

7.2 Quality Aspects of Human-Human Interaction over the Phone

In this and in the following section, taxonomies are developed which describe relevant factors affecting the quality of a service. The factors comprise quality elements which are related to speech transmission and communication for HHI, to dialogues and tasks in the context of HMI, and to the service and the context of use for both HHI and HMI. The resulting quality features can be categorized into quality aspects, i. e. classes of quality features such as effectiveness, efficiency, usability, utility, user satisfaction and acceptability of the overall service. These notions will be illustrated in the following by means of taxonomies which, to be sure, can be helpful in three different ways. (1) System elements which are in the hands of developers and which are responsible for specific user perceptions can be identified. (2) The dimensions underlying the overall impression of the user can be described, together with adequate auditory assessment methods. (3) Prediction models can be developed to estimate quality, as it would be perceived by the user, from purely instrumental or expert-derived measurements. Some of these application examples are presented in Sect. 7.4.

7.2.1 Taxonomy of Quality Aspects in Human-Human Interaction

The taxonomy discussed in this section refers to telecommunication services which permit human-to-human communication over the phone, e. g., wire-line telephony, wireless or mobile telephony, or a speech-communication service operated from an internet terminal. Services which are not within its scope are, e. g., responding machines or dialogue systems for information retrieval, see next section. The taxonomy of Fig. 7.4 has been developed by the author in another publication [36], and some labels have been slightly modified to be in line with the HMI case. It has proved to be useful for classifying quality features, assessment methods and prediction models.

The quality of service, QoS, is thought to be divided into three types of factors. The major category subsumes speech-communication factors, i. e. all factors which can be directly associated with the communication between the human partners over the speech-transmission system. A second category comprises the service-related influences, which will be called the "service factors" here. It includes service support, service operability, serveability, and service security – see the definition cited above. A third category comprises the contextual factors which are not directly linked to the physical realization of the service, but have, nevertheless, an influence on the overall quality, such as investment and operational costs and account conditions.

The given factors can be further subdivided. Speech-communication factors relate to the aspects of the auditory event itself, resulting in one-way voice-transmission quality, to quality aspects resulting from the system's conversational capabilities – conversation effectiveness, and to communication-partner-related aspects – ease of communication. Further, they are related to both the speech-link performance and to environmental factors, e. g., ambient noise. At this level of the taxonomy, the quality factors, and thus also the composing quality elements as being in the hands of the quality planners, are related to the quality aspects which, in turn, are composed from different quality features – the latter being perceived by the user.

In the lower half of the taxonomy, the user-related quality categories are presented in relation to each other. All speech-communication-quality constituents, i. e. one-way voice transmission quality, conversation effectiveness, and ease of communication, contribute to communication efficiency. Subsequently, communication efficiency and the service efficiency form the construct of usability, which is the appropriateness of a system or service to fulfill a defined task. Finally, usability in relation to the – financial – costs results in the utility of a system or service. The users may accept or refuse the utility which they experience when using the telecommunication service, depending on the costs. How readily users actually use a service is called the acceptability of this service. It can be measured by counting the users of a service and comparing this number to the number of potential users. Thus, acceptability is a purely economic concept, as reflected in the following definition given in [9].

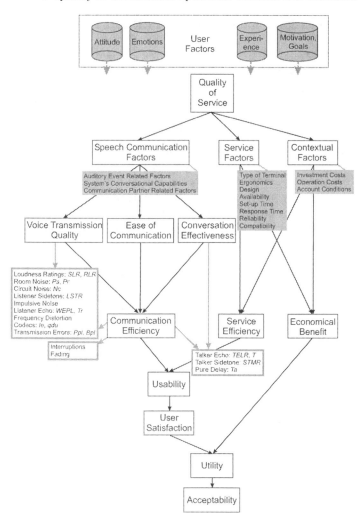

Fig. 7.4. Quality-of-service taxonomy for a human-to-human telephone service. In the **upper half**, factors affecting different aspects of quality perceived by the user, *white boxes*, and their relationships, *arrows*, are illustrated. As the user is the decision point for each quality aspect, user factors are distributed over the whole diagram. This fact has tentatively been illustrated in the **cover panel**, *gray cans*

> "*Acceptability of Service, AoS ...* Multidimensional property of a service, describing how readily a customer will use the service, represented as the ratio of the number of potential users to the quantity of the target group."

To be sure, concepts, such as quality, usability, user satisfaction and utility, are relative, as they represent congruence or trade-off between performance and user objectives. Thus, they are all subject to user factors.

7.2.2 Transmission Channel Impact in Human-Human Interaction

The transmission parameters of the reference connection in Fig. 7.3 can be classified according to their effect on the quality aspects as illustrated in the lower part of Fig. 7.4. Most parameters of the main voice-transmission path have an influence primarily on one-way voice-transmission quality. Amongst these are, e.g., parameters related to the loudness of the transmitted signal, namely, the loudness ratings, SLR and RLR, the different types of noise, i.e. ambient noise or continuous circuit noise, distortions due to the frequency shape of the transmission channel and codecs, as well as transmission errors, such as bit errors, packet loss or frame loss, and listener echo. In a second step, however, severe problems in one-way transmission will influence the conversational behavior as well. A too loud or too quiet connection, for example, will force the talkers to adapt their speech levels. This has an influence on ease of communication. Parts of the discourse which are lost because of short interruptions from lost packets lead to a request for confirmation – "Did you say that...?" – and influence the flow of the dialogue. Therefore, all such parameters also exert an influence on communication efficiency.

Communication efficiency is directly coupled to factors which occur for talking *and* listening, like talker echo, talker side-tone, absolute delay, or longer interruptions of the transmission flow. Such interruptions or fading channels predominantly occur in modern, i.e. mobile and packet-based networks. In a first step, they have an influence on one-way voice-transmission quality, but in a second step also on communication efficiency. Talker echo, talker side-tone and absolute delay are closely related to the conversation effectiveness, rather than to voice-transmission quality.

7.2.3 Influence of Service and Contextual Factors in Human-Human Interaction

Apart from the quality elements which are directly related to the physical connection set-up, service- and context-related factors will carry an influence on global QoS aspects in terms of usability, user satisfaction, utility and acceptability. The service efficiency is, e.g., co-determined by the type of terminal, including its ergonomic design, as well as by factors like availability, reliability, set-up times and response times. The advantage in terms of access that a mobile service may offer over a standard wire-bound service can be classified as an advantage in service comfort. Costs include the investment costs, e.g., for the terminal equipment, operational costs for each connection, as well as the account conditions. After all, they may also have a significant and even dominant effect on the utility and acceptability of a service.

As an example, auditory test results are presented which show the differences in overall-quality judgments in two simulated conversation situations, namely, for the standard wire-line telephone service, and for a simulated mobile service. Users with and without experience in using mobile phones carried

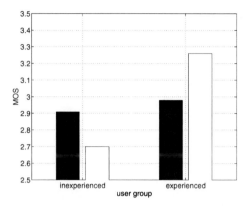

Fig. 7.5. Conversation-test results – average MOS over 14 circuit conditions – for users with and without experience of mobile phones. Wire-line scenario, *black*, mobile scenario, *white*. For details see [36]

out 14 conversations over connections with different amounts of degradation. They rated each connection individually on a 5-point category scale, as defined in [25]. The circuit characteristics were controlled to be identical for the two scenarios with the help of a simulation environment, which is described in detail in [36]. For the user, the main difference was the use of a standard wire-line handset in one case, and a simulated mobile handset, showing the same acoustic-electric transfer characteristics as in the wire-line case, in the other one.

In Fig. 7.5, the mean user judgments, average mean opinion scores, MOS, of all circuit conditions are presented. It can be seen that the wire-line scenario is rated nearly identical in both cases. On the other hand, the mobile scenario is rated better than the wire-line one by the experienced users, and worse by the inexperienced users. This shows that the users who know the service and contextual characteristics of mobile services take these factors into account in their judgment on overall quality. In particular, the advantage of access – which may be linked to the fact of being reachable at every location and time – seems to have a positive influence on their judgment. Users who are not experienced in these services cannot take these factors into account. On the contrary, they may have a negative attitude towards mobile services in general, a hypothesis which was supported by the results of a subsequent interview in the reported experiment.

The type of service also carries an influence on the price which users are willing to pay for it. In two comparative experiments, the relationships were analyzed between user judgments on overall quality, in terms of MOS, and the price they would be willing to pay for a time unit. The experiments were carried out in the simulation environment reported above, in order to guarantee identical transmission conditions. Two comparisons are depicted in Fig. 7.6, namely, between a simulated mobile scenario and a standard wire-line

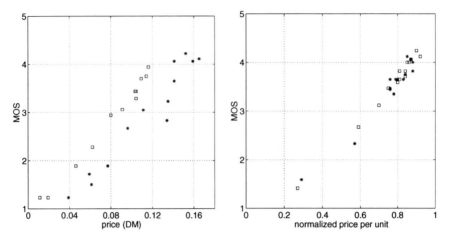

Fig. 7.6. Price that users are willing to pay for the service vs. MOS quality rating. Test with wire-line and mobile scenario, *left panel,* and with wire-line scenario and Voice-over-IP scenario, *right panel.* □ ... standard handset condition, ∗ ... portable handset, *left,* or computer terminal, *right.* Details of the experiments are described in [36]

scenario, and between a communication scenario from an internet terminal as typical for "voice-over-internet protocol", VoIP, and a standard wire-line telephone service. In both cases, a more or less linear relationship between both judgments was observed. This high correlation may indicate two things, i. e. that users mainly judge upon the observed transmission quality, and that a general inadequacy of laboratory experiments exists when it comes to obtaining realistic price judgments. However, an offset can be observed between the relationships of wire-line and mobile services, see Fig. 7.6 left. This indicates that the price which users are willing to pay depends strongly on service and contextual factors. For the VoIP service as simulated here, these factors, for the time being, do not yet seem to be stable enough to establish a reference different from that for the standard wire-line service.

7.3 Quality Aspects of Human-Machine Interaction over the Phone

The quality of interactions with spoken-dialogue systems is difficult to determine. Whereas structured approaches have been documented on how to design spoken-dialogue systems in such a way that would adequately meet the requirements of their users, compare [2], the quality which is perceived when interacting with SDSs is often addressed in an intuitive way. In [17] efforts are described to determine the underlying dimensions in user quality judgments by performing a multi-dimensional analysis on subjective ratings as obtained from a large number of different scales. It turned out that the problem was obviously multi-dimensional indeed. Nevertheless, many other researchers still try to estimate "overall system quality", "usability" or "user satisfaction" by simply calculating the arithmetic mean over several user ratings on topics as different as perceived text-to-speech quality, perceived system understanding, and expected future use of the system. The reason is the lack of adequate descriptions of quality dimensions, both with respect to system design and to the perceptions of the users.

The second taxonomy which is developed in this section addresses a specific type of interaction with a spoken-dialogue system, namely, a task-oriented interaction through a telephone network. Currently, most dialogue systems which are publicly available are only designed for task-oriented inter-action. Thus, this restriction is not a severe one. The limitation to telephone interfaces makes this scenario similar to the HHI case, which will be reflected in the similarity of the resulting taxonomies. In fact, when calling a service which is based on a spoken-dialogue system, users might initially not be aware of this and, hence, expect the characteristics of a human operator – with respect to speech input and output capabilities, but also with respect to the task and domain coverage. Thus, HHI with a human operator may actually form *one* reference against which quality is judged. Other references may evolve from alternative ways to perform the same task, e. g., by using a graphical user interface accessed through the internet, instead of speech.

7.3.1 Taxonomy of Quality Aspects in Human-Machine Interaction

In order to adequately describe QoS in these cases, a taxonomy for quality of service aspects has been developed by the author [33, 35], as presented in Fig. 7.7. The taxonomy illustrates the categories or factors, which can be sub-divided into aspects, and their mutual relationships. Again, user factors are distributed over the whole diagram. However, for the case of HMI, as discussed here, user factors will not be addressed further.

In [48] three factors have been identified to be considered with regard to the performance of SDSs. Consequently, they may contribute to its quality as perceived by the user. These factors are agent factors – mainly related to the dialogue and the system itself, task factors – related to how the SDS captures the task it has been developed for, and environmental factors – e. g., factors re-lated to the acoustic environment and the transmission channel. Since the tax-onomy refers to the service as a whole, a fourth point is added here, namely, contextual factors – such as costs, type of access and availability. All four fac-tor types subsume quality elements which can be expected to exert an influ-ence on the quality as perceived by the users. The corresponding quality fea-tures are summarized into aspects and categories in the lower part of Fig. 7.7.

Each type of factor may have a severe or even dominant influence on the overall quality as experienced by the user. For example, a system which is

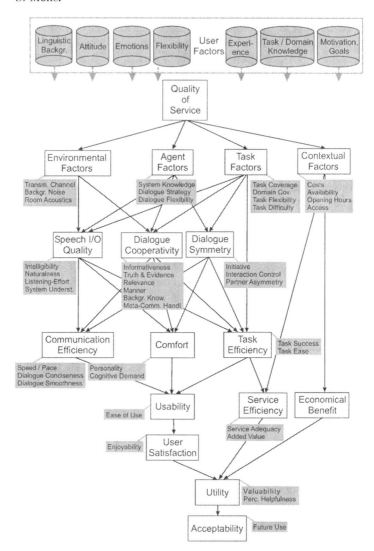

Fig. 7.7. Quality-of-service taxonomy for task-oriented human-machine interaction over the phone. Factor categories, *white boxes*, as subdivided into aspects, *gray boxes*, and their mutual relationships, *arrows*. User factors are distributed over the whole diagram

otherwise working well may nevertheless be unusable because of failure of a speech recognizer due to bad transmission lines or high background-noise levels. The machine agent has obviously a high impact on the overall quality, as its system components mainly determine the course of the interaction with the users. Regarding the task factors, a system may not be able to deliver the required information to the user, which will result in low quality for the user.

With respect to the contextual factors, costs and system availability may be decisive. For instance, a user might prefer the automated service only in cases where a comparable human-operated service is not available, e. g., owing to restricted opening hours.

The agent factors relate to three categories of quality aspects. On the speech level, input and output quality will exert a major influence. Quality features for speech output have been largely investigated in the literature and include well known issues, such as intelligibility, naturalness and listening-effort. They will further depend on the whole system set-up, and on the situation and task that the user is confronted with. Quality features related to the speech input from the user, and thus to the system's recognition and understanding capabilities, are far less obvious. They are, in addition, much more difficult to investigate as the users only receive indirect feedback on the system's capabilities, namely, from system reactions which are influenced by the dialogue as a whole. Both, speech input and output, are strongly co-determined by environmental factors.

On the language and dialogue level, dialogue cooperativity has been identified as a key requirement for high-quality services [2]. Classification of cooperativity into aspects, as proposed in [2] and being related to *Grice*'s maxims of cooperative behavior in HHI [13], is mainly adopted here, yet, with one exception. The partner-asymmetry aspect is covered under a separate category called dialogue symmetry, together with initiative and interaction-control aspects. Dialogue cooperativity will thus cover the following aspects: informativeness, truth and evidence, relevance, manner, the users' background knowledge, and meta-communication handling strategies.

Adopting a notion of efficiency as used by the European Telecommunications Standards Institute, ETSI, and the International Organization for Standardization, ISO [7], efficiency denotes the effort and resources expanded in relation to the accuracy and completeness with which users can reach specified goals. For HMI, it is proposed to distinguish the following three categories of efficiency. Communication efficiency relates to the efficiency of the dialogic interaction, and includes, besides the aspects speed and conciseness, the smoothness of the dialogue, which is sometimes called "dialogue quality" in the HMI literature. Please note, at this point, the significant difference to some other notions of efficiency which only address the efforts and resources, but not the accuracy and completeness of the goals to be reached. The second category, task efficiency, is related to the success of the system in accomplishing the task, thereby it covers task success as well as task ease. Service efficiency, the third category in this context, is the adequacy of the service as a whole for the purpose as defined by the user. It also includes the "added value" which is attributed to the service, e. g., in comparison to alternative means to obtain the desired information, e. g., a comparable graphical user interface or a human operator.

In addition to efficiency aspects, other aspects exist which relate to the agent itself, as well as to the perception of the agent by the user in a dialogic interaction. These aspects are subsumed under a category labelled "comfort", covering the agent's "social personality", i. e., perceived friendliness, politeness, as well as the cognitive demand on the user.

Depending on the area of interest, several notions of usability are common. In congruence with the HHI case, usability is defined here as the suitability of a system or service to fulfill the users' requirements. It mainly considers the ease of using the system and may result in user satisfaction. However, it does not cover service efficiency or economical benefits which exert an influence on the utility, i. e. usability in relation to financial costs and other contextual factors of the service. In [47] it is also stated that "user satisfaction ratings [...] have frequently been used in the literature as an external indicator of the usability of an agent." As in [32], it is assumed that user satisfaction is predictive of other system-designer objectives, e. g., the willingness to use or pay for a service. Acceptability, which is defined on this more or less "economic" level, can therefore be seen in a direct relationship to usability and utility.

The taxonomy as presented here for the HMI case has not been fully verified yet in the sense that it could guarantee to fully cover all relevant quality dimensions which are experienced by the users of the respective service. Nevertheless, it is in keeping with several less-structured analysis approaches reported in literature. As stated above, the separation of environmental, agent and task factors was motivated by [48]. The same categories appear in the characterization of spoken-dialogue systems given in [11] – plus an additional user factor which is found in the quality aspects owing to the fact that it is the user who decides on quality. The context factor is also recognized in [6]. Dialogue cooperativity is a category which is based on a relatively sophisticated theoretical as well as empirical background. It has proven useful especially in the system design and set-up phase. First results in evaluation have also been reported [2]. The dialogue-symmetry category captures the partner asymmetry aspect and has been designed separately to cover initiative and interaction-control aspects as well. To this author's knowledge, no similar category has been reported before.

A comparison between the two taxonomies discussed in this chapter shows a quite similar structure in both cases. In the HHI case, speech communication factors replace the environmental and agent factors of the HMI case. This category is divided into a one-way voice-transmission category, a conversational category and a user-related category – ease of communication, comparable to the category "comfort" in the HMI case. The task and service categories of interaction with SDSs are replaced by the service category of the HHI taxonomy since HHI over the phone is not restricted to task-oriented dialogues. The rest of the taxonomy is congruent in both cases, although there are obvious differences in the particular aspects which are covered by each category.

7.3.2 Application Examples

The new taxonomy can be used to classify quality features which have been observed in different experiments investigating QoS of task-oriented HMI over the phone. As discussed in another publication of this author [35], twelve investigations as reported in the literature have been analyzed, and the quality features have been identified which have been addressed by the questionnaires used in these investigations. For each aspect at least two questions could be found which address just this aspect. Thus, the taxonomy seems to cover most of the QoS aspects which have been observed in these experiments.

With the help of the taxonomy, it is also possible to identify dialogue and system measures which are supposed to be related to the individual quality aspects. An analysis of well-known dialogue and system measures is also reported in [33,35]. For most of the aspects, several measures can be identified which seem to be directly related to the particular aspect at stake. Only for "high level" aspects, such as speech-output quality, agent personality, service efficiency, usability, and user satisfaction, relevant measures are not obvious. It is expected that suitable combinations of dialogue and system measures addressing the individual influencing factors may be able to predict some of these aspects. However, this remains still to be proved. Experiments along these lines are well underway.

7.4 Quality Assessment and Prediction

As was pointed out in Sect. 7.1.2, formation of quality requires a human perception and judgment process. Thus, auditory assessment methods are needed in order to recognize and quantify the quality features as experienced by the user as well as the contribution of these to overall-quality aspects like user satisfaction or system usability. In the following section, the most important assessment methods will be reviewed separately for the HHI and HMI cases – see Sects. 7.4.1 and 7.4.2. Where appropriate, the aspect in the QoS taxonomy is indicated that the respective methods refer to.

Auditory assessment methods, however, have a reputation of being time-consuming and expensive. This is in contradiction to the demands of service providers, who would preferably like to get an indication of the quality to be reached with specific services, network and/or system configurations even before the respective networks and systems have even been set up in practice. Various prediction models have been developed to this end for HHI. While still being limited regarding potential scope of application and validity of the predicted results, they have already become important tools for network planning and optimization. A number of different modelling approaches are presented in Sect. 7.4.3 and it will be shown how some of the models can be used for predicting specific quality aspects of HMI services as well – see Sect. 7.4.4.

7.4.1 Assessment Methods for Human-Human Interaction

There are several ways of characterizing assessment methods, e. g., regarding the modality of the test – listening-only or conversational; the rating procedure – single stimulus or paired comparison, adjustment or constant method; or even the test environment – laboratory test vs. surveys or field tests. These classification schemes are well documented in the psychophysics literature. As a general rule, it is not possible to define a universal "best" assessment method or methodology independently of the test objective. In contrast, an optimum method has to be engineered by the test designer for each specific objective. This includes the choice of test stimuli, subject groups, number of objects to be tested, and other test-environment factors. Where no standardized test methods are available, new approaches have to be developed. Yet, first of all, these approaches have to be tested with respect to the fulfillment of general requirements for test methods [29], such as validity and reliability.

Telecommunication services are usually used in a bi-directional, conversational mode. As a consequence, in order to obtain valid integral quality judgments, conversation methods are generally preferred for telephone-quality assessment. If a detailed analysis of quality dimensions or test economy requires the use of listening-only tests, paired comparison, PC, is generally recommended for stimuli that are almost equal, because differential sensitivity of the auditory system is much larger than absolute sensitivity. In this way ordinal or interval judgments are predominantly obtained – see the definitions of scale levels as proposed by *Stevens* [44]. Absolute category rating, ACR, is often used for more varied test stimuli and for obtaining more "absolute" judgments. If valid results for high-quality conditions are required, comparison techniques such as degradation-category rating, DCR, or comparison category rating, CCR, may be more appropriate in differentiating between conditions. The differential sensitivity of the ear can further be increased by forced-choice comparison, i. e. the "equal" judgment being excluded. An alternative method to ACR and PC is the paired-rating method, PR, as recommended, e. g., for listening tests on loudspeakers [18]. A description of different test methods can be found, e. g., in the respective ITU-T literature, e. g., [19, 25, 36] – see also [10] this volume.

If more analytical information than just an integral quality judgment is required, multi-dimensional analysis can yield highly informative test results. Two main possibilities exist, namely, either no quality dimensions are given to the test subjects and distance or similarity scaling is required, or a number of carefully selected quality dimensions is selected beforehand, represented by different scales with semantic descriptors, and then a polarity profile can be obtained for the test samples. It is also possible to ask experienced test subjects analytical questions in a standard listening-only experiment, and sometimes even in a conversation experiment. The drawback, however, is the potentially low validity of the results obtained from the experts as compared to a naïve telephone-user population.

Laboratory tests always lack some naturalness in the sense that the assessing subjects are in an artificial situation when reporting their opinions on the test items. Especially in telephony, the motivation for simulated test calls is different from real life. These differences influence the test results, especially when user- and service-directed aspects are investigated, such as utility, acceptability and cost-related factors. The same problem occurs when subjects are asked to carry out test conversations with a spoken-dialogue system. In order to obtain more realistic results for such quality aspects, it might be more appropriate to perform user surveys or to use usability measures – see, e. g., [12].

The assessment methods described in this section can be classified based on the QoS taxonomy. Apparently, all the methods which make use of a listening-only situation can only investigate features of one-way voice-transmission quality. Amongst these are articulation and intelligibility tests, listening-only tests using absolute rating, ACR, multi-dimensional analysis with a polarity profile, as well as listening-only tests using paired-comparison techniques, e. g., DCR or CCR. Communication efficiency can only be investigated by performing conversation tests. Service efficiency and costs are mainly investigated by means of user surveys and usability-evaluation methods, in a natural – i. e. field test – environment. Real measures of acceptability, however, can only be obtained by counting the actual number of users in a test run of the prototype service.

7.4.2 Assessment Methods for Human-Machine Interaction

In the HMI case, we identified the factors which result from the environment – including those stemming from the transmission channel, the machine agent, the task and the context of use. Thus, when assessing quality in this scenario, it first has to be decided whether the object of investigation is the speech-technology device itself, i. e. the ASR system, the speech-synthesis system, the spoken-dialogue system, the transmission channel, or the service as offered to the users at large.

A large amount of work has been performed in the past which addresses the performance of speech technology. The investigations relate to individual components or to whole systems and include both the identification of respective dialogue and system measures as well as the definition of auditory assessment methods. Useful information on this topic can be found, e. g., in the EAGLES handbook, particularly on the assessment of speech recognition [46] and speaker recognition [3], of synthetic speech [45], and of interactive systems as an entity [11].

In contrast to this, the influence of the transmission channel on system-component performance and, subsequently, on overall service quality has hardly been investigated. An example is given below, and a more diagnostic analysis of the transmission-channel impact on ASR is presented in [38], and on synthesized speech in [34]. To this author's knowledge, comparable investigations regarding the effects of transmission channels on the overall quality and usability of services have not been reported yet.

In order to assess QoS in general, laboratory or field tests with users are commonly performed – see, e. g. [33]. Depending on the test situation, the users are able to carry out real tasks – field tests – or have to carry out tasks which are given to them in terms of scenarios. The latter leads to a loss in realism, and it is still not clear how strong the impact on the test results will be. Questionnaires are given to the test subjects before and after each conversation, addressing different aspects of the interaction and the system. Although several prototype questionnaires are reported for different tasks, there is still no well-established method defining which questions to ask in order to adequately cover the perceptual space of quality dimensions. Further information is needed showing how the different aspects relate to each other and how they are influenced by the quality elements which may be optimized by the service provider.

During experimental interactions with a prototype service, it is common to log a number of system and dialogue measures. By comparing these measures to subjects' responses on different quality aspects, relationships between both may be established. In this way, initial indicators of quality can be obtained – see Sect. 7.4.4.

7.4.3 Quality Prediction for Human-Human Interaction

It is the aim of quality prediction models to establish relationships between signals or parameters describing the quality elements of networks (system elements) on the one hand, and perceptive quality features on the other hand. The solution may be considered successful when a high correlation between auditory test results and input signals or parameters is reached. However, a high correlation does not necessarily mean that valid quality predictions can also be obtained for scenarios which have not been covered by the test databases as used for model definition. It is thus important not to extrapolate over the verified range of validity. Otherwise, the quality predictions will be useless or may mislead the service provider.

According to a classification of prediction models proposed in [39], five criteria have to be taken into account, namely,

- the application scope of the model, e. g., network planning, optimization, monitoring,
- the considered network components or configurations, e. g., mouth-to-ear, codecs only,

- the predicted quality features, e.g., integral quality, intelligibility, listening-effort, conversational aspects,
- the model's input parameters,
- the psycho-acoustically or perceptively motivated content vs. empirically motivated knowledge.

These criteria will be briefly discussed in the following section.

Prediction models can serve different purposes, depending on the phase of planning, set-up, operation and optimization, which the network under consideration finds itself in. In the planning phase, it is possible to predict the quality that the user of the future connection may encounter, provided that all or most of the planning parameters shown in Fig. 7.3 are known beforehand. In such a case, comparative calculations of different set-ups can help to find weaknesses in the network set-up or to decide whether to apply an additional piece of equipment, e.g., a codec or an echo canceller. When parts of the network equipment are available separately it is also possible to perform measurements with artificial or typical speech signals, e.g., for equipment optimization. Signal-based measures, see below, are a good means for this purpose. However, it is also possible to measure specific signals, speech/noise levels, or other parameters such as delay times, in operating networks. From these measurements, quality can be estimated for the specific network configuration, and potential weaknesses and network malfunctions can be identified.

Depending on which parts of the network are already available, different types of input parameters can be used for predicting quality. These parameters may refer to one or several network elements or to the whole transmission channel, mouth-to-ear. As a result, the predictions based on the parameters/signals will only be valid for this or a comparable type of equipment with identical degradations.

Most models aim at predicting integral quality, in terms of a mean opinion score, MOS, or a transmission rating, R, see below. The predicted quality index may express user opinion – which has been collected in specific situations, e.g., in a laboratory listening-only test environment, or they can be associated with quality of service classes as defined for network design [22]. Quality indices will only be valid for the specified range of impairments which were part of the tests for which the prediction model has been optimized. For example, if speech-sound quality has not been regarded as a quality dimension in the tests, the model cannot be expected to provide reasonable predictions for this dimension. Still, such a dimension may become dominant in specific network configurations. In this case, the validity of the predicted results is questionable.

The mapping of input parameters on output indices can be performed incorporating different degrees of psycho-acoustic knowledge. Either a pure curve-fitting to the results of auditory tests is performed, or certain known characteristics of human auditory perception are taken into account and are modelled, e.g., masking or loudness perception. Usually, both ways are chosen, and the models contain some perceptively motivated parts as well as some purely empirically determined parts – strictly speaking, however, all psycho-acoustically motivated knowledge is also empirically determined. Following these classification criteria, three major types of models can be distinguished, namely,

– signal-based comparative measures,
– network planning models, and
– monitoring models.

Signal-based comparative measures predict one-way voice-transmission quality for single network components. They have mainly been developed for the effects of low-bit-rate codecs, but newer versions also take background noise, transmission errors, or lost frames or packets into account. The quality prediction is performed on the basis of the speech signals which are available at the input and output side of the network element under consideration. These signals are first pre-processed, e. g., for loudness and delay compensation, and then transformed into an internal representation, e. g., a loudness representation - see, e. g., [15] for potential representations. Examples show that it is not necessary to choose a highly sophisticated perceptual model at this stage.

Between the two internal representations, a similarity or distance can be computed for each block, i. e. frame, of the signal. Alternatively, it is possible to decompose the degraded speech signal into a *clean* source part and a degradation-*noise* part. Subsequent temporal integration of the distance or similarity values, like simple temporal mean or more sophisticated time-scale remembering modelling, and subsequent transformation to the desired scale, e. g., MOS or R, finally yields a one-dimensional quality prediction index. This index is mainly fitted to the results of listening-only tests and reaches a high degree of correlation for most well-established speech codecs. Examples of such models include the PESQ as standardized in [27, 43], the TOSQA model [1], or the measures reported in [14, 15].

In contrast to signal-based measures, network planning models refer to the whole transmission channel from the mouth of the talker to the ear of the listener. They also take conversational aspects, e. g., the effects of pure delay and echo, into account. The models cover most of the traditional analogue and digital transmission equipment, handset telephones, low-bit-rate codecs, and partly also the effects of transmission errors. Input parameters are scalar values or frequency responses which can be measured instrumentally off-line, or for which realistic planning values are available. For example, the so-called E-model [21] uses the parameters depicted in Fig. 7.3 as an input. With realistic planning values being available, it is already possible to predict quality in the network-planning phase, before the network has been fully set up.

Network planning models aim at predicting overall quality, including conversational aspects and, to a certain extent, contextual factors, e. g., effects of user expectation towards different types of services, see Sect. 7.2.3. In order to integrate a multitude of perceptively different quality dimensions, a multi-step transformation has to be performed. For example, the E-model first performs a transformation on an intermediate "psychological" scale and then assumes additivity of different types of impairment on this scale. The model output is a one-dimensional quality index, the so-called transmission rating, R, defined on a scale from 0 for lowest possible to 100 for highest possible quality. This quality index can be related to mean users' opinion in terms of MOS or to network-planning quality classes as defined in [22]. The most important network planning model is the E-model, developed mainly by *Johannesson* [31] and now being standardized by ITU-T and ETSI [8,21]. Another example is the SUBMOD model [28] which is based on ideas from [42].

If the network has already been set up, it is possible to obtain realistic measurements of all or a major part of the network equipment. The measurements can be performed either off-line – i. e. intrusively when the equipment is put out of network operation, or on-line in operating networks – non-intrusive measurement. In operating networks, however, it might be difficult to access the user interfaces. Therefore, standard values are taken for this part of the transmission chain. The measured input parameters or signals can be used as an input to the signal-based or network planning models. In this way, it becomes possible to monitor quality in the specific network under consideration. Different models and model combinations can be envisaged. Details can be found in the literature [39].

Already from the principles being used by a model, it becomes obvious which quality aspects it may predict at best. When signal-based measures are used, only one-way voice-transmission quality can be predicted, for the specific parts of the transmission channel for which the particular method has been optimized. Such predictions usually reach a high accuracy since adequate input information is available. In contrast to this situation, network planning models, like the E-model, base their predictions on simplified and perhaps imprecisely estimated planning values. Yet, in addition to one-way voice-transmission quality, they cover conversational aspects and, to a certain extent, also effects caused by service- and context-related factors.

7.4.4 Quality Prediction for Human-Machine Interaction

Developers of telephone services based on spoken-dialogue systems are interested in identifying quality elements which are under their control, with the aim of using them for enhancing the quality for the user. Unfortunately, few such elements are known, and their influence on service quality is only partly understood so far.

Owing to the lack of accessible quality elements, modelling approaches for the HMI case are mainly based on dialogue and system measures. These measures relate to a certain extent to the quality elements which have to be optimized for a successful system design. They can be determined during the users' experimental interaction with the help of log-files, either instrumentally, e. g., dialogue duration, or by an expert evaluator, e. g., contextual appropriateness. Although they provide useful information on the perceived quality of the service, there is no general relationship between one or several such measures and specific quality features. A large number of measures are reported in the literature – for an overview see, e. g., [5, 11, 33].

The most-widely applied modelling approach for HMI – not particularly restricted to the telephone situation – is the PARAdigm for DIalogue System Evaluation, PARADISE, [47, 48]. The aim was to support the comparison of multiple systems (or system settings) doing the same task by developing a predictive model of "usability" of a system as a function of a range of system properties, i. e. dialogue measures. By doing this, a technique is, hopefully, provided for making generalizations across systems and for getting an indication of which system properties really impact usability.

PARADISE uses methods from decision theory to combine a set of system and dialogue measures and "user satisfaction" ratings into a single performance-evaluation function. It performs a multi-variate linear regression, using user satisfaction as an external validation criterion. Unfortunately, user satisfaction is calculated as a simple arithmetic mean across a set of user judgments on different quality aspects such as the intelligibility of the synthetic voice, perceived system understanding, stress or fluster when using the system, or system speed. Bearing in mind the different aspects which are addressed by these judgments, the calculation of a simple mean seems to be highly questionable. As an underlying assumption of PARADISE [47], it is postulated that

> "performance can be correlated with a meaningful external criterion such as usability, and thus that the overall goal of a spoken-dialogue agent is to maximize an objective related to usability. User satisfaction ratings [...] have been frequently used in the literature as an external indicator of the usability of an agent."

This assumption, and the use of user satisfaction as an external validation criterion for the prediction, are in line with the QoS taxonomy described above. According to [32], user satisfaction may also be predictive of other objectives, e. g., of the willingness to use or pay for a service – which may be seen as an indicator of service acceptability.

Although the PARADISE model has proved successful in predicting an amalgam of user-satisfaction judgments – see above – it does not really help to identify the individual factors contributing to the different quality features as perceived by the user. In fact, much more analytic investigations are necessary in order to quantify the correlation between individual quality elements – or dialogue and system measures which can be measured instead – and the quality features as experienced by the user.

In the following, a brief example is given for such an analytic investigation. It shows the dependence of the recognition accuracy of two prototyp-

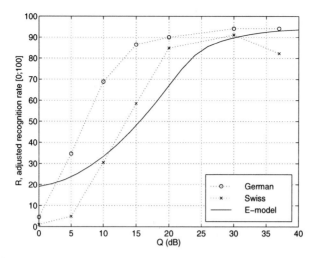

Fig. 7.8. Comparison of E-model predictions for speech quality in HHI, transmission rating factor R, and normalized recognition rates for two speech recognizers, for different levels of signal-correlated noise with a signal-to-noise ratio of Q. Recognition rates have been normalized to a range of 0–100 to conform with E-model quality predictions

ical speech recognizers, one for the German and one for the Swiss-French language, on the characteristics of the transmission channel. The relevant transmission channel characteristics shown in Fig. 7.3 have been generated in a controlled way by using the simulation model described above. In this case, the amount of signal-correlated noise, generated with the help of a modulated-noise reference unit [26] and inserted into the transmission channel at the position of the codec, is used as the variable parameter. This type of noise is perceptively similar to quantizing noise as resulting from PCM coding, and it can be scaled to a desired signal-to-noise ratio, Q. Speech files have been collected at the input of the transmission channel and been then transmitted through different channel settings. The files processed in this way can be used as an input to speech recognizers, and the recognition rate can be determined as a function of the channel characteristics. Details of the experiment are reported in [38].

In order to be able to compare the recognition rates to the quality predictions provided by network planning models for HHI, only the relative recognition rates are shown in Fig. 7.8. They have been obtained via a linear transformation and normalization to their maximum values as can be observed for a clean channel, resulting in the highest recognition rate corresponding to $R = 100$, i.e. the highest transmission rating obtained from the E-model, and the lowest recognition rate corresponding to $R = 0$. The comparison shows a similar behavior of recognition performance and human-to-human speech quality as predicted by the E-model. In both cases, an S-shaped curve describes the degradations as caused by signal-correlated noise. However, the E-model prediction is more pessimistic in that it predicts a significant decrease of speech quality already at higher signal-to-noise ratios.

It has to be noted, however, that the relationship between transmission rating, R, and required – normalized – recognition performance had been chosen somehow arbitrarily. Provided that a fixed relationship between these entities can be established, it may become possible to use quality prediction models which were originally designed for predicting quality in HHI, also for estimating the transmission-channel influences on dialogue measures which are related to speech-input and -output quality. Such approaches, however, will have to be validated against the set of transmission parameters which are described in Sect. 7.1.1. In particular, the entity which might be predicted is recognition accuracy and not the speech-input quality perceived by the user. It will be necessary to establish further relationships between dialogue measures and quality features perceived by the user in order to obtain more general indicators of service usability, user satisfaction, and even acceptability.

7.5 Conclusions and Outlook

In this chapter, the quality of transmitted speech was addressed in two application scenarios which are common to modern telecommunication networks, namely, human-to-human telephone conversation and task-oriented interaction of a human user with a spoken-dialogue system over the phone. For both scenarios, a taxonomy has been introduced which identifies those factors of the service which are relevant for the users' quality perception. The taxonomies show many similarities but also obvious differences, which result from the fact that the HMI service comprises both the transmission channel and the machine interaction agent. It is useful to treat both scenarios in parallel, since the transmission networks under consideration will have to satisfy the requirements of human users as well as the requirements of speech-technology devices. Provided that adequate assessment methods or prediction models are available for one of the scenarios, it does not seem to be far off the mark to attempt transfering these to the other scenario as well. An example for such a transfer was presented in the last section.

Following the presented taxonomies, quality of telecommunication services turns out to be a multi-dimensional and multi-layered problem. In this context, the quality of transmitted speech is just one aspect of the quality of the service at large. Apart from the obvious transmission-channel- and environment-related factors, service and contextual factors have been shown to have influence on how the user judges quality. In the case of HMI over the phone, task and agent factors are important for communication and task efficiency as well. When services are to be planned according to quality considerations, all factors which affect quality have to be considered.

The taxonomies help identifying those quality aspects which are affected by specific system properties, i.e. the quality elements. They explicitly show interrelations between the characteristics of the service and the quality dimensions on which they exert an influence. Such a structured view avoids establishing non-meaningful relationships as may result from an undue mixing of different quality aspects. So far, the taxonomies presented here have been used for classifying quality features as perceived by the user and for classifying instrumentally or expert-derived dialogue and system measures, auditory assessment methods, and quality estimations that can be obtained from current prediction models [33].

An overview was given on auditory assessment methods which are well-established for judging impairments that occur in traditional and modern networks. These assessment methods primarily refer to time-invariant impairments affecting the listening-only as well as the conversational situation. Assessment methods for time-variant impairments, such as packet loss, fading channels, etc. are still under study [40]. There are several indications that – apart from the loss rate – the "burstiness" and the temporal distribution of the lost packets may also be important and, consequently, these aspects have to be taken into account in the assessment method. For HMI over the phone, most assessment methods involve the distribution of questionnaires and the collection of dialogue and system measures during subjects' interaction with a prototype system. Unfortunately, there is still no structured approach of how to set up questionnaires in such a way that they would adequately cover all relevant quality aspects for the service under investigation. It is hoped that the taxonomy developed in Sect. 7.3.1 will help provide relevant information to this end.

For HHI over the phone, a number of quality prediction models have been developed which aim at predicting part of the quality aspects which contribute to overall quality of the service. Depending on the application purpose, the models estimate indices of integral quality in the network planning, optimization, or operation phase. Although there is a continuous need to investigate the quality-estimation results of these models regarding their validity and reliability, these models generally reach a relatively high correlation with auditory test results, though only for the specific purposes for which they have been designed. Thus, it turns out be very dangerous to extrapolate the predictions to scenarios which have not been used for optimizing the models. Consequently, substantial extensions to the current models are needed – in particular with respect to describing the effects of user interfaces other than handsets, time-variant degradations [40], as well as new dimensions of quality which might not have been considered in the networks that the models have been optimized for, e.g., speech-sound quality, see [41]. The combination of different types of impairments to a one-dimensional quality index is a topic which needs further study. Work is underway in this respect within the ITU-T and elsewhere.

In the case of task-oriented HMI over the phone, no comparable prediction models are available. The only approach in this context is the PARADISE framework, which attempts to estimate an integral "user-satisfaction" index via a multi-variate regression analysis from dialogue and system measures.

Although this framework certainly provides some valuable information about relationships between overall-quality estimates and different system and dialogue measures, the user-satisfaction index which it predicts is questionable, as it is calculated as a simple non-weighted mean across different quality features which have shown to be not independent of each other.

A better solution might be to start modelling single quality aspects on a local level. For example, it might be possible to predict speech-input quality from several measures relating to the system's recognition performance, the feedback provided by the system, etc. This prediction of speech input quality could then be combined with other indicators to calculate an estimation of communication efficiency. In this way, it might be possible to establish relationships between the quality aspects gradually, and thus obtain information about the weighting of different quality aspects for the overall quality of the service under investigation. Models set up in this way might have a higher potential for being generic, i.e. providing valid predictions also for other application scenarios and tasks. In the end, the information obtained from such modelling approaches will help to clarify underlying mechanisms of quality perception and judgment.

Acknowledgement

The presented work was performed at the Institute of Communication Acoustics, Ruhr-University Bochum. The author would like to thank *A. Raake, S. Schaden* and *U. Jekosch* for stimulating discussions and comments on this work.

References

1. Berger J (1998) Instrumentelle Verfahren zur Sprachqualitätsschätzung – Modelle auditiver Tests (Instrumental methods for speech-quality estimation – models of auditory tests). Shaker, Aachen
2. Bernsen N O, Dybkjær H, Dybkjær L (1998) Designing interactive speech systems: from first ideas to user testing. Springer, Berlin
3. Bimbot F, Chollet G (1997) Assessment of speaker verification systems. In: Gibbon D, Moore R, Winski R (eds) Handbook of standards and resources for spoken language systems 408–480. Mouton de Gruyter, Berlin
4. Blauert J (1997) Spatial hearing: the psychophysics of human sound localization, 2nd rev edn. MIT Press, Cambridge
5. Danieli M, Gerbino E (1995) Metrics for evaluating dialogue strategies in a spoken language system. In: Empirical methods in discourse interpretation and generation. 1995 AAAI Symp, Stanford CA 34–39. AAAI Press, Menlo Park
6. Dybkjær L, Bernsen N O (2000) Usability issues in spoken dialogue systems. Nat Lang Engr 6:243–271
7. ETSI Techn Rep ETR 095 (1993) Human Factors (HF); guide for usability evaluations of telecommunication systems and services. Europ Telecom Standards Inst, Sophia Antipolis

8. ETSI Techn Rep ETR 250 (1996) Transmission and Multiplexing (TM); Speech communication quality from mouth to ear for 3,1 kHz handset telephony across networks. Europ Telecom Standards Inst, Sophia Antipolis
9. EURESCOM Proj P.807 Deliverable 1 (1998) Jupiter II – Usability, performability and interoperability trials in Europe. Europ Inst Res Strategic Stud Telecom, Heidelberg
10. Fastl H (2005) Psychoacoustics and sound quality. Chap 6 this vol
11. Fraser N (1997) Assessment of interactive systems. In: Gibbon D, Moore R, Winski R (eds) Handbook of standards and resources for spoken language systems 564–615. Mouton de Gruyter, Berlin
12. Gleiss N (1992) Usability – concepts and evaluation. TELE 2/92:24–30, Engl edn, Swedish Telecom Adm, Stockholm
13. Grice H P (1975) Logic and conversation. In: Cole P, Morgan J L (eds) Syntax and semantics, Speech Acts 3:41–58. Academic Press, New York
14. Hansen M (1998) Assessment and prediction of speech transmission quality with an auditory processing model. Doct diss, Univ Oldenburg, Oldenburg
15. Hauenstein M (1997) Psychoakustisch motivierte Maße zur instrumentellen Sprachgütebeurteilung (Psycho-acoustically motivated measures for instrumental speech-quality assessment). Doct diss. Christian-Albrechts-Univ Kiel, Kiel. Shaker, Aachen
16. Heute U (2003) Speech and audio coding: aiming at high quality and low data rates. Chap 14 this vol
17. Hone K S, Graham R (2001) Subjective assessment of speech-system interface usability. In: Proc 7th Europ Conf on Speech Comm Techn, EUROSPEECH 2001, Aalborg 3:2083–2086
18. IEC Technical Rep 60268–13 (1998) Listening tests on loudspeakers. Int Electrotechn Com, IEC, Geneva
19. ITU-T Handbook on telephonometry (1992) Int Telecom Un, Geneva
20. ITU-T Recomm E.800 (1994) Terms and definitions related to quality of service and network performance including dependability. Int Telecom Un, Geneva
21. ITU-T Recomm G.107 (2003) The E-model, a computational model for use in transmission planning. Int Telecom Un, Geneva
22. ITU-T Recomm G.109 (1999) Definition of categories of speech transmission quality. Int Telecom Un, Geneva
23. ITU-T Recomm G.113 (2001) Transmission impairments due to speech processing. Int Telecom Un, Geneva
24. ITU-T Recomm P.79 (1999) Calculation of loudness ratings for telephone sets. Int Telecom Un, Geneva
25. ITU-T Recomm P.800 (1996) Methods for subjective determination of transmission quality. Int Telecom Un, Geneva
26. ITU-T Recomm P.810 (1996) Modulated noise reference unit (MNRU). Int Telecom Un, Geneva
27. ITU-T Recomm P.862 (2001) Perceptual evaluation of speech quality (PESQ), an objective method for end-to-end speech quality assessment of narrowband telephone networks and speech codecs. Int Telecom Un, Geneva
28. ITU-T Suppl. 3 to P-Series Recomm (1993) Models for predicting transmission quality from objective measurements. Int Telecom Un, Geneva
29. Jekosch, U (2001) Projektionsmodell zu Sprachqualitätsmessungen (Projection model for speech-quality measurements). In: Fortschr Akust, DAGA'01, 596–597. Dtsch Ges Akust, Oldenburg

30. Jekosch U (2000) Sprache hören und beurteilen: Ein Ansatz zur Grundlegung der Sprachqualitätsbeurteilung. Inaugural diss (habilitation), Univ Essen, Essen. Engl vers: Foundations of voice- and speech-quality perception, Springer, Berlin Heidelberg New York, in press

31. Johannesson N O (1997) The ETSI computational model: a tool for transmission planning of telephone networks. IEEE Comm Mag 70–79

32. Kamm C A, Walker M A (1997) Design and evaluation of spoken dialogue systems. In: Proc 1997 IEEE Worksh Autom Speech Recogn Understanding, Santa Barbara. 14–17

33. Möller S (2005) Quality of telephone-based spoken dialogue systems. Springer, New York

34. Möller S (2004) Telephone transmission impact on synthesized speech: quality assessment and prediction. Acta Acustica united with Acustica 90:121-136

35. Möller S (2002) A new taxonomy for the quality of telephone services based on spoken dialogue systems. In: Proc 3rd SIGdial Worksh Discourse Dialogue, Philadelphia, 142–153. Ass Comp Linguistics, New Brunswick

36. Möller S (2000) Assessment and prediction of speech quality in telecommunications. Kluwer Academic Publ, Boston

37. Möller S, Berger J (2002) Describing telephone speech codec quality degradations by means of impairment factors. J Audio Engr Soc 50:667–680

38. Möller S, Bourlard H (2002) Analytic assessment of telephone transmission impact on ASR performance using a simulation model. Speech Comm 38:441–459

39. Möller S, Raake A (2002) Telephone speech quality prediction: towards network planning and monitoring models for modern network scenarios. Speech Comm 38:47–75

40. Raake A (2004) Assessment and parametric modelling of speech quality in voice-over-IP networks. Unpubl doct diss, Ruhr-Univ Bochum, Bochum

41. Raake A (2000) Perceptual dimensions of speech-sound quality in modern transmission systems. In: Proc 6th Int Conf Spoken Language Processg, ICSLP 2000, Beijing, 4:744–747

42. Richards D L (1973) Telecommunication by speech. Butterworths, London

43. Rix A W, Beerends J G, Hollier M P, Hekstra A P (2000) PESQ – the new ITU standard for end-to-end speech quality assessment. 109th Conv Audio Engr Soc, Los Angeles. Preprint 83

44. Stevens S S (1946) On the theory of scales of measurement. Science 103:677–680

45. van Bezooijen R, van Heuven V (1997) Assessment of synthesis systems. In: Gibbon D, Moore R, Winski R (eds) Handbook of standards and resources for spoken language systems 481–563. Mouton de Gruyter, Berlin

46. van Leeuwen D, Steeneken H (1997) Assessment of recognition systems. In: Gibbon D, Moore R, Winski R (eds) Handbook of standards and resources for spoken language systems 381–407. Mouton de Gruyter, Berlin

47. Walker M A, Litman D J, Kamm C A, Abella A (1998) Evaluating spoken dialogue agents with PARADISE: two case studies. Comp Speech Language 12:317–347

48. Walker M A, Litman D J, Kamm C A, Abella A (1997) PARADISE: a framework for evaluating spoken dialogue agents. In: Proc ACL/EACL 35th Ann Meeting Ass Compl Linguistics 271–280. Morgan Kaufman, San Francisco

8 Assigning Meaning to Sounds – Semiotics in the Context of Product-Sound Design

Ute Jekosch

School of Architecture, Rensselaer Polytechnic Institute, Troy NY

Summary. Sound design constructs audibility of the world. Sounds carry information about the world. When listening to sounds, communication takes place. These are well-known facts for speech sounds, but it is also true for other types of sounds such as music or product sounds. In principle, each acoustic event can be perceived as a sign carrier through which information about the world is communicated. In its ultimate sense, sound designers are *engineers of communication*. To be successful, they have to take design decisions on the basis of how listeners perceive sounds and of what kind of communication takes place during this event. Suitable sound design requires a special view on acoustic/auditory communication. Among other sciences, *semiotics* deals with this field.

8.1 Introduction

Semiotics is a scientific discipline which deals with signs. Generally speaking, a sign is a mental unit which is processed as standing for something other than itself, as pointing at something else. Well-known items which are treated as signs are, e. g., traffic lights, number codes, national flags and speech. In each of these cases recipients need background knowledge of the perceived physical objects, i. e. they must know the relation between the primary object of perception and the concept that it is going to denote.

From a functional point of view, the sign is an element of communication. Using signs means to communicate, it means to send or receive information about "the world". There are highly complex sign systems such as oral speech which utilize acoustic waves to communicate thoughts, ideas, and facts. Such sign systems have to be learned, as the relation between the acoustic forms and their content is arbitrary to a high extent, i. e. it is based on conventions. This means that communication will fail if the relation between the primary perceptual object and the intended concept is not known to the recipients. This holds also true for simpler sign systems, e. g., traffic signs. If the convention is not known the colour red, e. g., will not lead to the action to stop.

Apart from signs where there is an arbitrary relation between perceived form and denoted content, i. e. the case of a "symbol", there are other types where the relation is based either on similarity, "icon", or on causality, "index". These sign systems do not have to be learned in the same way as arbitrary systems, yet their meaning is acquired through experience, i. e. through perception and operation [29].

Basically, all physical/perceptual events can be approached from the perspective of semiotics. Consequently, the perception of product sounds can be analyzed from the point of view of sign theory as well. Moreover, from what has been said so far it also follows that the perception of product sounds is a *communicative event*. This perspective is taken in this chapter. Acoustic/auditory items taken from the field of engineering acoustics will be discussed – with special emphasis being put on product sounds.

8.2 Product Sound and Semiotics

During the last decade, product-sound design has become an important issue. In the traditional understanding of product quality – products are exclusively related to industrial function products here, e. g., to cars or electrical household appliances – the sound was mostly seen as something secondary, if not equated with unwanted noise. In other words, the sound was treated as an issue regarded to be of minor relevance to the product as such.

Yet, in more general terms, product quality has been understood as the result of technical, ergonomic and aesthetic functionality in product use. But even in this scope, the sound has usually only then been viewed at as a feature of product quality when it happened to come negatively to the fore. Nevertheless, there have always also been cases where product sounds have supported product quality, though primarily in the sense of an incidental supporting background event. There are also examples where product sounds have become a positive foreground event, and even some where the sound stands for the whole product. Although initially caused by mechanically-functional processes, in these cases sounds are more than just a desired accompanying auditory event. They can indeed be seen as the "identity card" of the product – *product identity* is a related keyword. It is one of the tasks of sound design to understand why such sounds are product-quality supporting and, consequently, to have the basic knowledge to design such sounds on purpose in a systematic way. Semiotics is a supporting science in this context.

8.3 The Sound as a Sign

Every sound – be it wanted or unwanted – can be regarded as a sign carrier. *Von Uexküll* [56] states that every activity that consists of perception and operation imprints its meaning on an otherwise meaningless object and, thereby, turns it into a subject-related carrier of meaning in the respective "Umwelt", i. e. perceptual environment, subjective universe [56]. In this view, each perceived sound in product use is a carrier of information and can hence be analyzed as a sign carrier.

A product sound carries information about the product. With the sound contributing to a positive meaning of the product, users get attracted, access to the product is simplified and its use is supported. That, in turn, improves the perceived product quality. Consequently, the sound is a central determinant of product quality. Even products of which the functional qualities of elements other than sound have already been rated very high, can get an additional quality push by a suitable and supportive product sound. Obviously, however, there is a strong relationship between types of sounds and their qualification to function as a quality-supporting sign carrier. Experience shows that not every sound is qualified to fulfil the required conditions.

To be sure, industries has discovered the huge potential of sound design already some time ago. But, although there are powerful tools available to administer natural sounds and to create artificial sounds, and although there are often numerous sound designers involved in product planning and specification, many aspects remain anonymous to a certain extent. Sound design seems to be more like an artistic task where intuition is asked for rather than a task that can be approached in a rational and systematic way.

It is the aim of this chapter to discuss product sound design from the perspective of the sounds being carriers of information. This is a typical top-down approach, i. e. a deductive attempt, which offers valuable impulses for the design process. In this sense, semiotics is functionalized here. It is an instance of applied semiotics.

8.4 What is Semiotics?

Usually, when introducing a scientific discipline such as semiotics, there is the need to sketch not only the scope and the objects that the respective field is concentrating on, but also to standardize terms, perspectives, objectives, common methods and methodologies – always keeping in mind the demands of both theory and practice. In this chapter, we cannot accomplish this task to any amount of completion but shall restrict ourselves to the absolutely necessary, predominantly based on *classical* semiotics.

As already said, semiotics is the theory of signs [15]. As to its scope, it comprises quite different research fields such as text semiotics, semiotics of the literature, the theater, of poetry, film, paintings and music [18]. Particularly specialized areas are, e. g., zoo-semiotics, architectural semiotics, computer semiotics – even semiotics in mathematics has recently become a research field [27]. As far as scientific terminology within the area of semiotics is concerned, there is ample literature which deals with the definition of fundamental terms such as "sign", "signal", "communication", "information", "code", "codification", "aesthetics", to mention only some.

This list indicates that a definition of semiotics free of contradictions is a task on its own. In this paper, the scope of semiotics will be outlined in a more implicit way. In discussing several questions related to product-sound design with examples taken from other fields as well, the research subject/object of semiotics will hopefully become clear. Anyhow, despite of all different streams within the field there is consensus as to the following understanding of semiotics.

> "Semiotics, being the theory of signs, investigates cultural processes as communicative processes. It is the aim of semiotics to show that and – what is even more important – how cultural processes are based on systems." [14]

In other words, a key term in semiotics is communication. As is well-known, it is not only semiotic theory which deals with communication but information theory and communication theory do so as well. Yet, these disciplines look at communication from different perspectives as discussed in the following.

Information theory: In information theory the term "communication" is understood as an exchange of information between systems, such as humans, machines, societies, animals, plants, physical and biological environments. The focus is put on the physical signals, the code and on channel characteristics.

Communication theory: Communication theory can be regarded as both a special case and an extension of information theory. It is a special case concerning systems in so far as it (a) concentrates exclusively on the system "human being", both as source and recipient, and (b) solely on spoken and written language as the medium for communication. Nevertheless, it exceeds the scope of information theory by investigating the question of how human beings process codes. Thus, the scope of communication theory comprises the entire process during which spoken or written language is perceived as auditory or visual sign carriers. It are indeed these sign carriers which enable the formation of meaning. Consequently, communication theory takes into account the nature of human systems as well as the events which take place during the processes of communication.

Semiotic theory: Semiotics as applied in this chapter is inspired by the understanding of the term "communication" as it appears in both information and communication theory as well as in other communication-related fields. *Eco* [13] is a well-known proponent of this view. In semiotics, communication is not reduced to spoken and written language, but also applies to other modes of communication such as gestures, mimics, illustrations, and paintings, to mention only some. This view stems already from classical semiotics, as, e. g., represented by *Peirce* [42] or *Morris* [38,39] – for an overview see [37].

Further, communication is neither a phenomenon that solely takes place between different systems, as it is treated by information theory, nor between humans alone, as seen by communication theory. It is, in general, an event that takes place between a human being and "the world". So the general scope of semiotics as understood here is in essence human perception and interpretation of the world [55]. Along these lines of thinking, the specific scope of semiotics in the context of product-sound design comprises the question of how human perception of a product is affected by product sounds.

In this respect semiotics is a basic science. All of its research objects are related to communication. Its traditional areas overlap with fields of other scientific disciplines related to communication, such as information and communication theory – as already discussed. Further disciplines such as linguistics, psychology, epistemology and sociology are touched upon as well. Each of these fields is directed towards its specific object area, but they all are concerned with the phenomenon of *communication*.

Semiotics overlaps with the scope of all these disciplines in that it is the theory of structuring and describing objects in a systematic way, which themselves are research objects of individual disciplines. Thus semiotics is a systemic science. It is not directed towards identifying elements, structures, or patterns as such, but it focuses on the question of what makes an element an element, a structure a structure, a pattern a pattern – always seen from the perspective of a perceiving and operating human being. It aims at identifying basic principles that rule the processes of building structures from elements and combination rules, always caring for a stable but restricted analogy with the entirety of different individual fields of knowledge at large, i. e. all the individual scientific disciplines. It is these basic principles which motivate a semiotic analysis of product sound perception.

Summary

– Semiotics is a basis theory for sciences which deal with phenomena that are perceived as sign carriers. As such it can be applied to sound design and to product-sound design, respectively.
– Communication is a key term in semiotics. Communication, as introduced here, is an event that takes place between a perceiving human being and "the world". Communication is based on the processing of perceptual events as sign events.

8.5 What is a Sign?

As semiotics is the theory of signs, it is mandatory to introduce the concept of sign in this context. In everyday-language use we speak of signs as being physical objects which are given a specific meaning by conventions. Typical examples are spoken language, written texts, *Morse* signals, flags and traffic signs. When listening to or looking at them we associate a meaning with them – provided that we know the conventions, the code. If we do not know the code, these "signs" cannot be decoded and understood appropriately.

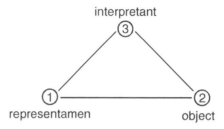

Fig. 8.1. The sign triad of *Peirce* as a semiotic triangle, modified by *Bense* [40]. *Peirce*'s terminology roughly relates to the one used in this article as follows: representamen → sign carrier, object → object of experience, interpretant → meaning

However, semiotics as introduced here speaks of signs in a different way. In semiotics a sign is something that happens in our brain. It is a mental phenomenon. Signs are the results of a specific mode of processing perceptual events. These processes are mainly stimulated by the perception of physical objects as sign carriers, e. g., acoustic speech signals. When processing a signal as a sign carrier, perceived features of the object, e. g., sequences of phones like in the spoken word /mIlk/, are functionalized in that they signify objects of experience – the concept of milk, its taste, colour, a cow, etc. In signifying objects of experience, associations and connotations are triggered, e. g., "I do not like hot milk!". Finally a large quantity and great variety of associations is interpreted and, as a result, meaning is assigned to the object of perception. This is roughly what we mean when we state: The object of perception has been processed as a sign.

In many traditional theories the sign is graphically symbolized as a triangle, the "semiotic triangle". An overview of sign models is given in [40], pp. 137–141. In Fig. 8.1 the sign triad of *Peirce* is depicted [40] – please beware not to mistake the term "interpretant" for the interpreting subject.

Summary
- The term "sign" denotes the processing of a perceptual object.
- The sign is a mental phenomenon.
- Constituents of signs are (a) the sign carrier, (b) the object of prior experience to which it refers and (c) the meaning which evolves.

8.6 What is Semiosis and How Does Semiosis Take Place?

The exemplary introduction of the concept of sign as illustrated in Fig. 8.1 suggests that the processing of a perceptual object as a sign carrier were a simple stepwise-predictable, i. e. transversal, sequence of events as follows. You perceive fractions of the world through your sensory organs, then you treat them as sign carriers. Finally you associate objects with them and, consequently, meaning evolves. Or, to put it into semiotic terms, *"semiosis"* takes place.

In fact, this is an abstraction and a simplification of what actually happens. Such a description of sign processing would not be explicit enough to be of help for sound design. Actually, what we have described so far is *that* certain processes do take place, but not *how* they take place. To treat the latter issue it is helpful to look at the perceptual processes even more analytically and to investigate further how perceptual phenomena are selected, how sign carriers are identified, how they are organized and how meaning finally actually evolves – in other words, to pose the question of how *semiosis* operates in detail.

In *semiosis*, perceptual events, i. e. *items of perception* are processed in view of what is already known, of what the current situation is, of what is anticipated and experienced, namely, in view of *objects of prior experience*. *Semiosis* is a process of selection, organization, coordination and structuring, not only for the items of perception, but also for the objects of experience. When an auditory event appears in your perceptual world, i. e., when you are listening to something, the object of prior experience – to which items of perception are directed – does not become mentally "visible" immediately. The object of prior experience is not simply *there*, but it has to be spotted, selected and identified within the vastness of all experiences. Obviously, there is a close interaction between processing items of perception and data of experience: A sound cannot be classified as a speech sound unless the code is known, and the code itself is an object of experience and of "world knowledge", which has to be activated as well.

At this point, among other things, the degree of salience of the prior experience plays an important role. As is known, expectation guides perception, and the more the expected is in line with the perceived, the easier the object of experience can be identified.

Within this scope, the processes of selection and identification can analytically be described as follows. Upon being triggered by the perceptual event, both items of perception and items of experience are processed in a mutual process of extracting data from perception and extracting data from experience. The process of organization, coordination and structuring is only completed when the following mental state is reached: Extracted data of perception carry certain characteristics of data of experience – or, seen from the other perspective, data of experience carry certain characteristics of data of perception. Consequently, in merging these two data types an object of experience is signified. The object of perception becomes a signifier for an object of experience, i. e., *semiosis* takes place. These processes are schematically simplified in Fig. 8.2.

What is important now is the fact that the mutual interplay between extracting data of perception and data of experience, which both are processes

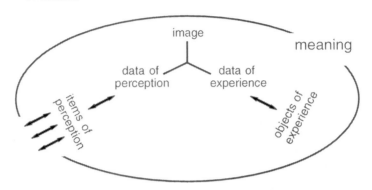

Fig. 8.2. Model of *semiosis*

for organization, coordination and structuring, are organized and structured processes themselves. The processes of extracting data are ruled. This is one of the reasons that although much of what has been discussed so far is highly individual, yet inter-individually comparable at the same time.

The processing rules as discussed before are called "schemas". Among other things, these schematic processes have been emphasized by Gestalt theorists as follows.

> The range of various signals does not gradually lead to individual perceptual objects themselves, but only enables perception to be autonomously organized. However, this organization obviously has its own set of rules in which the various performances of the perceptual system are integrated and abstracted. In genetic epistemological theory autonomous systems like these are called schemas [43, 45].

Schemas can be described as systems of experience [20]. They provide techniques to process objects of perception as signs. They provide what semiotics calls codes. Please note that the term "code" is used in a different way than in information theory. In semiotics and Gestalt theory codes are understood as mechanisms that rule perception [49]. On the basis of schemas as well as on the basis of anticipation and expectation, data of perception and data of experience are identified. Bringing them together is a dynamic process of building and verifying hypotheses. Treating the object of perception as a sign carrier leads to associations, to denotations, connotations and interpretations, i. e. to the assignment of meaning. The object is processed as a sign, *semiosis* occurs and something is communicated.

8.6.1 Rules of Semiosis

It was stated above that *semiosis* is a ruled, self-organizing process. But what are the rules? In fact, rules of *semiosis* can only be inferred from observations of human behaviour.

To get an idea of how *semiosis* is ruled, it is, e. g., helpful to depict and discuss extreme cases of communication. For instance, to understand ruling processes in *semiosis*, it is interesting to analyze behavioural strategies of listeners who are confronted with a signal they have never encountered before or, to mention a point which will be discussed in detail later, observe and analyze cases where no appropriate schemas are available. In such a situation an incompatibility, also called imbalance or disequilibrium, between data of perception and data of experience may occur between the hypothetical objects of perception on offer and the hypothetical objects of experience – as seen from a higher processing level. In fact, perceiving systems tend to correct this imbalance and "adaptation" occurs. According to *Piaget* [46] two strategies are conceivable to this end, namely, "assimilation" and "accommodation" [1].

Assimilation: The "new offer" is to be aligned with the existing schema. It is put "into a different code" and integrated into the existing system. In relation to audition this means that *we tend to hear what we want to hear*. To this end we actively modify data until they are compatible with what we have learned already, with what we know and with our objects of experience.

Accommodation: The existing schema is changed such as to suit the conditions of the arriving signal. The schema remains flexible and adaptable until a possibility for integrating the "new offer" has developed [2].

In the course of many investigations, *Piaget* and others could prove that, usually, both processes control the formation of an equilibrium. Whenever perception is only assimilating or only accomodating, this has to be considered to be a pathological case – see, e. g. [19].

Synthetic speech may be mentioned as a typical example of being confronted with a "new offer" which causes a perceptual disequilibrium. When listening to a synthetic voice for the first time, the "automatism" which identifies, organizes and structures data of perception as well as of experience may, depending on the quality of the synthetic voice, be severely disrupted. Naïve listeners may experience a new auditory event which, on the one hand, points them to the well-known sign carrier speech, but, on the other hand, the well-established processing schema for speech cannot be applied. Thus a disequilibrium evolves. Data of perception, then, cannot be coupled to data of experience. What happens in such a case? Most likely assimilation occurs, i. e. the perceived synthetic signal will be "corrected" mentally – it will be aligned with features of natural speech. At the same time listeners will learn the peculiarities of synthesized speech and, consecutively, develop a more appropriate processing schema. Also, after having being exposed to synthetic signals for some time, learning will take place and ease of communication will improve.

It goes without saying that the new auditory events must of course qualify for being processed as sign carriers which denote speech at large. If they do not do so, the data of perception and those of experience cannot be aligned to each other, which means that adaptation and, thus, communication would fail completely.

Still, the question remains of how the ability to process signs develops at all. For example, how do newborns learn to identify and process sign carriers? The genesis of semiotic systems and the development of *semiosis* is one object of research that contributes to answering the question of how schemas develop and of how assimilation and accommodation processes are ruled and controlled. The following brief look into child development, i. e. into ontogeny, may exemplify the approach:

Assuming that *semiosis* takes place as described above, how do new-born children cope with the tremendous richness of "the world", particularly bearing in mind that their sensory and cognitive abilities are yet in a pre-mature state? How do children manage to live and develop under these circumstances? Do they recognize an imbalance between what is on offer and what their reduced processing abilities can perform? If so, how do they solve disequilibria? And how do processing schemas develop, what are stimulating inputs to the sensory apparatus and to the cognitive system? What is the ratio of assimilating and accommodating?

Actually, these questions are posed from the perspective of an adult observer, and hypothesizing answers can only be given by observing newborns and interpreting their behaviour. There are indeed clear hints which support the assumption that there are genuine rules according to which schemata develop. As an example, there are comparable performances of children of a specific age, not only within one specific culture, but also across different cultures. Observations support the hypothesis that these performances are all based on well-ordered and ruled processes. *Piaget* [46], e. g., identifies so-called sensory-motor stages. The important fact in our context is that these stages occur in a specific order which cannot be reversed. In other words, a child has first to go through an sensory-motor process before being able to proceed to formal thinking [52].

And what are the underlying principles of development? With regard to auditory perception, it can be observed that small children are basically respondent to two types of sounds, (a) to those ones which "possess" a certain "Gestalt", i. e. a general form, and are thus perceived in a holistic way – think, e. g., of a baby being rocked in mother's arms while she is humming a soothing tune, and (b) those ones which do not possess a Gestalt, such as a sudden acoustic/auditory event which catches attention. The underlying strategy used by the small children is to unify, "syncretism", and to diversify, "discretism", at the same time, and to differentiate by establishing opposi-tions [28]. Among other things, it is a matter of age, experience, attention, mood, emotion, abilities which give dominance to one or the other behaviour in a specific situation. Nevertheless, establishing oppositions seems to be a fundamental principle of establishing schemas.

Another finding that should be mentioned in this context is the development of the loss of abilities due to old age or for pathological reasons, e. g., aphasia, language loss. A basic principle there is as follows. What has been learned last is lost first. Of course there are exceptions to this rule, but generally it shows that schemas which have been established at an early age show highest stability.

To put it into different terms, a schema is an organizing and structuring entity which in itself is organized and structured. Schemas become effective when repeated objects of perception are to be processed. Schemas allow for an "automated" mental processing. This saves time and effort. Further, it supports efficient communication, i. e. an efficient *semiosis* and, consequently, efficient understanding of the environment in general.

8.6.2 Types of Schemas

The types of schemas available seem to be quite different. With regard to the role of learning and adaptation they cover a broad range from archetypical ones, e. g., the ones for processing baby crying, to others which are highly conventionalized, e. g., schemas for understanding speech, schemas for interpreting the strike of a church bell as an indication of the hour, or schemas for a blind person to understand the sound attached to traffic lights as a prompt to stop or go. The availability of a schema to process baby crying has an advantage for survival. By *automatically* identifying and bringing together data of perception and data of experience, a time-consuming *semiosis* with hypotheses to be formulated and to be verified can be passed by. This schema is specialized in identifying highly specific features of items of perception, and, as such, it is an extremely stable processing system. There are supporting examples along these lines from the animal world. For instance, the schemas for fleeing, as activated by a specific object of perception, is extremely specialized – i. e. specialized with regard to the degree of freedom of processing and reacting.

Without doubt, schemas for processing speech or other stimuli whose perception and interpretation is based on arbitrary conventions are specialized as well, but they are specialized on different aspects. As to the mutual interaction when extracting data of perception and data of experience there are schemas which are specialized in establishing similarities and analogies. These schemas become effective when we are recognizing objects which we do have seen before, yet under very different circumstances, e. g., regarding light, distance or position. Further, this happens when we identify people who have aged considerably or when we understand voices we have never heard before. Here the schemas allow an active and individual processing of perceived items, of course based on what has been learned before, i. e. on objects of experience. This shows that these schemas are systems of experience.

Summary

- Processing of perceived objects as signs requires prior experience. It is a ruled procedure.
- The processing systems of experience are called schemas. Schemas are techniques to select, identify and correlate data of perception and data of experience in a fast, *automatic* way.
- "Assimilation" and "accommodation" are strategies to bring together non-fitting data of perception and experience.
- Different types of schemas can be distinguished. There are schemas which are highly specialized in recognizing objects of perception and do not require complex processing, and there are others which administer mental activities.

8.7 What are the Functions of Signs?

Signs can have different functions. *Bühler* 1879–1963 [10] describes signs in their communicative relations. In this context, the sign is viewed at as a tool. *Bühler*'s famous ORGANON Model [11], is related to the sign carrier speech, but can as well be applied to sounds in general – see Fig. 8.3 and Fig. 8.4.

Generally, when being asked of what the function of a sign carrier such as speech is, one will refer to the fact that it carries information encoded in the vocabulary, i. e. in language use. However, when we listen to speech we do not only try to understand what is being said, but we also, among other things, take into account who the speaker is and why he/she addresses us. In other words, a sign carrier has different semantic functions. When *semiosis* takes place, the sign is always a tool which points to the speaker who produces the sign carrier, to the listener who perceives it and to reference objects. To give an example, the speech sign gives an indication of the speaker's sex, age, emotional state. At the same time it is directed towards the listener in that it may provoke an action or a change in attitude. Finally, the sign refers to *third* objects which it informs about.

It is a central fact that all these different semantic functions of signs are generally present when *semiosis* occurs. When one listens to the sound of a product such as a car, one gets information about the product with regard to its general functionality, its current functional state, with regard to product material, safety, location, spaciousness, etc. At the same time, what is heard has an effect on one's present affective and cognitive state, i. e. it may be stimulating, appealing, one's confidence into the product may be consolidated, one may be animated to buy and use the product. Further, one receives information about the product and may get an idea about the product's credibility, reliability and durability, about the degree of innovation that it represents, etc.

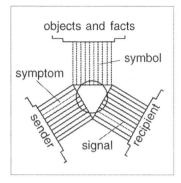

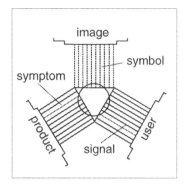

Fig. 8.3. *Bühler*'s ORGANON Model of speech – shown according to *Nöth* [40]

Fig. 8.4. Semantic functions of product sounds in analogy to *Bühler*'s model

Of course, it is not only audition on which *semiosis* is based. Auditory perception provides only one set of cues among many others one has available in a specific situation in life. In fact, *semiosis* is a multi-sensory action where each individual cue carrier may play a relevant role. For product quality it is thus important to take care that different sign carriers such as auditory, visual, haptic or olfactory objects do not provide inconsistent informational cues which the perceiving product user cannot or can only hardly bring together. The more harmonized the different signals on offer are for the perceiving and operating subject, the more convinced of the product quality the users tend to be.

Figures 8.3 and 8.4 are meant to be interpreted as follows. The circle in the middle symbolizes the acoustic event. Three aspects of it hint at it being a sign. These aspects are symbolized by three legs of the triangle. On the one hand the triangle encloses less than the circle does, i. e. denotes selection, on the other hand it encompasses more than the circle itself to indicate that perception is always also a result of anticipation and expectation. The assemblage of lines symbolizes the semantic function of the sign [40].

8.8 Links Between Semiotics and Product-Sound Design

Being asked about the object of industrial product design, a common answer is as follows. The object is the industrial product, of course. This is not completely wrong, but rather simplified. More precisely one must state that the industrial product is not the object of design, but rather its product. Although this differentiation might appear to be unusual and perhaps even far-fetched, it is important in this context, because it looks at product design from a different perspective in a sense as discussed in the following.

The object of product design is to create physical facts which – when giving rise to an object of perception in the context of product use – lead to a positive meaning of the product. Typically, a positive meaning is established when the perceived product performance and its perceived aesthetic characteristics satisfy user expectations. In this sense the product of design is a tool that helps to create a mental image. The industrial product is an object of function with regard to its purpose to stimulate certain objects of perception, imaginations and fantasies when it is used.

Among other things, acoustic events are a means of accomplishing this purpose. Looking at product-sound design from this perspective, the product sound is an object of function. It is a tool to create an image which supports product quality.

Product use is a communicative act where the sound is one possible carrier of information with regard to the product as a whole. When using the product, *semiosis* takes place in the listening user. Data of perception are, then, correlated with data of experience, and, finally, end meaning is associated with the auditory event specifically, and with the product generally. All this happens in a communicative context.

Modern design of industrial products takes exactly this view in that the mental image created through the process of product use is emphasized. Today many designers of industrial products take the perspective of designing a *process* through the design of a product. In the end, the object of design is not simply the industrial product as such, but it is the process in which the product influences the way of how it is used. Through the design of this process the product as such influences man-machine interaction as well as the way how people interact. In other words, product design is always process design [21], namely, it is *design of communication*.

Accordingly, it is the objective of product-sound design to convey specific information for a specific purpose by means of a specific code, in the semiotic sense, using specific media for specific user groups. As for the type of the product, specific parameters for design have to be aligned with each other, taking into account that the sound is only one design element of the product. There are a number of others, such as technical functionality, visual design or material, just to mention some.

These design elements are not necessarily orthogonal, i.e. independent from each other. Having in mind that the design efforts aim at an industrial product of high quality, the following question comes up. What quality does an individual element, such as an acoustic signal, has to have in order to contribute as much as possible to the perceived quality of the product in its use? Following from what we have mentioned above, it would certainly not be adequate in this context to regard every element of quality independently, i.e. as a unit that has to have the highest quality of its own to the end of creating a top quality product by putting all the quality elements together. To be sure, the sound of a product is also functional with regard to other product components, as well as with regard to requirements imposed by the actual use of the product.

As is well-known in this field, design decisions – when a specific product is on the construction board – are usually based on goals and specifications with regard to, e. g., usability, utility, information, purpose, code, medium and users. The task is to construct a physical unit, namely, a product which when being used, creates a unified and coherent mental concept that is qualified to fulfil or even exceed user expectations [50]. Consequently, a successful product-sound design is achieved when there is no incongruency between user expectation and the perceived auditory image of the product in the sense as stated in the following.

> "Elements make their mark in the quality shaping process of a product in that acoustic forms are audible by product users. When product users perceive and evaluate product-sound quality, they assign quality to auditorily perceived features of the product in such a way that they are identical to those they regard as desirable or positively expected in the sense of individual expectations and/or social demands." [30]

A key term in this context is expectation. Quality, of which object soever, is always the result of a judging comparison of data of perception with data of expectation. This may be recalled by mentioning the following definition of product-sound quality:

> "Product-sound quality is a descriptor of the adequacy of the sound attached to a product. It results from judgements upon the totality of auditory characteristics of the said sound – the judgements being performed with reference to the set of those desired features of the product which are apparent to the users in their actual cognitive, actional and emotional situation." [6]

In order to be able to specify a product sound in a rational way it is essential to know what desired features are in fact apparent to the users in their actual cognitive, actional and emotional situation. Some design experts doubt that expectation with regard to product sounds can at all be broken down analytically. Nevertheless, in the following paragraphs an attempt is made to pave the way in the sense that some classical theoretic semiotic considerations as introduced above are applied to sound and product-sound quality. [1]

---

[1] It should be mentioned here again that we only deal with *classical* semiotics in this chapter. Modern semiotics has reached a high degree of differentiation and specialization which to discuss would exceed the scope of this introductory chapter

8.9 Semiosis of Product Sounds

In the context of product use the perceived sound of a product is a sign carrier. It causes imaginations of the product, it can auralize specific quality-supporting aspects and it can even bring something to the ear which would not be audible otherwise. In order to construct a high-quality product it is obligatory to meet user expectations with regard to the product in general and to its sound in particular. Further, it is indispensable to know what potential user groups would like or dislike, what they prefer and what they neglect. This is indeed an important issue in the engineering context. But what do users of industrial products actually expect with regard to product sounds?

8.9.1 Ruling Schemas and Gestalt Perception

As outlined earlier, expectation plays a role in the process of *semiosis* in the following way. Listening users expect items of perception which can be processed with schemas in such a way that meaning is associated to them which supports product quality. But what are these schemas like? Generally, schemas are directed towards the processing of characteristics of the auditory event. An exemplary set of questions which arise in the context of the concept of schemas is listed below.

How does a human listener manage to select certain characteristic features from the auditory event and take them as information carrying units, i. e. as sign carriers? What are invariant types of behaviour as opposed to individual or spontaneous ones? What is the role of the context? Why are humans able to recognize auditory objects which they have heard before, even under very different circumstances, such as regarding distance and direction? Why are they able to identify types of acoustic sources although individual sources generate different sounds? Why are they able to recognize the identity of different shapes? Why can they predict the way well-known but transposed melodies will continue, why can they understand voices they have never heard before?

The answer to these and related questions is as follows. This is so because the perceptual process is based on schemas. When analyzing schemas more analytically, one very soon comes across the theory of "Gestalt" perception. In fact, schemas are processing units which are closely related to the concept of *Gestalt*, German plural→*Gestalten*. *Piaget* [44] points out that objects are recognized and perceived by us not because we have analyzed them in detail, but because of the *general forms*, i. e. Gestalten, which are as much constructed by ourselves as they are given by the perceived items – refer to the discussion of assimilation and accommodation in Sect. 8.6.1.

Gestalten have first been thoroughly investigated, e. g., by *Ch. von Ehrenfels* 1859–1932 [16], *M. Wertheimer* 1880–1943) [35], *W. Köhler* 1887–1967 [26], and *K. Koffka* 1886–1941 [32]. Generally spoken,

Gestalt theorists take up the question of why we perceive and organize objects the way we do. *Von Ehrenfels* defines the following criteria for perceptual organization.

Ehrenfels criteria

- "Übersummenhaftigkeit" ... A Gestalt is more than the sum of its parts.
- "Transponierbarkeit", transposability ... A Gestalt does not change even when its elements are substituted by other ones. The requirement is that the structural whole of the Gestalt must be kept the same.
- Characteristics of structure, material and qualities of representation

The main rules of Gestalt theory are [3]

- Gestalten are perceptually primary and appear before their so-called perceptual parts or elements.
- To perceive and to react to a Gestalt is more natural, easier, and occurs earlier than the perception of and reaction to its parts.
- Gestalten tend to be as complete, symmetrical, simple, and good as possible under prevailing conditions – *law of salience*, German → Prägnanz.
- Gestalten tend to be governed by internal rather than by external factors – *law of autonomy.*
- Parts derive their properties from their place and function within the Gestalt.

Elements tend to be grouped into one organized Gestalt when the following criteria are fulfilled – the so-called *laws of Gestalt*, which, in fact, are rules rather than laws.

- Proximity ...   Elements are close together.
- Similarity ...   Elements appear to be alike.
- Common fate ...   Elements move according to a pattern, e. g., a wave.
- Good continuation ...   Elements are aligned in the same direction.
- Closure ...   Elements are connected to each other although they form an incomplete structure.
- Simplicity ...   Elements are "simple"
- Experience or habit ...   Elements form familiar and regular shapes.
- Persistence or inertia of initial groups ...   Elements as initially grouped recur over time.

A major part of literature on Gestalt perception is related to visual perception, much less can be found on auditory perception so far, e. g., [9,36,48]. Anyhow, there is ample proof that schemas, directed to which sensory modality soever, are built according to Gestalt principles. In terms of Gestalt theory, the main questions in the scope of product-sound design are as follows. What are the general forms that our auditory system is reacting to? How can these be described, systematically grouped and labelled? In fact, having in mind the sparse literature which is as yet available in the field of auditory perception, the rich results from the area of visual perception may give valuable hints for further auditory research [53].

8.9.2 Patterns of Semiosis that Create Context

While the previous discussion of schemas and Gestalt perception describe *processes* which rule *semiosis*, it is rather the *context* of *semiosis* that we aim to concentrate on in this chapter. As elaborated upon above, schemas and Gestalt perception are related to rules that enable *semiosis*. However, by applying these rules, listeners may take different perspectives. They may, e. g., concentrate on different aspects when associating meaning with a specific product sound. As for their reaction, sounds can be classified according to *cognitive activity, affect,* and *motor activity* [57, 58].

Indeed, it is this variance in individual behaviour which is one of the reasons why in the context of product-sound engineering the construction of meaning is often seen as a black box which cannot yet be broken up and analytically processed. Nevertheless, in this paragraph an attempt will be made to model the different aspects of content in *semiosis*, i. e. in the course from the auditory event to the association of meaning in product use. The approach is based on semiotic theory, which has, however, been modified partly here so that it is applicable to the field of product-sound perception.

The question is: What are the criteria that the mental process of extracting and comparing data of perception and experience are based upon – with the auditory events being processed as sign-carriers? The model of *semiosis* as depicted in Fig. 8.2 may serve as an anchor point for the following discussion. In principle, with regard to semantic functions, the listening product user can concentrate on three different aspects during *semiosis*, namely,

- on auditory features of the sound itself, i. e. on *form characteristics* of the auditory event $\rightarrow$ mental image, I_1, of the form, F,
- on the person who is listening to the sound $\rightarrow$ mental image, I_2, of the listener, L,
- on the product $\rightarrow$ mental image, I_3, of the product, P.

For the resulting meaning, M, as associated with a sign carrier, the following certainly holds: $M = f\{I_1, I_2, I_3\}$. Consequently, *semantic functions* of product sound can be modelled as in Fig. 8.5, where the *perceived form characteristics* point at the form, namely, the *sound*, further, at the *listener* and at the *product*, such initiating the formation of meaning.

Sign carrier, related to the form: When analyzing aspects of *semiosis*, form characteristics of the product sound itself can be of different levels of abstraction as regards signification:

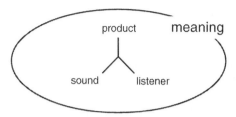

Fig. 8.5. Model of meaning as a function of the sound, the listener and the product

– "Symptom" ... The form of the sound indicates that *something happens.*
– "Signal" ... The form of the auditory event provokes a reactional behaviour; it catches the listener's attention.
– "Symbol" ... The sound signifies objects.

Sign carrier, related to the listener: When analyzing the product sound from the perspective of the listener, again three differentiations can be made. The focal point is the listener's ambition to extract information via the perceived form characteristics of the product sound. The listener's cues for understanding can be related to the source, to the recipient and to the code – as summarized in the following listing.

source	recipient	code
aesthetic cues	imperative (re)-action	index (causality)
technical cues	indicative (re)-action	icon (similarity)
functional cues	suggestive (re)-action	symbol (arbitrarity)

For the sake of explication some analytic questions are put forward in the following.

– *As to the source*: Are all auditory cues compatible, i. e. aesthetic, technical and functional ones? Are they congruent, i. e. do they all refer to the same product type as the source? Which of these three has highest, which has lowest priority?
– *As to the recipient*: Is the sound directed towards the recipient in that there is the need to react directly – like with a warning signal, i. e. imperative cues? Does the sound indicate a specific state of the sound source, e. g., such that its state of operation can easily be sensed, i. e. indicative cues? Is the sound directed towards the listener in such a way that a specific reaction is provoked indirectly, i. e. suggestive cues?
– *As to the code*: Can a specific source or type of source be associated with the perceived sound? Is the sound referring to an object of experience where the listener has already processed comparable cues in the past, e. g., as delivered by schemas or Gestalt perception? What is the degree of abstraction from the sign carrier, the product sound, to its content? There are three categories of abstractions as follows.

- "Index" ... The sound refers the listener directly to its physical source, e. g., the sound of a hammer banging on wooden material. If this event is an object of experience to the listener, wooden material will be associated with the sound.
- "Icon" ... The relation between the sound and its source is based on similarity, e. g., an artificial sound which contains reduced characteristics of natural sounds – preferably the most relevant ones – and avoids redundancy. Most relevant criteria are configuration, continuity and analogy. As to making use of a reduced set of basic design elements refer to auditory icons or visual ones, e. g., pictograms [12].
- "Symbol" ... The connection between a sound and its *signifié* is based on conventions, e. g., a church bell indicating the hour, a warning signal in a car pointing the listener to not having fastened the seat belts.

Sign carrier, related to the product as such: Still another perspective is taken when the product sound is processed from the point of view of its being emitted by a product-in-use – what has been referred to as the communicative process before. Again, differentiation of three aspects helps to analyze the process of understanding, see the following listing.

source	user	sign
direct observation	emotional (re)-action	cultural values
transformation	rational (re)-action	
interpretation	actional (re)-action	

Some examples are itemized as follows.

- *As to the source*: Regarding the product-in-use profile, can the sound as single-sensory input among input from other senses be instantly associated with its source, namely, the product-in-use – such acting as an index? Does the sound possess all relevant features that are necessary to associate a supporting product image, for example, an artificially-designed sportive sound of a car? Is the percept of the product communicated in such a way that the encoded information can be extracted, e. g., telephone speech?
- *As to the user*: What is the reaction of the product users? Do they respond affectively, e. g., with satisfaction/dissatisfaction, sympathy/antipathy. Do the product users, whilst listening, deduce that the product is applicable and suitable for use, promising high performance, economy, etc.? As to the actional response, does the sound support, e. g., operational control of the product-in-use, its ease of use, interactivity?
- *As to the sign carrier*: The sound of a product-in-use may be a sign carrier for different cultural values, such as tradition, luxury, originality.

Meaning: Based on all these different perspectives with priority put on one or another criterion, meaning is associated after all. Meaning itself as well as processes which lead to the association of meaning can only be understood appropriately in this comprehensive context, i. e. in the "semiosphere" where product sounds are perceived and cognitively processed. The concept of semiosphere has been introduced by *Lotman* [34]. It is the semiotic space which is necessary for the existence and functioning of sign systems. *Hoffmeyer* [34] calls it the autonomous sphere of communication. Semiosphere is the presently used term for all patterns that generate context [51]. It is created by the structural coupling of cognitive systems, and ensures that every perceptual object becomes meaningful [22]. In other words, depending on the particular semiosphere, i. e. attention, activity, the listener's general state and others, meaning can in principle either be the result of the overall image, e. g., the sound is supportive with regard to the general product image, it is adequate to requirements of product-in-use, or it can relate to very specific features of the product sound, e. g., the form of the sound is too weak, the sound is a secondary perceptual event only, or it is inadequate with regard to the product function. In general, specific features of the product sound tend to come to the fore when expectations are not met.

8.10 Conclusions to be Drawn for Product-Sound Design

Based on what has been discussed so far, various criteria for product-sound design can be itemized. The following may serve as an example:

- Motivation for product use: What are the reasons for product use? In which respect does the listener gain when using the product? The product may, e. g., compensate for deficiencies as a hearing-aid does, support mobility as cars do, make work much easier as household appliances do, or the product may just be a status symbol.
- Function of the acoustic component in the product context: Is the acoustic signal a key quality element of the product when the product is used or is it a secondary phenomenon in the background?
- Function of the acoustic component in relation to other product components: Is it suitable as regards the sound in product use, are the sounds adequate with regard to other modalities, to adverse conditions, etc.?
- Meaning associated in product use: e. g., applicability, suitability for use, saturation of demands, meeting of requirements, product performance, practicability, functionality, security.
- Dominant quality items of product sounds: How is the product sound judged upon – e. g., extraordinary, independent, classical, innovative, simple, complex, practical, distinctive, convincing, typical, comfortable?
- Provoked reactive behaviour of the user as to product quality, e. g., the product is reliable, safe, of high performance, as to affect, e. g., reactions of disgust, contempt, indignation, dissatisfaction/satisfaction, desire, empathy, and as to motor activity, e. g., ease of handling, ergonomy, flexibility, operational control.

–   User specification: Can a typical user be specified? If so, how can the group of users be typified, e. g., according to age, education, experience of techniques and expectations of quality, socio-cultural background?

The know-how of product-sound designers is, among other things, related to form and function criteria of acoustic/auditory events as follows.

syntactic form	*syntactic function*
– configurative order	– accentuation
– structure and rhythm	– foreground vs. background event
– symmetry or asymmetry	

semantic form	*semantic function*
– simplification	– product: symptom, e. g., illustration
– alienation	– user: signal, e. g., attention
– exaggeration	– image: symbol, e. g., product identity
– underdetermination	
– ironization	

pragmatic form	*pragmatic function*
– clarity	– rationality and aesthetics
– appearance, German→Anmutung	– appropriateness for use
– usability	– desirability
– ergonomy	

The design of product sounds can be understood as an illustration of a specific product idea. The sound should be supportive to the product, and the product itself should provide an answer to a question, meet a demand. In this way, the principle of acoustic design is a functionalistic one. It is the functional forming of acoustic/auditory events in such a way that they meet specific goals. In this respect sound design is to a certain degree in line with the well-known *Bauhaus* tradition – refer to *Sullivan's* product design slogan "Form Follows Function" of the twenties in Germany [21]. There is, however, the following major difference between product-sound design as discussed here and the Bauhaus tradition. Bauhaus does not only stand for the form-follows-function principle, but also for the method of how to satisfy this design objective. The Bauhaus method is to reduce the conceivable richness of forms to simplicity, try to avoid redundancy and ambiguity – and at the same time support product function. This is a typical iconic approach. Product-sound design as discussed in this chapter does in a way follow the Bauhaus tradition, however, it is not restricted in scope by suggesting only iconic solutions. The "Form Follows Function" principle is mentioned here as it is one specific approach out of various semiotically motivated solutions conceivable. In principle, many more solutions regarding the construction of an acoustic form object are feasible – provided that the chosen acoustic form supports the idea of the product.

In order to be able to take design decisions with regard to how the sound as a sign carrier is processed in isolation as well as in the context of other types of sign carriers, i. e. in the semiosphere, an analysis of the processes of *semiosis* as described in the previous chapter, but in reverse order, is required. However, the auralization of product features is never 100% determined by the mere acoustic signals, as there is, for instance, always room for individual imagination and interpretation at the product user's side. This fact, by the way, justifies a transient area between rational and intuitive, artistic sound design. Yet, this area will not be elaborated further in this paper. What counts here in the end is the output of overall design competence, that is the question of whether the acoustic event does finally auralize salient product features in a valid and appropriate way.

8.11 Acoustic Design Between Innovation and Convention

The aim of product-sound design typically is, among other things, to create acoustic events that will be perceived as both novel and already known at the same time. They should be novel to avoid monotony and uniformity and, at the same time, they should in a way be known already – in order to enable the listener to quickly understand the meaning of the sound and its function when using the product. The sound should further be an active stimulant for imagination and invite the listener to use the product. But what does the dichotomy "novel" and "known" mean? In fact "novel" and "known" are relative terms, and to explain what is meant here, the concepts of schemata, assimilation and accommodation, as discussed above, can be applied. Indeed, what is novel cannot immediately be processed with available schemata, but only via mediation after assimilating and accommodating.

As described before, assimilation and accommodation are ruling principles in data alignment. One of the tasks of the sound designer might be to construct an acoustic event which makes the auditory event unfamiliar to a certain degree – without, however, violating basic relations and/or patterns of experience. To this end, a good product-sound design may well break the habitual view and take a new perspective. The change in perspective can be initiated through "ear-catching" features of the sound which are not immediately expected in the communication situation the listeners are in. Nevertheless, the listeners should be able to treat the auditory event as a token of a specific known type. This is basically important when they are confronted with artificial signals which they have never listened to before. Please recall, the objects of perception are being analyzed, structured, organized, and compared with data of prior experience, and during this process analogies are established.

Among other things, such analogies can be based on patterns and on structural characteristics which by their very nature have no reality of their own in the physical world. As outlined before, a pattern becomes a pattern because someone states a sequence of items to be meaningful or cohesive [4]. It is Gestalt theory which gives extremely valuable explanations for an analytic understanding of what listeners base their statement on, namely, Gestalt perception paradigms as proximity, common fate, good continuation, or simplicity – see above.

This way of looking at product-sound design sets a particular requirement, namely that a target product sound is either explicitly or implicitly available and has "only" to be adapted to a new context in an inventive way. That seems to be an unrealistic requirement at first sight, especially for the sound design of new industrial products. If there is no best target sound available, then there is no sound that can be creatively modified.

Nevertheless, having in mind that one of the design maxims is that the object of design has in one way or another to dock on what is known and experienced, this is exactly this point where sound design starts. It is not necessarily a specific target sound which has to be known, but it is more the product context in general, i. e. the configurative battery of items which in their sum make up the semiosphere. All these items are not incidental events with regard to their qualitative form. They are identical or at least similar with regard to their perceived persisting patterns. These patterns are identified as sign carriers, namely, sign carriers of different nature. Only when these contexts can be classified, they confer meaning [4]. In other words, perceptual stimuli which are processed by other sensory systems are patterned as well – we say that they are systematic.

It follows that one approach which could be pursued is to establish criteria of reference in order to understand the various common features of product sounds, the patterns and structures of other acoustic sign carriers as, e. g., speech, or those of different sensory modalities such as vision or haptics. All of these can be understood as mutable systems which, among other things, can be characterized by invariance and variation. Consequently, it would be of profound advantage to learn more about the principles of invariance and variation.

In order to be able to take design decisions in a structured and rationally motivated way, it is also enriching to have a closer look at some further paradigms of semiotics. These paradigms are related to different systems in which semioses happen. Although they seem to be of minor importance for product-sound design at first glance, they are mentioned here as they hint indirectly at invariances. If the task is to extract qualitative or structurally relevant features of systems on which *semiosis* is based, it may be an option to take the following perspectives.

– *Natural vs. artificial sign systems:* What are typical natural sign systems? *Von Uexküll* [56] mentions such which are related to the physical medium, to food, enemy and sex as the most important ones. Are these systems patterned? Which items contribute to the concept of a pattern?

Can specific genres be identified on the basis of such patterns? How can they be described, both in terms of qualitative and structural features? Which objects of natural sign systems qualify for being used in different contexts, e. g., for product-sound design? Which kinds of associations are stimulated when listening to natural sounds? Can new artificial sounds be created on the basis of natural sounds? Is learning required or can such artificially created sign systems be dealt with intuitively? How stable are natural sign systems? How distinguished are they? How complex are they?

- *Animal vs. human communication systems:* How do animals communicate with each other? Which signs do they use for communication? Which cues do they concentrate on? Can these signs be misunderstood by other species? Are there signs common to all species or at least to groups of species? Which animals are able to learn new functions of signs [23]?

- *Diachronic vs. synchronic view* – seen either from the perspective of ontogeny, i. e. child development, phylogeny i. e. the evolution of species, or pathogeny, i. e. loss of abilities.
  - Diachrony: How do sign systems develop over time [54]? What are the origins of sign systems? How do these systems look? How do they change? What are influencing factors [31]?
  - Synchrony: What are the characteristics of a specific sign in relation to other signs at the same time?

- *Syntagmatic vs. paradigmatic view:* Can different signs and types of signs be combined? According to which criteria can a specific sign be chosen from a number of different candidates? An analysis of these different systems shows that they are made up of invariant and of variant features. Looking at them might indeed offer specific answers and solutions to the design of product sounds.

Summary

- The object of product design is the engineering of a tool for creating a mental image.
- A sign carrier is not processed as an "unbiased" perceptual phenomenon [60], but as one which carries information.
- The assignment of meaning to a sound is dependent on the listener's perspective, on the semiosphere in which the auditory event happens.
- Product use and product perception are not independent processes as product use is an aspect of product perception. Auditory product perception is one aspect of perception in general, yet, it is closely related to percepts of other modalities.

- When applied to product sound, a possible design philosophy might be that the product sounds *follow the function* of the product as expected by the user.

8.12 Conclusions

Today designers work in quite diverse fields, and they have tools at hand which open up really new possibilities. This leads, among other things, to a diversification in the field of product-sound design. However, in comparison to other design areas – such as visual design – sound design, as a rule, has been rather timid and modest so far. One of the reasons for this is that sound design was mostly decoupled from the product-design task in the past, and even today it is in a way still hidden behind the facade of the industrial product as such. Consequently, the future of product-sound design is open for more radical design solutions. There are many ways to reach this goal. In this chapter, sign theory, or semiotics, has been introduced, as it offers a number of different perspectives and paradigms for describing and understanding the various processes of product-sound design. In this way, knowledge and competence will, hopefully, be added to this multi-fold and multi-disciplinary area. For sound designers, the primary benefit of semiotics is not so much that it will offer specific helps to take particular design decisions. The advantage of semiotics, as an additional source of knowledge, is primarily that this theory provides a specific overview and offers new ways to structure the problems. Without doubt, this will prove to be of substantial benefit for the sound-design process at large.

Sound design takes place in a broad range between sensuality and rationality, and this makes sound design an extremely interesting task. Consequently, there are purely application-oriented persons as well as theorists approaching this field. A theoretic approach, as offered here, may be more exerting than a practical view. Yet, there is no doubt that the theory of semiotics offers perspectives which render interesting hints that practitioners of product-sound design may well profit from.

Acknowledgement

Part of this work was carried out at the Institute of Communication Acoustics at Bochum, headed by *J. Blauert*. Among other things, it is based on semiotic theory as taught by *W. A. Koch* and *K. Eimermacher* – also Ruhr-University Bochum, and by *A. Eschbach* – University of Duisburg–Essen. I would like to thank all of them for fruitful discussions – sometimes taking quite unconventional perspectives. My thanks are also directed towards my former colleagues *S. Möller, A. Raake, S. Schaden, R. Pegam* and *M. Hanisch*, and towards my colleagues *G. Canévet* and *S. Meunier* at CNRS, Marseille, with whom we co-operated under the PROCOPE project, funded by the German Academic Exchange Service, DAAD.

References

1. Atherton J S (2003) Learning and teaching: Piaget's developmental psychology. http://www.dmu.ac.uk/ jamesa/learning/piaget.htm, accessed Oct 21, 2004
2. Ausubel D P, Sullivan E V (1978) Historischer Überblick über die theoretischen Ansätze. In: Steiner G (ed) Die Psychologie des 20. Jahrhunderts. Piaget und die Folgen. vol VII:547–567. Kindler, Zürich
3. Avant L, Helson H (1973) Theories of perception. In Wolman B B (ed): Handbook of general psychology. Prentice Hall, Englewood Cliffs, 419–448
4. Bateson G (1979) Mind and nature. E P Dutton, New York
5. Bateson G (1985) Ökologie des Geistes. Anthropologische, psychologische, biologische und epistemologische Perspektiven (Steps to an ecology of mind). Suhrkamp, Frankfurt/Main
6. Blauert J, Jekosch U (1996) Sound-quality evaluation – a multi-layered problem. ACUSTICA/acta acustica 83:747–753
7. Blauert J, Jekosch U (2003) Concepts behind sound quality: some basic considerations. Proc. Internoise'03, Seogwipo, Korea, CD-ROM
8. Bednarzyk M (1999) Qualitätsbeurteilung der Geräusche industrieller Produkte. Der Stand der Forschung, abgehandelt am Beispiel der Kfz-Innenraumgeräusche. Doc diss Ruhr-Univ Bochum, VDI Verlag, Düsseldorf
9. Bregman A S (1990) Auditory scene analysis. The perceptual organization of sound. MIT Press, Cambridge MA
10. Bühler K (1933) Die Axiomatik der Sprachwissenschaft. Klostermann, Frankfurt/Main
11. Bühler K (1934) Sprachtheorie. Fischer, Stuttgart
12. Dürrer B (2001) Untersuchungen zum Entwurf von Auditory Displays. Doct diss Ruhr-Univ Bochum, dissertation.de Verlag im Internet, Berlin
13. Eco U (1976) A theory of semiotics. Indiana University Press, Bloomington
14. Eco U (1994) Einführung in die Semiotik. 38. Authorized German ed by J Trabant. Fink. München,
15. Eschbach A, Trabant J (eds)(1983) History of semiotics. Benjamins, Amsterdam
16. Fabian R (ed)(1990) Christian von Ehrenfels. Philosophische Schriften: 4. Metaphysik. Philosphia Verlag, München
17. Fastl H (2003) Psychoacoustics and sound quality. Chap 6 this vol
18. Fricke J P (1998) Systemische Musikwissenschaft. In: Niemöller K W (ed) Perspektiven und Methoden einer Systemischen Musikwissenschaft. Peter Lang, Frankfurt/Main, 13–23
19. Ginsburg H, Opper S (1998) Piaget's Theorie der geistigen Entwicklung. 8th ed. Klett-Cotta, Stuttgart
20. von Glasersfeld E (2000) Scheme theory as a key to the learning paradox. In: Tryphon A, Vonèche J (eds) Working with Piaget. Essays in Honour of Bärbel Inhelder. Psychology P, Hove, 141–148
21. Godau M (2003) Produktdesign. Eine Einführung mit Beispielen aus der Praxis. Birkhäuser, Basel

22. Goppold A (2003) A spatio-temporal perspective vision of neuronal reso-
    nance technology, of technological ars memoriae and of the aesthetics and
    architectonics of tensegrity structures in hypertext node networks. www.uni-
    ulm.de/uni/intgruppen/memosys/persp.htm, accessed Oct 21, 2004
23. Gottlieb G (2002) On the epigenetic evolution of species-specific perception:
    the developmental manifold concept. Cogn Developm 17:1287–1300
24. Götz M (1989) Das grafische Zeichen - Kommunikation und Irritation.
    In: Stankowski A, Duschek K (eds) Visuelle Kommunikation. Ein Design-
    Handbuch. 2nd ed (1994) Dietrich Reimer, Berlin, 53–76
25. Habermann H (2003) Kompendium des Industrie-Design. Von der Idee zum
    Produkt. Grundlagen der Gestaltung. Springer, Berlin
26. Henle M ed. (1971) The selected papers of Wolfgang Köhler. Livernight, New
    York
27. Hoffmann M H G (ed)(2003) Mathematik verstehen? Semiotische Perspektiven.
    Franzbecker, Hildesheim
28. Jekosch U (1982) Syncretism and Gestalt in Psychogenesis. In: Koch W A
    (ed) Semiogenesis. Essays on the analysis of the genesis of language, art and
    literature. Lang, Frankfurt/Main, 252–273
29. Jekosch U (1999) Meaning in the context of sound quality assessment. ACUS-
    TICA/acta acustica 85:681–684
30. Jekosch U (2000) Sprache hören und beurteilen: Ein Ansatz zur Grundlegung
    der Sprachqualitätsbeurteilung. Inaugural diss (habilitation), University of Es-
    sen, Essen
31. Koch W A (1982) Semiogenesis. Some perspectives for its analysis. In: Koch
    W A (ed) Semiogenesis. Lang, Frankfurt/Main, 15–104
32. Koffka K (1935) Principles of Gestalt psychology. Lund Humphries, London
33. Kuwano S, Namba S, Schick A, Höge H, Fastl H, Filippou T, Florentine M,
    Muesch H (2000) The timbre and annoyance of auditory warning signals in
    different countries. Proc. Internoise 2000, Nice, CD-ROM
34. Lotman J M (1990) Universe of the mind: A semiotic theory of culture. 125.
    Indiana University Press, Bloomington
35. Luchins A S, Luchins E H (1982) An introduction to the origins of Wertheimer's
    Gestalt psychology. Gestalt Theory 4:145–171
36. McAdams S, Bigand E (eds) (1993) Thinking in sound. The cognitive psychol-
    ogy of human audition. Clarendon, Oxford
37. Mersch D (ed)(1998) Zeichen über Zeichen. Texte zur Semiotik von Charles
    Sanders Peirce zu Umberto Eco und Jacques Derrida. Deutscher Taschen-
    buchverlag, München
38. Morris C W (ed)(1939) Writings in the general theory of signs. Mouton, The
    Hague
39. Morris C W (1946) Signs, language, and behavior. Braziller, New York, 1971
40. Nöth W (2000) Handbuch der Semiotik. 2nd rev ed. Metzler, Stuttgart
41. Patsouras C, Filippou T G, Fastl H (2002) Influences of colour on the loudness
    judgement. Proc. Forum Acusticum 2002, Sevilla, PSY-05-002-IP, CD-ROM
42. Peirce C S (1982-1989) Writings of Charles S. Peirce: A chronological edition.
    Vol 1:1857-66, Vol. 2:1867-71, Vol. 3:1872-78, Vol. 4:1879-84. Indiana University
    Press, Bloomington
43. Piaget J (1924) Le jugement et le raisonnement chez l'enfant (The judgement
    and reasoning in the child). Engl transl (1962) Routledge & Kegan, London

44. Piaget J (1926) La représentation du monde chez l'enfant (The child's conception of the world) Engl transl (1962) Routledge & Kegan, London
45. Piaget J (1937) La construction du réel chez l'enfant (Der Aufbau der Wirklichkeit beim Kinde) Germ transl (1974) Klett, Stuttgart
46. Piaget J (1952) The origins of intelligence in children. International Universities Press, New York, 6–7
47. Piaget J (1969) The mechanisms of perception. Routledge & Kegan London
48. Purwins H, Blankertz B, Obermayer K (2000) Computing auditory perception. Organised Sound 5:159-171
49. Quinn N, Holland D (1987) Cultural models of language and thought. Cambridge Univ Press, New York
50. Reitzle W (1993) Qualität - Wichtige Komponente der F- und E-Strategie eines Automobilherstellers. In: Seghezzi H D, Hansen J R (eds) Qualitätsstrategien. Anforderungen an das Management der Zukunft. Hanser, München, 94–108
51. Santaella L (1996) Semiosphere: The growth of signs. Semiotica 109:173–186
52. Schröder E (1988) Vom konkreten zum formalen Denken: Individuelle Entwicklungsverläufe von der Kindheit zum Jugendalter. Doct diss Techn Univ Berlin, Berlin
53. Stankowski A, Duschek K (eds) (1989) Visuelle Kommunikation. Ein Design-Handbuch. 2ndedition (1994) Dietrich Reimer, Berlin,
54. Szagun G (1983) Bedeutungsentwicklung beim Kind. Wie Kinder Wörter entdecken. Urban & Schwarzenberg, München
55. Trabant J (1996) Elemente der Semiotik. Francke, Tübingen
56. von Uexküll J (1982) The Theory of Meaning. Semiotica 42:25–82 and 31
57. Västfjäll D (2002) Mood and preference for anticipated emotions. Doct diss Chalmers Univ, Gothenborg
58. Västfjäll D, Gulbol M A, Kleiner M, Gärling T (2002) Affective evaluations of and reactions to exterior and interior vehicle auditory quality. J Sound Vibr 255:501-518
59. Vygotskij L S (1992) Geschichte der höheren psychischen Funktionen. Lit Verlag, Münster
60. Zwicker E, Fastl H (1990) Psychoacoustics. Facts and models. Springer, Berlin

9 Binaural Technique – Basic Methods for Recording, Synthesis, and Reproduction

Dorte Hammershøi and Henrik Møller

Department of Acoustics, Aalborg University, Aalborg

Summary. The term "binaural technique" is used as a cover label here for methods of sound recording, synthesis and reproduction, where the signals in focus are the acoustic signals at the eardrums. If these are presented authentically to listeners, the listeners will obtain acoustic cues which are deemed sufficient for authentic auditory experience – including its spatial aspects. This chapter reviews the basic principles of binaural technique - putting a special focus on results of investigations which have been performed at Aalborg University. These basic principles form the foundation for current utilization of binaural technique at large. They include basic theory, investigations on sound transmission in the ear canal, measurements and post-processing of head-related transfer functions, HRTFs, transfer functions of headphones and their adequate equalization, and results from localization experiments in real life as well as with binaural recordings from real heads and artificial heads. Numerous applications to these methods exist. Some of them will be introduced exemplarily.

9.1 Introduction

Binaural technique starts from the concept that our auditory percepts are predominantly formed on the basis of only two inputs, namely the sound-pressure signals at our two eardrums. If these are recorded – literally – in the ears of listeners, and reproduced authentically when played back, then all acoustic cues are available to the listeners for forming authentic replicas of the original auditory percepts – including all spatial aspects.

The signals to the two ears, called binaural signals, may, instead of being picked up in the ears of listeners, also be synthesized by computer – once the sound transmission from the source to the listeners' ears has been mapped for all possible positions of the sound source. It can then be controlled by computer *where* the listener will hear the sound, and sound sources can be placed anywhere in the listeners' perceptual space. This technique is obvious for virtual-reality application, but many further areas of applications exist.

The definition of *binaural technique* may be extended more generally to *binaural technology* as follows.

"Binaural technology is a body of methods that involves the acoustic input signals to both ears of the listener for achieving practical purposes, for example, by recording, analyzing, synthesizing, processing, presenting, and evaluating such signals." [8]

This chapter presents methods for obtaining or generating binaural signals and reproducing them adequately. For a wider view on spatial hearing and binaural technology, see e. g., [9, 35].

9.1.1 Structure

With binaural recording the sound is recorded in the ear canals of listeners. For practical reasons the listener is often replaced by a manikin with the relevant acoustical properties similar to those of a human. The technique is therefore commonly referred to as "artificial-head" recording technique or "dummy head" recording technique. When binaural recordings are made, the spatial properties of the exposing sound field are transformed into two electrical signals representing the input to the left and right ears, respectively.

Play-back is most often done by headphones, as they offer complete channel separation and acoustical isolation from the surroundings, thus enabling optimal control of the reproduction situation. Authentic reproduction requires correct calibration of the complete recording and reproduction chain. Section 9.2 addresses principles of calibration from a theoretical point of view.

The headphone transfer function is often given very little attention. Often, headphones at hand are used without electronic compensation, disregarding the intended use and design goals of the headphones and their typically non-flat frequency responses. In Sect. 9.3 examples of headphone transfer functions and equalization filters are shown.

When an artificial head is used during recording – or a human different from the listener – then the recorded signals do not exactly correspond to the ear signals that the listener would have been exposed to in the recording situation. Localization performance with binaural recordings from humans and artificial heads is addressed in Sect. 9.4. This section also includes a comparison of head-related transfer functions, HRTFs, of some artificial heads and humans.

Binaural signals may alternatively be generated electronically by the use of filters representing HRTFs, in which case the method will be denoted "binaural synthesis". The success of binaural synthesis depends strongly on details of the procedures applied for determining and realising HRTFs, such as physical aspects of the measurement situation, post-processing of data, and implementation as digital filters. These issues are addressed in Sect. 9.5.

Finally, in Sect. 9.6, some of the numerous applications of binaural technology are reviewed. The list is in no way exhaustive, but offers some input for inspiration and useful links to players in the field. Applications of binaural technology are also described in other chapters of this book [10, 19, 84].

9.2 Theory

9.2.1 Sound-Transmission Model

When the binaural signals are recorded – whether in the ears of a human or an artificial head – the spatial sound field is transformed into only two signals, one for each side. A prerequisite for the successful reproduction is thus that the full spatial information is maintained in these signals.

This issue has been frequently addressed in the literature, e. g., [37, 66–68, 99–101, 110], the latter containing a summary. It is generally agreed that there are other places in the ear canal than at the eardrum where the full spatial information is available. This has been demonstrated for points on the center line of the ear canal, including the point on the entrance plane. It has also been shown that the spatial information is maintained, even if the ear canal is blocked at the recording point or further towards the eardrum. This is of significant practical relevance, since larger microphones can then be used in human ear canals, and artificial heads can be made without ear canals. In addition, as will be seen in Sect. 9.2.3, individual effects from the ear-canal and eardrum impedance are removed, whereby the signals become more general with respect to the spatial information contained in it.

A sound-transmission model as introduced in [71] splits the sound transmission up into direction-dependent and direction-*independent* parts. The model, and a similar one for the reproduction situation – see Sect. 9.2.2, is used for determining the correct calibration of the complete recording and play-back chain. Further, it enables assessment of other relevant properties, such as inter-individual variation, see Sect. 9.2.3. The model is depicted in Fig. 9.1[1]

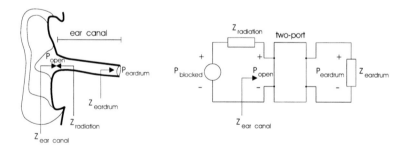

Fig. 9.1. Model of free-field sound transmission to the human external ear: Sketch of anatomy, *left*, and analogue model, *right* – adapted from [71]

---

[1] In the present chapter, capital letters denote terms given in the frequency domain. The symbols thus stand for complex functions of frequency. Time domain representations are denoted by the corresponding lower-case symbols. The connection between the two domains can, e. g., be given by the *Fourier* transform, $\mathcal{F}$, and its inverse, $\mathcal{F}^{-1}$, thus $P(f) = \mathcal{F}(p(t))$ and $p(t) = \mathcal{F}^{-1}(P(f))$

In the model, the complete sound field outside the ear canal – whatever the source and the sound field – is described by a *Thévenin* equivalent, consisting of the open-circuit sound pressure, P_{blocked}, and a generator impedance. The generator impedance is identical to the radiation impedance, $Z_{\text{radiation}}$, as seen from the ear canal into the free air. P_{blocked} does not exist during normal listening conditions, but if the ear canal is blocked to make the volume velocity zero, P_{blocked} can be found just outside the blockage. P_{eardrum} denotes the sound pressure at the eardrum and P_{open} denotes the sound pressure at the entrance to the ear canal. The ear canal itself is modelled by a two-port which is loaded by the eardrum impedance, Z_{eardrum}. The input impedance of the two-port as seen from the entrance of the canal, is denoted $Z_{\text{ear canal}}$. The total sound transmission from the free field to the eardrum can be described by the ratio of (i) the sound pressure at the eardrum and (ii) a reference sound pressure which is found at the position corresponding to the center of the subject's head with the subject absent, i.e.

$$\frac{P_{\text{eardrum}}}{P_{\text{ref}}}(\angle). \tag{9.1}$$

This ratio, $P_{\text{eardrum}}/P_{\text{ref}}$, depends on the angle of incidence of the sound wave, as indicated by the symbol $\angle$. It defines a head-related transfer function, usually abbreviated HRTF[2]. The sound transmission can be separated in consecutive parts in the following way,

$$\frac{P_{\text{eardrum}}}{P_{\text{ref}}}(\angle) = \frac{P_{\text{blocked}}}{P_{\text{ref}}}(\angle) \cdot \frac{P_{\text{open}}}{P_{\text{blocked}}} \cdot \frac{P_{\text{eardrum}}}{P_{\text{open}}}. \tag{9.2}$$

$P_{\text{eardrum}}/P_{\text{open}}$ describes the sound transmission along the ear canal and does not depend on the angle of incidence – due to one-dimensional wave propagation inside the ear canal. Nor does $P_{\text{open}}/P_{\text{blocked}}$ depend on the angle of incidence. This ratio denotes the pressure division at the entrance of the ear canal between $Z_{\text{ear canal}}$ and $Z_{\text{radiation}}$ as given in the equation below. In essence, only $P_{\text{blocked}}/P_{\text{ref}}$ depends on direction.

$$\frac{P_{\text{open}}}{P_{\text{blocked}}} = \frac{Z_{\text{ear canal}}}{Z_{\text{radiation}} + Z_{\text{ear canal}}}. \tag{9.3}$$

An example of the transfer functions for three directions of sound incidence which illustrates directional dependence and independence as discussed above is given in Fig. 9.2.

The consequence of this is that both sound pressures at the eardrum, at the ear-canal entrance, blocked or open, or elsewhere on the center line of the ear canal, will have the full spatial information and may be used for binaural recording. Likewise, it suggests a wider definition of the head-related transfer

---

[2] Note that the term HRTF is commonly understood as standing for a set of two transfer functions, namely, one for the left and one for the right ear

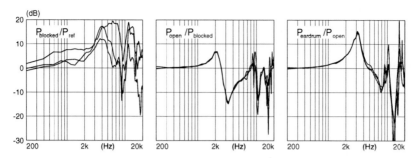

Fig. 9.2. $P_{\text{blocked}}/P_{\text{ref}}$, *left*, $P_{\text{open}}/P_{\text{blocked}}$, *center*, and $P_{\text{eardrum}}/P_{\text{open}}$, *right*. Measurements with sound coming from the front, from the left side and from the back are overlaid in each frame – left ear of Subject ML, from [37]

function, so that it covers all points in the ear canal that have the full spatial information and not only the point at the eardrum – as was denoted in (9.1). An HRTF, then, is, generally given by the ratio of the sound pressure at the ear of a listener and the reference sound pressure as follows:

$$HRTF(\angle) = \frac{P_{\text{ear}}}{P_{\text{ref}}}(\angle). \tag{9.4}$$

9.2.2 Calibration of the Complete Recording and Play-Back Chain

In order to determine the full transmission of a binaural recording and play-back system, a model similar to that described in Sect. 9.2.1 for the free-field transmission is used for the play-back situation. The superscript "hp" is used in the following for signifying sound pressures that refer to the reproduction situation over headphones. Again, the input is described by a *Thévenin* generator, i. e. the blocked-entrance sound pressure, $P_{\text{blocked}}^{\text{hp}}$, and a generator impedance, $Z_{\text{headphone}}$. The complete transmission from voltage at the headphone terminals, $E_{\text{headphone}}$, to sound pressure at the eardrum, is then

$$\frac{P_{\text{eardrum}}^{\text{hp}}}{E_{\text{headphone}}} = \frac{P_{\text{blocked}}^{\text{hp}}}{E_{\text{headphone}}} \cdot \frac{P_{\text{open}}^{\text{hp}}}{P_{\text{blocked}}^{\text{hp}}} \cdot \frac{P_{\text{eardrum}}^{\text{hp}}}{P_{\text{open}}^{\text{hp}}}. \tag{9.5}$$

The sound transmission along the ear canal is the same whatever the source, thus $P_{\text{eardrum}}^{\text{hp}}/P_{\text{open}}^{\text{hp}} = P_{\text{eardrum}}/P_{\text{open}}$ holds. However, the *Thévenin* impedance is not the same. Consequently, the pressure division at the entrance to the ear canal differs between the two situations. The pressure division during reproduction is

$$\frac{P_{\text{open}}^{\text{hp}}}{P_{\text{blocked}}^{\text{hp}}} = \frac{Z_{\text{ear canal}}}{Z_{\text{headphone}} + Z_{\text{ear canal}}}. \tag{9.6}$$

If recording is made at the eardrum, M describes the transfer function of the recording microphone, and G_{eardrum} denotes the electrical gain from the output of the recording microphone to the input of the headphone, the total transmission from recording to reproduction can be written as

$$\frac{P_{\text{eardrum}}}{P_{\text{ref}}} \cdot M \cdot G_{\text{eardrum}} \cdot \frac{P_{\text{eardrum}}^{\text{hp}}}{E_{\text{headphone}}}. \tag{9.7}$$

The headphone produces the *correct* sound pressure at the eardrum when the transmission in (9.7) equals $P_{\text{eardrum}}/P_{\text{ref}}$, which means that the design target for the equalizer is given as follows – see [71] for calculations,

$$G_{\text{eardrum}} = \frac{1}{M \cdot (P_{\text{eardrum}}^{\text{hp}}/E_{\text{headphone}})}. \tag{9.8}$$

Similarly, when the recording is made at the entrance, open or blocked, respectively, the following equations apply,

$$G_{\text{open}} = \frac{1}{M \cdot (P_{\text{open}}^{\text{hp}}/E_{\text{headphone}})} \quad \text{and} \tag{9.9}$$

$$G_{\text{blocked}} = \frac{1}{M \cdot [P_{\text{blocked}}^{\text{hp}}/E_{\text{headphone}}]} \cdot \frac{Z_{\text{ear canal}} + Z_{\text{headphone}}}{Z_{\text{ear canal}} + Z_{\text{radiation}}}. \tag{9.10}$$

The last term in (9.10) is the ratio of the pressure division during recording (9.3) and the pressure division during reproduction (9.6), which is denoted the pressure-division ratio, PDR. If the PDR is ignored – at least at this point, it can be seen that the exercise for finding the correct equalization is the same whatever the recording point is. In other words, the transfer function of the equalizer, G, shall equal the inverse of the microphone transfer function times the inverse of the headphone transfer function as measured at the same physical point in the ear where the recording was made.

Many of modern miniature and probe microphones have fairly flat frequency responses, in which case they may simply be represented by their overall sensitivities in the design target for G. Thus the main component of G is the inverse of the headphone transfer function measured at the position in the ear canal where the recording was made. For this reason, the equalizer is often referred to as the *headphone* equalizer, although, by definition, it serves to ensure that the entire reproduction chain is adequately calibrated. Headphone equalization will be further discussed in Sect. 9.3.

9.2.3 Inter-Individual Variation

In the previous sections it was inherently assumed that the recordings and the determination of the headphone transfer function were done with the

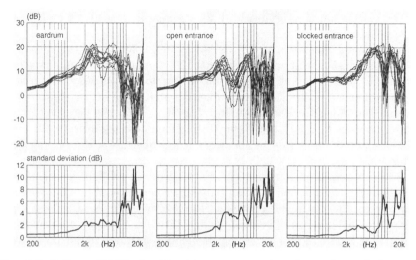

Fig. 9.3. Left-ear HRTFs for 12 subjects, measured at the eardrum, **left**, at the open entrance, **center**, and at the blocked entrance, **right**, to the ear canal – adapted from [37]. Sound from the left, standard deviations computed frequency by frequency in dB

same person, and that the listener was identical to this person. This is rarely the case, and since there are considerable differences in the shape of humans, including their outer ears, it must be anticipated that there will be individual differences in the binaural signals and in the headphone transfer functions. The consequence is that the binaural technique may result in correct signals only for those particular individuals that the recordings and the headphone-transfer-function measurement were made with. Thus, only these individuals may achieve the authentic listening experience.

The separation of free-field sound transmission into cascaded components, such as described in Sect. 9.2.1, paves the road to objective assessment of inter-individual variations. Data for inter-individual variation in sound transmissions have been presented previously, see, e. g., [9,30,37,39,76,97,110,111, 117]. Comparisons as performed in [37,76] provide the basis for the following discussion.

Examples of HRTFs measured at the eardrum, the open entrance and the blocked entrance are given in Fig. 9.3. If the general characteristics of the three types of HRTFs are compared, there are clear differences. This does not come by surprise, given the characteristics of the pressure division, $P_{open}/P_{blocked}$, and the transmission along the ear canal, $P_{eardrum}/P_{open}$, see e. g., Fig. 9.2. These elements are additively included in the transfer functions in the frames of Fig. 9.3 – in the order mentioned, i. e. from right to left. Individual characteristics of the elements will also add, but since the elements may be correlated, the variance will not necessarily sum up for each step in the transmission.

In the lower panels of Fig. 9.3, the standard deviations for the data in the top panels are shown. It can be seen that the standard deviation at the open entrance is higher than at the eardrum or the blocked entrance. An analysis of correlation [37] shows that this is due to a high negative correlation of the transmission along the ear canal, $P_{\text{eardrum}}/P_{\text{open}}$, and the pressure division at the entrance, $P_{\text{open}}/P_{\text{blocked}}$. This correlation does not surprise, since both terms are highly influenced by the acoustical properties of the ear canal and its termination.

It also becomes evident that the blocked-entrance HRTF has the lowest variation. This means that it is not only convenient from a practical point of view to record at the blocked entrance, but it also results in recordings which may offer a wider range of general applicability. An example of a convenient microphone technique for recording of blocked-entrance sound pressures is shown in Fig. 9.4.

Also the headphone transfer functions vary across subjects, as will be shown by examples in Sect. 9.3. The significance of inter-individual differences will thus also depend on the properties of these transfer functions, the procedures for determining them and the design of adequate equalization filters. An analysis – using terms from the model – is given in Møller et al. [77].

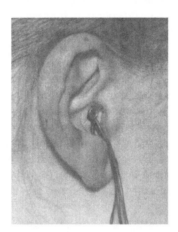

Fig. 9.4. Example of a microphone technique for blocked-entrance binaural recordings. A hole of suitable size is made in an EAR earplug – for instance by a soldering iron – and a Sennheiser KE 4-211-2 miniature microphone is inserted in the hole. The earplug, with the microphone inside, is then compressed and, once well compressed, earplug and microphone are inserted into the ear canal. Earplug and microphone should be held in position while the earplug decompresses, so that the microphone is ensured the correct position flush with the entrance – after [76]

9.3 Headphone Reproduction

As stated earlier, authentic reproduction of binaural recordings is aimed at through one-to-one transmission from the recorded sound pressure to the reproduced one. In the following, examples of headphone transfer functions, pressure divisions and equalization filters [73, 77] are given. For further examples check, e. g., [40, 55].

9.3.1 Headphone Transfer Functions

In [73] the transfer characteristics of a selection of headphones were studied, mostly of Hi-Fi type – for examples see Fig. 9.5. It became obvious that none of the headphones had the flat frequency response required for authentic reproduction of binaural signals, compare Sect. 9.2.2, and that several of them have complicated structures that are difficult to compensate with low-order filters.

For frequencies above 7 kHz the structures are also highly individual, so that individual equalization has to be considered. It was also noted that the headphones have very different frequency responses, even those that are claimed to be designed according to the same design goal, e. g., free-field or diffuse-field – see Sect. 9.3.4. Further, the sensitivity of the headphones varied considerably between subjects for some headphones. This may be considered a minor problem, but the accurate calibration of level may be important for the generation of perceptually-authentic auditory scenes, particularly for scientific experiments. Thus, generally speaking, the transfer function of the headphones used for the reproduction of binaural signals must be known and compensated for.

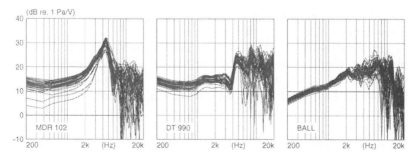

Fig. 9.5. Headphone transfer-function examples for Sony MDR-102, **left**, Beyer-Dynamic DT-990 headphones, **center**, and for a free-standing ball loudspeaker, **right**, measured at the blocked-ear-canal entrance of 40 human subjects [73]

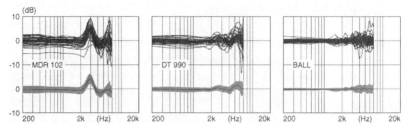

Fig. 9.6. Examples of pressure-division ratios, PDRs, for the Sony MDR 102, **left**, the BeyerDynamic DT 990 Professional headphones, **center**, and for a free-standing ball loudspeaker, **right**. Data from 40 human subjects [73], individual curves, **above**, mean values ± one standard deviation, **below**. Each PDR is computed on basis of four measurements, i.e. open and blocked entrance measured in the free field and open and blocked entrance measured with the headphone as the source. The validity of the results for higher frequencies is doubtful, thus no results are shown for frequencies above 7 kHz

9.3.2 Pressure Division

If the pressure division in the headphone situation equals the pressure division in a free-air situation, PDR reduces to unity – refer to Sect. 9.2.2. The term "free-air-equivalent coupling", FEC, was introduced in [73] to describe headphones for which this is the case. For headphones that have FEC properties, the equalization filter thus only needs to compensate for the headphone transfer function itself, even if recordings are made at the blocked ear canal. For headphones that do not have FEC properties, the equalization filter also needs to compensate for the PDR.

Results from [73], see the examples in Fig. 9.6, indicate that the pressure divisions in the free-air and headphone situations differ only slightly. Hence, the pressure division ratio can be assumed negligible for some headphones. Not surprisingly, this is the case for small, free-standing ball loudspeakers – Fig. 9.6, right frame – but also some of the more traditional headphones, including the BeyerDynamic DT 990 Professional – center frame – have pressure division ratios close to unity. Even for headphones for which pressure division ratios differ most clearly from unity, e.g., the Sony MDR-102 – left frame – deviations are, although significant in magnitude, limited to a rather narrow frequency range. In any case, the equalization needed to compensate for the pressure division ratio is much smaller than the one needed to compensate for the headphone transfer function itself. This is particularly relieving from a practical point of view, as determination of the PDR is complicated and requires a total of four measurements, i.e. in the free field, P_{open} and P_{blocked}, and with headphones on, $P_{\text{open}}^{\text{hp}}$ and $P_{\text{blocked}}^{\text{hp}}$ – see Sect. 9.2.2.

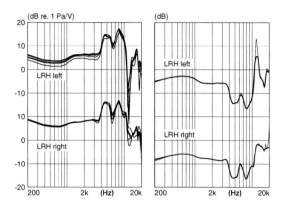

Fig. 9.7. $P_{\text{blocked}}^{\text{hp}}/E_{\text{headphone}}$ for BeyerDynamic-DT-990-Professional headphone, **left**, subject LRH, five repetitions. 32nd-order-IIR equalization filter, **right** [77]. Target function depicted as *thin line*

9.3.3 Headphone Equalization

Headphone equalization requires a successful inversion, i. e. de-convolution, of the headphone transfer function – see (9.8), (9.9), (9.10). Numerous mathematical methods exist for this inversion, and – considering the variation in characteristics of headphone transfer functions – it may be that one method may be ideal for one headphone type, and another ideal for another type.

Whatever the method used for inverting the headphone transfer function, the problem generally becomes easier to handle if the transfer function to be inverted represents a minimum-phase system. Generally, all-pass sections do exist in headphone transfer functions, but they are often ignored in the design of the equalization filter – see [69] for further discussion.

An example of measurements of a headphone transfer function and a designed filter is shown in Fig. 9.7. A target function for the filter was derived from five measurements. For each measurement, the subjects themselves arranged the headphone for most comfortable fit. It can be seen that the variations between measurements are much less than the variations between subjects, compare with Fig. 9.5. The low variation in the repeated measurements means that an individual headphone filter can be reliably designed. The high variation in transfer functions across subjects means that it probably should be designed individually, at least for critical purposes.

It can further be seen that a very deep dip exists in the left-ear headphone transfer function around 12–13 kHz which, when inverted for the equalization filter, gives a strong peak. Since it *is* slightly dependent on the exact positioning of the headphone, it should also be conservatively compensated for. This is done by reducing the amplitude of the corresponding high peak in the equalization filter, so that excessive amplification is avoided.

9.3.4 Pre-Equalization of Binaural Recordings

The possibility of applying a standard equalization for the use with binaural recordings is vital for the wider applicability of the methods. It has been suggested to pre-filter or pre-equalize binaural recordings with a weighting function that corresponds to the inverse of the design goal for standard stereo headphones, i. e. either a free-field or a diffuse-field equalization. This would, in the ideal case, lead to compatibility of the binaural recordings with existing standard-stereo headphones. Unfortunately, the variation across headphones and the general lack of success in reaching these design goals have compromised this idea. See [74] for a comparison of headphone transfer functions with traditional design goals, and [9, 102] for discussions of design philosophies.

9.4 Performance with Binaural Recordings

The listening experience obtained with reproduced binaural signals has been discussed recurrently, among other things, questioning the fundamentals principles – particularly regarding the issue of whether the frontal direction can be properly reproduced with human-like transfer functions. This was the background for a series of localization experiments with binaural recordings using real [77, 78] and artificial [70, 79] heads. The results are summarized in the following.

9.4.1 Localization Experiments

A series of listening experiments were carried out in which the localization with binaural recordings was compared to real-life localization. A set-up including 19 loudspeakers at various directions and distances was made in a standard listening room built according to IEC 268–13 [43] – see [59] and Fig. 9.8. The subjects were asked to indicate on a tablet the loudspeaker where they perceived the sound. In other words, although only loudspeaker positions were possible responses, the subjects were asked to report on the position of "what they heard", i. e. the *auditory event*, and not in any way encouraged to speculate on sound-source positions.

The subjects listened in consecutive experimental series to (a) the stimulus sound played back in the real-life set-up, (b) their own binaural recordings, (c) binaural recordings from other humans and (d) binaural recordings from artificial heads. Responses which did not correspond to the stimulus position, were subdivided into four main *error* categories as inspired by the nature of our hearing, namely

- Distance error ... when the direction was correct but stimulus and response differed in distance. There were four loudspeakers in the front

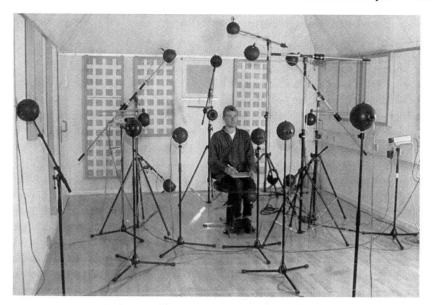

Fig. 9.8. Photo of set-up in standard listening room [70]

and three loudspeakers at 45° azimuth, right, that could be confused in distance.

– Median-plane errors ... when stimulus and response differed in direction but were both within the median-plane. There were ten source positions in the median plane.

– Within-cone errors ... confusions between source positions that were on "cones" with the same inter-aural time difference. There were two such source positions on the left side, 45° elevated up and down, and five on the right, 45° azimuth in the horizontal plane at three different distances, and 45° elevated up and down.

– Out-of-cone errors ... all other mistakes, indicating severe directional errors.

Directional errors were considered more *severe* than distance errors, thus, if a response represented both a distance *and* directional error, it was counted as a directional error only.

The results of these localization experiments do not depict spatial hearing in every detail or give a full story of human sound localization. They do, however, provide a fairly representative performance map for common and natural listening situations. Authentic reproduction would result in a performance map comparable to the performance map for real-life listening. This, indeed, held true for the experiments with the subjects' own recordings, whereas a statistically-significant degradation was seen for all experiments with reproduction of non-individual and artificial-head recordings.

9.4.2 Non-Individual Human Recordings

In [78] a panel of 20 listeners participated in localization experiments including binaural recordings from a panel of 30 human *recording heads*. There was considerable variation in the percentage of errors obtained with the different sets of non-individual recordings. The heads were subsequently "ranked" according to the number of median-plane errors, and the head which rendered least median-plane errors in total was "selected". This procedure will always as a result point to a "best" head, even if the low number of errors occurred by chance. An additional experiment was therefore carried out to verify that the recordings from the "selected" head consistently rendered the lower number of errors. The result of this experiment confirmed that it actually did so. This recording head thus incorporates the most common and/or salient features required for good sound localization. Consequently, it was denoted a "typical" human head.

Also others have studied the performance with non-individual binaural signals, see, e. g., [3,5,6,22,49,50,61,80,105–108,112], and other philosophies have been put forward for the selection of "typical" or "optimal" heads. One suggestion is that the heads of persons who localize well themselves, "good localizers", will also provide the best cues for other listeners. This idea can be assessed from Fig. 9.9. If the ears or head of a good localizer would also provide good cues for other listeners, then the points should follow a monotically increasing trend. Such pattern is not seen – thus this hypothesis is not supported by the current data.

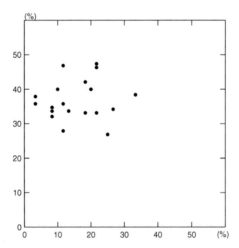

Fig. 9.9. Average median-plane errors of all listener of the panel, listening to the recording made from specific listeners, *ordinate*, plotted against the specific listeners' own performance in real life, *abcissa* [78]

9.4.3 Artificial Heads

The performance of artificial heads was first investigated in the advent of artificial-head technique, e. g., [16, 27, 28, 30, 31, 33, 61, 62, 86–88, 103, 114] – reviewed in [79]. All investigations pointed at imperfect performances, but the differences in experimental conditions, way of specifying and presenting results, computing statistics – only few have employed statistical comparisons at all, and, particularly, the very limited number of subjects in most of these investigations, disable further comparison.

A series of listening tests similar to the localization experiments reported in Sect. 9.4.1 and 9.4.2 have been carried out with recordings from artificial heads as well. The same group of listeners participated and the experimental design was identical in every other way. The results, median-plane errors only, of both human and artificial recording heads are shown in Fig. 9.10.

Figure 9.10 shows the average percentage of median-plane errors for a given recording head, human as well as artificial. The mentioned variation in error percentages for the human recording heads is clearly seen. The hu-

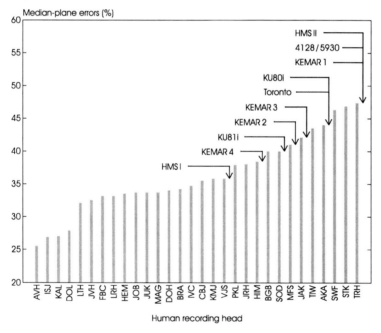

Fig. 9.10. Average median-plane error percentages for a panel of 20 listeners for the recording head indicated on the abscissa. Arrows indicate the performance with artificial head recordings: KEMAR X from Knowles Electronics with ear 'X', i. e. four types of ears included, KU80 and KU81i, HMS I and HMS II from Head Acoustics, 4128/5930 from Brüel and Kjær, types 4128 and 5930, respectively, and the TORONTO experimental head from Toronto University – adapted from [79]

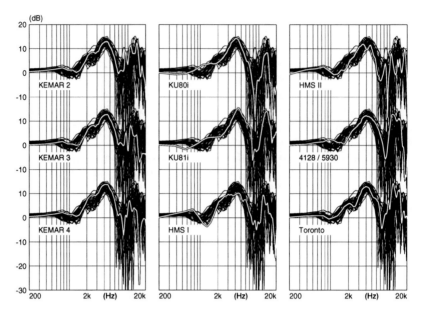

Fig. 9.11. Blocked-entrance HRTFs for frontal sound incidence for 40 humans, *thin lines* and manikins, *white line*, [36]

man recording head, AVH, was the one that gave the lowest median-plane error percentage, i. e. 26%. The performances for several artificial heads are indicated with arrows inserted in between the columns for humans. It is seen that even the better artificial heads compare only to the human recording heads in the poorer end of the scale, namely, 60% of the human recording heads are better. This is disappointing as artificial heads should in principle represent the "average" or "typical" listener.

The differences between human and artificial heads may be assessed objectively by comparison of their HRTFs. Frontal HRTFs for the artificial heads are given in Fig. 9.11. It can be seen that the human HRTFs are well grouped, though with higher variations the higher the frequency is, as can also be seen in Fig. 9.3. It can further be seen that the HRTFs of the artificial heads do generally not well represent the human HRTFs. The artificial-head HRTFs deviate in several cases considerable from the "average" or "typical" structure, and often do not even fall within the range of humans. This is disappointing, since the general design goal for the artificial heads is to replicate humans.

The deviations in Fig. 9.11 do, however, not directly predict the ranking shown in Fig. 9.10. The TORONTO head has a very human-like transfer function, but it has a relatively high percentage of median-plane errors even so. The HMS I, on the contrary, has a fairly non-human-like transfer function, but has a lower percentage of median-plane errors, at least when compared to the other artificial heads. If more directions are inspected – not shown here, it is seen that whereas the front direction may reveal certain types of shortcomings of the heads, other directions reveal other types of shortcomings that result in other types of localization errors [70, 79]. Generally, both the results from the listening experiments and the analysis of the transfer functions indicate that the differences between artificial heads and humans are larger than differences between humans.

Successive experiments were carried out including more artificial heads, among others an in-house experimental recording head, VALDEMAR. These experiments were carried out over several years with a varying group of listeners, thus requiring more caution when comparing across investigations. It was shown that, generally, the various experimental series gave results of good validity, and that artificial heads provide the better performances the more human-like they are [70].

9.5 Binaural Synthesis

A human recording head, or alternatively an artificial head, serves to perform the alteration of the sound field that the listener would have caused in the recording situation. When placed in a free sound field, the listener will obstruct an incoming sound wave, and ears, head and body will cause a linear filtering of the sound signal. This filtering is completely and uniquely described by the HRTFs.

Figure 9.12 shows the HRTF in the time and frequency domains for a single subject with sound coming from the side. It is obvious in both domains that the sound reaching the ear closest to the sound source, left ear, is stronger than the sound reaching the ear opposite to the sound source, right ear. In the frequency-domain representation it can also be seen that the sound transmission to the left ear exceeds 0 dB, which is mainly due to pressure

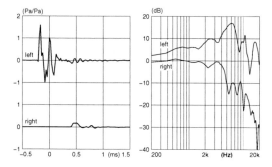

Fig. 9.12. Head-related transfer function, HRTF, in the time domain, **left**, and in the frequency domain, **right** – measured at the blocked entrance with sound from the left [76] – data from subject LRH

build-up and diffraction around ear, head and body. The frequency response for the right-ear sound transmission resembles the characteristics of a low-pass filter – due to the shadowing effect of the head. In the time domain it can be seen that the sound at the closest ear arrives some hundred microseconds earlier than the sound at the opposite ear.

It is the direction-dependent characteristics in the sound signals reaching our two ears which enable us to localize the sound source by our hearing. A very strong localization cue is the arrival-time difference between left and right ear signals, which *positions* the auditory event laterally on a given spatial cone of equal inter-aural time differences, also when other cues are ambiguous – thus the term "cone of confusion".

In the general definition of the HRTF all linear properties of the sound transmission are included. All proposed descriptors of localization cues, such as inter-aural differences in arrival-time, ITD, in phase, IPD, in level/intensity, ILD/IID as well as monaural cues, are contained in the HRTFs. They can thus be derived from the HRTFs, whereas the opposite is not generally the case.

9.5.1 Measurement

The ideal HRTF describes the transformation from a freely-propagating plane wave to sound pressures that occur when a listener *enters* the sound field and, thus, obstructs wave propagation. In the laboratory, plane waves exist only in approximation. When a source is set up for measurements in the anechoic chamber, it will be within a limited distance of the subject, and the wave that hits the subject will thus not be exactly plane, but marginally spherical.

The an-echoic chamber also presents an approximation in itself, since the absorption at the boundaries will fail to work at low frequencies, and possibly – depending on design and material – also for higher frequencies. Platforms, chairs, etc., that are needed for sources and subjects, also cause unintended reflections and, thus, are a source of measurement errors. Hence, it is preferable to minimize the number of loudspeakers present in the room, and to reduce the amount of equipment to a minimum needed carrying, supporting and/or fixating the subjects. One solution is to use an arc with one or more loudspeakers, and to move the arc [57, 58] or the listener [76]. An example is shown in Fig. 9.13.

In recent years, a number of measurement techniques have been developed that return the transfer function in the time-domain representation, i. e. for the present application, as head-related impulse responses, HRIRs. A particularly successive technique is based on a pseudo-random noise stimulus, known as a maximum-length sequence, e. g., [18, 90, 97, 116]. This technique enables a time gating of the measured impulse where later parts of the impulse can be rejected. In this way it is, e. g., possible to "gate out" an unavoidable reflection from an imperfectly absorbing wall, or from the platform carrying the subject.

Fig. 9.13. Set-up for measurement of full-sphere HRTFs. Eight loudspeakers are placed in an arc, and the subject is standing on a platform supported with a back-rest. The platform can be turned in discrete steps to render source directions for the whole sphere. The monitor shows the subject from above and enables the subject to center head position accurately [76]. For photo-technical reasons the monitor is closer to the subject than during measurements

Raw measurement of P_{ear} and P_{ref} will include the transfer functions of all elements of the transmission chain, e. g., power amplifier, loudspeaker, microphone, microphone amplifier, anti-aliasing filters. Thus, a prerequisite for cutting away unwanted reflections from, e. g., supporting platforms, is that the head-related impulse response convolved with the impulse responses of amplifier(s) and loudspeaker does not overlap in time with the unwanted reflections. This can lead to a peculiar contradiction of compromises, since, e. g., a good loudspeaker response at low frequencies is desirable for a good signal-to-noise ratio, SNR, but a such loudspeaker will often have a long impulse response, and will thus limit the possibility of gating out reflections. Also, band-limiting filters with steep slopes of their transfer functions can result in long impulse responses.

HRTFs are computed by a complex division in the frequency domain – see (9.4). For the highest frequencies, i. e. above the cut-off frequency of the anti-aliasing filters and other band-limiting elements, the result will be based on one set of invalid data divided by another set of invalid data. This results in a time-aliased impulse response with random high-frequency, i. e. "ringing" components which must be removed by appropriate low-pass filtering. For the lowest frequencies it is tempting to do the same, but as will be demonstrated in the following, it is important to control the low-frequency characteristics in a more direct way.

9.5.2 Low-Frequency Control

The two measurements used for the calculation of an HRTF are always carried out with equipment that has a lower limiting frequency. Thus, even if the measurements return values at DC, these values are not the results of true measurements, but they simply reflect the off-set-voltage properties of the amplifiers and A/D converters involved. As a consequence, the complex division carried out to obtain the HRTF gives a wrong and more or less random value at DC. One might think that this is not important as long as the signals do not have DC components. However, an incorrect DC value has consequences for the processing of signals in a wide frequency range as it will be seen in the following.

Fortunately, it is easy to correct for the DC value. At low frequencies, the presence of a person does not alter a propagating sound wave and an HRTF will thus asymptotically approach unity gain, i. e. 0 dB, when the frequency decreases toward DC. As a consequence, the calculated – and wrong – DC value of the HRTF can simply be replaced by unity in the frequency domain, just before taking the inverse *Fourier* transform to obtain the HRIR. Alternatively, the individual taps in the HRIR filter can be adjusted accordingly.

Figure 9.14 shows the consequence of slightly incorrect DC values. In the time-domain representation the errors certainly look modest, but in the frequency-domain representation it is seen that the magnitude is severely affected not only at DC, but also at frequencies well within the range of hearing.

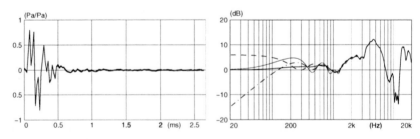

Fig. 9.14. Blocked-entrance HRTF in the time domain, 128 points, **left**, and in the frequency domain, **right**. Transfer function for a DC value corrected to unity, *medium solid line*, for a slightly-too-high DC value, *dashed line*, for a DC value of zero, *dash-dotted line*, and for a slightly-too-low DC value, *thin solid line* [72]

At the frequencies, which equal an integer times the frequency resolution of the original HRIR, in this case 375 Hz, all curves coincide. The concerning observation is the behaviour of the amplitude response between these points. These bumps are several dB in magnitude, and can be clearly audible as miscolouration of the lowest frequencies. If, by misfortune, the DC-value is slightly too low in the HRIR for the one ear and slightly too high in the HRIR for the opposite ear, then a large difference can exist. This does not normally occur in real life, and can give a quite unpleasant feeling of sub- or super-pressure in the ears. Hence, in the authors' experience, it is generally required to correct the DC-value to ensure a good sound quality.

9.5.3 Filter Length and Order

It appears from visual inspection that the length of the HRIR itself is on the order of 1 ms or slightly more, e. g., Fig. 9.12 and 9.14. The HRIR can be implemented directly as an FIR filter with an order corresponding to the number of taps. The possibility of reducing the order for minimizing computational resources has repeatedly been investigated, e. g., [3, 4, 14, 15, 38, 41, 46, 64, 93, 94] – for a review see [42].

Figure 9.15 shows the probability of detection in a three-alternative forced choice listening experiment, where finite-impulse-response filters, FIR filters, with reduced order were compared with the original HRIR, i. e. 256-point FIR. The sampling frequency was 48 kHz, so 48 taps correspond to a filter duration of 1 ms. From Fig. 9.15 it can be seen that for high filter orders, i. e. down to 72 taps, the difference is undetectable. For shorter filters the differences are heard, though still with a low detection at 48 taps and even at 24 taps. Thus, depending on the application at hand, filters of these or-

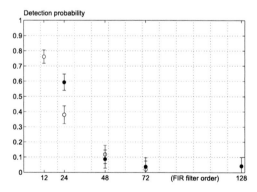

Fig. 9.15. Detection probability of reduced-order FIR implementations, pooled data for 15 subjects and 17 directions covering the whole sphere. Individual synthesis and individual headphone calibration was employed with speech, *open circles*, and noise signals, *solid circles* [94]

ders may in many cases work satisfactorily. In general, with current signal processing technology, FIR filters of 72 taps are considered relatively short.

The results match well with what one would conclude from visual inspection of HRIRs and from considering the physical dimensions of humans. Nevertheless, it is not an uncommon understanding that the length needs to be considerably longer, and that the quality of the filter is better the more taps are included. One reason for disagreements is the possible lack of low-frequency control. The longer the filter is, the lower is the frequency at which the "bumps" occur. This means that the errors can be indirectly shifted to such low frequencies that they become inaudible. An easily overlooked fact is that similar errors are likely to occur when the transfer functions are measured using frequency-domain methods or when the filtering is carried out in the frequency domain.

The possible advantages of using alternative, possibly computationally more efficient filter representations, e. g., infinite-impulse-response filters, IIR filters, has also been investigated, e. g., [3,13,15,38,41,44,51,52,64,91,93,94]. None of the results points to a convincingly superior computational performance. This may – at least partly – be due to the fact that IIR filters are often superior for describing low frequency spectral details, which are virtually non-existent in HRTFs.

It is a pre-requisite for many IIR-filter-design methods that the system has minimum-phase properties. This is, however, not the case for HRTFs, e. g., [76]. All HRTFs include a linear-phase component, i. e. pure delay, which is vital for maintaining the correct inter-aural time difference. HRTFs do also typically have one or more all-pass sections. Yet, it was demonstrated [69] that, without any audibly consequences, the all-pass sections may be replaced by an additional pure delay – determined as the low-frequency phase delay or group delay of the all-pass sections. Though the use of IIR-filter design may not be computationally favourable, the possibility of using minimum-phase approximations has many other practical advantages.

9.5.4 Performance of Binaural Synthesis

There have been numerous investigations testing or using binaural synthesis, e. g., [3,5,6,49,50,80,105–108,112]. Most have been made in an-echoic chambers with the subjects in fixed positions, thus simulating a single sound wave with a given direction, i. e. static free-field simulation. There may be several reasons for this. Technically, all that is required for a free-field simulation is the HRTFs, and the appropriate headphone equalizations. Second, if a simulation of more realistic environment should be made, a sound-transmission-analysis program is required. Such a program will also have its approximations and imperfections, which will influence and possibly deteriorate the overall result. Third, the first-arriving sound wave has a leading significance, when it comes to directional hearing, the so called "precedence effect", see e. g., [9] for a review. Thus, the static free-field simulation is in a sense a relevant and critical test of the success in rendering authentic *directional* perception.

A general conclusion from these investigations is that, when properly implemented, the binaural synthesis is successful in giving a near-real-life performance. The results have contradicted some of the myths associated with binaural synthesis, e. g., that the technique cannot render frontal localization or inevitably leads to inside-the-head locatedness. In the studies where these effects have been observed, often non-individual synthesis has been used, or the studies have had such diverting objectives that details of the synthesis technique have been given very little attention.

The binaural synthesis can, however, be considerably improved by making use of a single very important difference between recording and synthesis of binaural signals, namely, the possibility in the synthesis to change the HRTFs interactively when the direction of the sound incidence changes due to listeners' head movements. It is well known that in real life head turns are often used unconsciously, e. g., to resolve front/back confusions, e. g., [8].

With binaural recording, the listener has to sit still to have an authentic listening experience, since such binaural signals cannot respond to later dynamical changes, such as the listener's head movements. This is not an impediment of principle nature for the synthesis technique. The position and orientation of the listener can be tracked by so called head trackers, e. g., magnetic devices, and the synthesis can be made responsive to these changes by continuously updating signal processing.

The importance of correct updating of binaural synthesis in response to head-movements has been understood for years. More recently, a number of investigations, e. g., [7, 20, 92, 113] have demonstrated the proficiency of dynamic synthesis. Localization errors such as front/back confusion are almost non-existent, and cone-of-confusion errors are well resolved in general, also for non-individual recordings.

The very different properties of the listening experience in dynamic simulation calls for considerably revised experimental schemes for scientific assessments of the methods. Localization experiments are still fundamental in such assessments, but the continuously-changing signal processing can lead to artifacts, e. g., "clicks", that degrade the quality of the sound, without severely impeding localization.

9.6 Applications

In the following, the principles of some *classical* application areas of binaural technique, mostly synthesis, are introduced – see also [10,19,84], this volume.

Room Simulation

Room-simulation systems perform numerical analyses of the acoustical environment and, with the use of binaural synthesis, the analysis results may be rendered audible. Room-simulation systems are usually developed for assistance in architectural design and acoustical adjustment of concert halls, auditoria, theaters etc. Re-arrangements of walls, orchestras or audience can be listened to before alterations are made in the real world. Listening conditions at critical places in the concert hall and the speech intelligibility of the auditorium can be checked by actual listening. The simulation systems cannot only be used to suggest physical changes of the room, but also, by applying superposition, multiple sound sources can be simulated in the computer model and the system can be used for adjusting any multi-channel public address or electro-acoustical enhancement system prior to the costly set-up. The room-simulation system can also be used in various contexts where standardized listening conditions are needed. This could be used for instance for the evaluation of different loudspeakers. This way the loudspeaker producer can avoid building a rather costly listening environment. It can also be used in mobile monitoring environments for studio recording and mixing that lack space for a loudspeaker set-up. The potential of room-simulation systems has been, and is, a driving force in the research and development of the binaural-synthesis techniques. In later years the term binaural "auralization" has been used in general for rendering such numerical predictions audible – see, e. g., [2, 26, 53, 56, 60, 63, 81, 104, 115].

Binaural Mixing Consoles

A common way of making recordings in studios is to record each instrument or voice in its own channel, and then afterwards mix these down to the intended stereo perspective, using time and intensity differences. With a binaural mixing console such multi-channel recordings are transferred into binaural signals in the mixing phase. The binaural mixing console can thus be regarded an *electronic artificial head*. If an artificial head is used for the recording, then the physical set-up in the recording studio defines the position of the sound source for the play-back situation. These recordings, as well as ordinary recordings processed by means of the binaural mixing console, can be mixed together to binaural signals. Yet, the success of using binaural recordings in audio has been limited, possibly due to the lack of standardized methods for the processing procedures – for examples, see, e. g., [32, 34, 45, 48, 95].

Surround Sound by Headphones

The use of multi-channel sound-reproduction systems is increasing, e. g., in HDTV and DVD formats. The compatibility with headphone play-back is not trivial. Using binaural synthesis, the five channels can be transformed into a two-channel signal for headphone play-back. The idea is not at all new, e. g., [11], and has also been proposed for headphone play-back of traditional stereo recordings, where the same compatibility problem exists. Today, two approaches are seen. One that strives to simulate the optimal surroundings, and one that strives to simulate the end-user multi-channel set-up and environment, whatever the properties and qualities of this – for examples, see, e. g., [1, 65, 83, 89, 98].

Communication Systems

Binaural technique is often proposed for communication systems, e. g., [24, 25, 47, 109], for situations where the listener must pay attention to more than one communication channel at a time. The idea is to spatially arrange the communication channels around the listeners and, in this way, enable them to distinguish between the different channels as in daily life. The advantage of spatially arranged channels originates from the fact that our hearing can focus on one out of several sound sources in noisy environments with the sounds being heard at different directions, the so-called "cocktail-party effect". Implementations of this kind are, e. g., seen within military research institutions to be used between pilots and air control towers, e. g., [107]. More general kinds of tele-conferencing systems also exist.

3-D Auditory Displays

The idea of 3-D auditory displays is to create spatially arranged auditory icons for situations where it is either natural to have audible information passed or where, for instance, the visual information channel is not available or is fully occupied, e. g., [17, 21, 82, 85]. The sound of a fuel tank drying out could carry an important message for a pilot, and the sound of a printer could be a way to transfer the graphical interfaces of modern computers into something blinds can use. Auditory icons – sometimes called "earcons" – can also be used for transfer of information which has no origin in the auditive world. Examples are radar or ultrasonic information, where the distance information is passed and used, for instance, as a navigation tool for blind persons, or the association of sounds to stock exchange information, e. g., high- and low-pitched sounds for rates going up and down – e. g., [23]. Obviously, earcons serve a purpose even when they are not arranged spatially, yet, but 3-D auditory displays offer an environment where the user may effectively monitor more activities at the same time.

Virtual Reality

In Virtual-Reality applications, e. g., [12, 29, 84, 96] the ultimate objective is immersive simulation of a non-existing environment. The distinction between VR systems and other sound simulation systems can sometimes be difficult to see. VR systems aim at providing stimuli to a persons' senses, which are perceptually plausible to such an extent that these persons develop a persistent experience of actually being somewhere else – so called sense of "presence". The perceptual world that they are experiencing this way may only exist as a computer model which the computer uses to control various kinds of actuators that stimulate multiple modalities – besides the auditory one, one or more of the following, i. e. visual, tactile, olfactory, gustatory and proprioceptive ones.

Acknowledgement

This chapter reviews a series of studies carried out at the Department of Acoustics of Aalborg University, where all members of the staff have contributed to an inspiring atmosphere. In particular, *S. B. Nielsen, P. Ch. Frandsen, M. F. Sørensen, C. B. Larsen* – former *Jensen, J. Sandvad, F. Christensen, S. K. Olesen, P. Minnaar*, and *J. Plogsties* are thanked for their collaboration in different phases of the work. Various sources of financial support are also acknowledged, particularly the Danish Technical Research Council, the European Commission, the DANMON Research Fund, the *C. W. Obel*'s Fund, the National Agency of Trade and Industry, and the national Center for IT Research.

References

1. Aarts R (2003) Applications of DSP for Sound Reproduction Improvement. 23rd Audio Engr Soc Int Conf, Copenhagen, Denmark, paper 4
2. Ahnert W, Feistel R (1993) EARS auralization software. J Audio Engr Soc 41:894–904
3. Asano F, Suzuki Y, Sone T (1990) Role of spectral cues in median plane localization. J Acoust Soc Amer 88:159–168
4. Begault D R (1991) Challenges to the successful implementation of 3-D sound. J Audio Engr Soc 39:864–870
5. Begault D R (1992) Perceptual effects of synthetic reverberation on three-dimensional audio systems. J Audio Engr Soc 44:895–904
6. Begault D R, Wenzel E M (1991) Headphone localization of speech stimuli. Proc Human Factors Soc, 35th Ann Meet, Santa Monica, CA 82–86
7. Begault D R, Wenzel E M (2001) Direct comparison of the impact of head tracking, reverberation, and individualized head-related transfer functions on the spatial perception of a virtual speech source. J Audio Engr Soc 49:904–916
8. Blauert J (1997a) An introduction to binaural technology. In: Gilkey R H, Anderson T R (eds) Binaural and spatial hearing in real and auditory environments 593–609, Lawrence Erlbaum, Mahwah NJ
9. Blauert J (1997b) Spatial hearing: the psychophysics of human sound localization. 2nd rev edn. MIT Press, Cambridge MA
10. Blauert J (2005) Analysis and synthesis of auditory scenes. Chap 1 this vol
11. Blauert J, Laws P (1973) True simulation of loudspeaker sound reproduction while using headphones. Acustica 29:273–277
12. Blauert J, Lehnert H, Sahrhage J, Strauss H (2000) An interactive virtual-environment generator for psychoacoustic research. I: Architecture and implementation. ACUSTICA/acta acustica 86:94–102

13. Blommer M A (1996) Pole-zero modeling and principical component analysis of head-related transfer functions. Doct diss. University of Michigan, Ann Arbor, Michigan
14. Blommer M A, Wakefield G H (1994) On the design of pole-zero approximations using logarithmic error measure. IEEE Trans Sig Processg 42:3245–3248
15. Blommer M A, Wakefield G H (1997) Pole-zero approximations for head-related transfer functions using logarithmic error criterion. IEEE Trans Speech Audio Processg 5:278–287
16. Boerger G, Laws P, Blauert J (1977) Stereophone Kopfhörerwidergabe mit Steuerung bestimmter Übertragungsfaktoren durch Kopfdrehbewegungen (Stereophonic reproduction by earphones with control of special transfer functions through head movements). Acustica 39:22–26
17. Bolia R S, D'Angelo W R, McKinley R L (1999) Aurally aided visual search in three-dimensional space. Hum Fact 41:664–669
18. Borish J, Angell J B (1983) An efficient algorithm for measuring the impulse response using pseudorandom noise. J Audio Engr Soc 31:478–488
19. Braasch J (2005) Modelling of binaural hearing. Chap 4 this vol
20. Bronkhorst A M (1995) Localization of real and virtual sound sources. J Acoust Soc Amer 98:2542–2553
21. Bronkhorst A W, Veltman J A H, van Breda L (1996) Application of a three-dimensional auditory display in a flight task. Hum Fact 38:23–33
22. Butler R A, Belendiuk K (1977) Spectral cues utilized in the localization of sound on the median sagittal plane. J Acoust Soc Amer 61:1264–1269
23. Carlile S, Cohen M (Eds) (2002) Proceedings of the 8th International Conference on Auditory Display. Advanced Telecomm Res Inst (ATR), Kyoto, Japan (www.icad.org)
24. Cohen M, Koizumi N (1992) Exocentric control of audio imaging in binaural telecommunication. IEICE Trans Fund Electr E75A(2):164–170
25. Cohen M (2003) The internet chair. Int J Hum-Comput Interact 15:297–311
26. Dalenbäck B-I, Kleiner M, Svensson P (1993) Audibility of changes in geometric shape, source directivity, and absorptive treatment – experiments in auralization. J Audio Engr Soc 41:905–913
27. Damaske P, Wagener B (1969) Richtungshörenversuche über einen nachgebildeten Kopf (Directional hearing tests by aid of a dummy head). Acustica 21:30–35
28. Damaske P, Mellert V (1969/70) Ein Verfahren zur richtungstreuen Schallabbildung des oberen Halbraumes über zwei Lautsprecher (Sound reproduction of the upper semi-space with directional fidelity using two loudspeakers). Acustica 22:153–162
29. Djelani T, Pörschmann C, Sahrhage J, Blauert J (2000) An interactive virtual-environment generator for psychoacoustic research. II: Collection of head-related impulse responses and evaluation of auditory localization. ACUSTICA/acta acustica 86:1046–1053
30. Genuit K (1984) Ein Modell zur Beschreibung von Aussenohrübertragungseigenschaften (A model for description of outer-ear transfer functions). Doct diss. Techn Hochsch Aachen, Aachen
31. Genuit K (1988) Simulation des Freifeldes über Kopfhörer zur Untersuchung des Räumliche Hörens und der Sprachverständlichkeit (Simulation of the free field using headphones for investigation of the spatial hearing and speech intelligibility). Audiol Akust 6:202-221

32. Genuit K, Gierlich H-W, Künzli U (1992) Improved Possibilities of Binaural Recording and Playback Techniques. 92nd Audio Engr Soc Conv, Amsterdam, The Netherlands, preprint 3332. Abstr in: J Audio Engr Soc 40:444

33. Genuit K, Platte H-J (1981) Untersuchungen zur Realisation einer richtungstreuen Übertragung mit elektroakustischen Mitteln. Fortschr Akust, DAGA'81, Berlin, Germany 629–632

34. Gierlich, H-W, Genuit K (1987) Processing Artificial-Head Recordings. 82nd Audio Engr Soc Conv, London, UK, preprint 2460. Abstr in: J Audio Engr Soc 35:389–390

35. Gilkey R H, Anderson T R (1997) Binaural and spatial hearing in real and virtual environments. Lawrence Erlbaum, Mahwah N J

36. Hammershøi D, Møller H (1992) Artificial heads for free field recording: How well do they simulate real heads? Proc. 14th Int Congr Acoust, ICA, Beijing, paper H6-7

37. Hammershøi D, Møller H (1996) Sound transmission to and within the human ear canal. J Acoust Soc Amer 100:408–427

38. Hartung K, Raab A (1996) Efficient modelling of head-related transfer functions. Acta Acustica 82:suppl 1:88

39. Hellström P A, Axelsson A (1993) Miniature microphone probe tube measurements in the external auditory canal. J Acoust Soc Amer 93:907–919

40. Hirahara T (2004) Physical characteristics of headphones used in psychophysical experiments. Acoust Sci & Tech 4:276–285

41. Huopaniemi J, Karjalainen M (1997) Review of digital filter design and implementation methods for 3-D sound. 102nd Conv Audio Engr Soc, Munich, Germany, preprint 4461. Abstr in: J Audio Engr Soc 45:413

42. Huopaniemi J, Zacharov N, Karjalainen M (1999) Objective and subjective evaluation of head-related transfer function filter design. J Audio Engr Soc 47:218–240

43. IEC/TR 60268-13 (1981) Listening tests on loudspeakers. Publication 268-13, Int Electrotechn Comm, IEC, Geneva

44. Jenison R L (1995) A spherical basis function neural network for pole-zero modeling of head-related transfer functions. Proc IEEE worksh appl sig processg audio and acoust, New Paltz NY, IEEE catalogue number 95TH8144

45. Jot J-M, Wardle S, Larcher V (1998) Approaches to Binaural Synthesis. 105th Audio Engr Soc Conv, San Francisco, CA, preprint 4861. Abstr in: J Audio Engr Soc 46:1038

46. Jot J M, Warusfer O, Larcher V (1995) Digital signal processing issues in the context of binaural and transaural stereophony. 98th Conv Audio Engr Soc, Paris, France, preprint 3860. Abstr in: J Audio Engr Soc 43:396

47. Kang S H, Kim S H (1996) Realistic audio teleconferencing using binaural and auralization techniques. ETRI J 18:41–51

48. Karamustafaoglu A, Spikofski G (2001) Binaural Room Scanning and Binaural Room Modeling. 19th Audio Engr Soc Int Conf, Schloss Elmau, Germany, paper 1880

49. Kawaura J, Suzuki Y, Asano F, Sone T (1989) Sound localization in headphone reproduction by simulating transfer functions from the sound source to the external ear (in Japanese). J Acoust Soc Jpn (J) 45:756–766

50. Kawaura J, Suzuki Y, Asano F, Sone T (1991) Sound localization in headphone reproduction by simulating transfer functions from the sound source to the external ear. J Acoust Soc Jpn (E) 12(5):203–216

51. Kendall G S, Martens W L (1984) Simulating the cues of spatial hearing in natural environments. Proc Int Comp Music Conf, Paris, France, 111–125

52. Kendall G S, Rodgers A P (1982) The simulation of three-dimensional localization cues for headphone listening. Proc Int Comp Music Conf 225–243

53. Kleiner M, Dalenbäck B-I, Svensson P (1993) Auralization – an overview. J Audio Engr Soc 41:861–875

54. Kulkarni A, Colburn H S (1995) Efficient finite-impulse-response filter models of head-related transfer function. J Acoust Soc Amer 97:3278

55. Kulkarni A, Colburn S (2000) Variability in the characterization of the headphone transfer function. J Acoust Soc Amer 107:1071–1074

56. Kuttruff K H (1993) Auralization of impulse responses modelled on the basis of ray-tracing results. J Audio Engr Soc 41:876–880

57. Langendijk E H A, Bronkhorst A W (2000) Fidelity of three- dimensional-sound reproduction using a virtual auditory display. J Acoust Soc Amer 107:528–537

58. Langendijk E H A, Bronkhorst A W (2002) Contribution of spectral cues to human sound localization. J Acoust Soc Amer 112:1583–1596

59. Langvad B, Møller H, Budzynski G (1989) Testing a new listening room. Arch Acoust 14:45–60

60. Larsson P, Vastfjall D, Kleiner M (2002) Better Presence and Performance in Virtual Environments by Improved Binaural Sound Rendering. 22nd Audio Engr Soc Int Conf, Espoo Finland, paper 228

61. Laws P. Platte H-J (1975) Spezielle Experimente zur kopfbezogenen Stereophonie (Special experiments regarding head-related stereophony). Fortschr Akust, DAGA'75, Braunschweig, Germany 365–368

62. Laws P, Platte H-J (1978) Ein spezielles Konzept zur Realiserung eines Kunstkopfes für die kopfbezogenes stereofone Aufnahmetechnik (The design of a special artificial head for head-related stereophony). Rundfunktechn Mitt 22:28–31

63. Lehnert H, Blauert J (1993) Principles of binaural room simulation. Appl Acoust 36:259–291

64. Mackenzie J, Huopaniemi J, Välimäki V, Kale I (1997) Low order modelling of head-related transfer functions using balanced model truncation. IEEE Sig Processg Let 4:39–41

65. McKeeg A, McGrath D S (1997) Using Auralization Techniques to Render 5.1 Surround to Binaural and Transaural Playback. 102nd Audio Engr Soc Conv, Munich Germany, preprint 4458. Abstr in: J Audio Engr Soc 54:412

66. Mehrgardt S (1975) Die Übertragungsfunktion des menschlichen Aussenohres: Richtungsäbhangigkeit und genauere Bestimmung durch komplexe Strukturmittelung (The transfer function of the human external ear: dependence on direction and accurate determination by means of complex structural averaging). Fortschr Akust, DAGA'75, Braunschweig, Germany 357–361

67. Mehrgardt S, Mellert V (1977) Transformation characteristics of the external human ear. J Acoust Soc Amer 61:1567–1576

68. Middlebrooks J C, Makous J C, Green D M (1989) Directional sensitivity of sound-pressure levels in the human ear canal. J Acoust Soc Amer 86:89–108

69. Minnaar P, Plogsties J, Olesen S K, Christensen F, Møller H (200x) The audibility of all-pass components in binaural synthesis and reproduction. In preparation

70. Minnaar P, Olesen S K, Christensen F, Møller H (2000) Localization with binaural recordings from artificial and human heads. J Audio Engr Soc 49:323–336

71. Møller H (1992) Fundamentals of binaural technology. Appl Acoust 36:171–218

72. Møller H, Hammershøi D, Jensen C B, Sørensen M F (1995) Method of generating binaural signals with filter sets – uses head-related transfer functions with short time domain descriptions and low differences between individuals. Patent, PCT/DK 95/00089, publication No. WO 95/23493

73. Møller H, Hammershøi D, Jensen C B, Sørensen M F (1995) Transfer characteristics of headphones measured on human ears. J Audio Engr Soc 43:203–217

74. Møller H, Jensen C B, Hammershøi D, Sørensen M F (1995) Design criteria for headphones. J Audio Engr Soc 43:218–232

75. Møller H, Minnaar P, Plogsties J, Olesen S K, Christensen F (200x) The audibility of all-pass sections in electro-acoustical transfer functions. In preparation

76. Møller H, Sørensen M F, Hammershøi D, Jensen C B (1995) Head-related transfer functions measured on human subjects. J Audio Engr Soc 43:300–321

77. Møller H, Sørensen M F, Jensen C B, Hammershøi D (1996) Binaural technique: Do we need individual recordings? J Audio Engr Soc 44:451–469

78. Møller H, Jensen C B, Hammershøi D, Sørensen M F (1996) Using a typical human subject for binaural recording. 100th Conv Audio Engr Soc, Copenhagen, Denmark, preprint 4157. Abstr in: J Audio Engr Soc 44:632

79. Møller H, Hammershøi D, Jensen C B, Sørensen M F (1999) Evaluation of artificial heads in listening tests. J Audio Engr Soc 47:83–100

80. Morimoto M, Ando Y (1980) On the simulation of sound localization. J Acoust Soc Jpn (E) 1:167–174

81. Naylor G (1993) Computer modeling and auralization of sound fields in rooms. Appl Acoust 38:89–92

82. Nelson W T, Hettinger L J, Cunningham J A, Brickman B J, Haas M W, McKinley R L (1998) Effects of localized auditory information on visual target detection performance using a helmet-mounted display. Hum Fact 40:452–460

83. Nielsen S (2001) Multichannel Signal Processing Tools - differences to multiple single channel processing. 19th Audio Engr Soc Int Conf, Schloss Elmau, Germany, paper 1901

84. Novo P (2005) Auditory virtual environments. Chap 11 this vol

85. Perrott D R, Cisneros J, McKinley R L, D'Angelo W R (1996) Aurally aided visual search under virtual and free-field listening conditions. Hum Fact 38:702–715

86. Platte H J (1977) Zur Bedeutung der Aussenohrübertragungseigenschaften für den Nachrichtenempfänger "menschliches Gehör" (On the relevance of the transmissions characteristics for the information receiver "human hearing"). Doc diss. Techn Hochsch Aachen, Aachen

87. Platte H-J, Laws P (1976) Die Vorneortung bei der kopfbezogenen Stereophonie (Frontal localization in head-related stereophony). Radio Mentor Electron 42:97–10

88. Poulsen T (1978) Hörvergleich unterschiedlicher Kunstkopfsysteme (Auditory comparison of different dummy-head systems). Rundfunktechn Mitt 22:211–214

89. Richter G (1992) BAP Binaural Audio Processor. 92nd Audio Engr Soc Conv, Amsterdam, The Netherlands, preprint 3323. Abstr in: J Audio Engr Soc 40:444

90. Rife D D, Vanderkooy J (1989) Transfer-function measurements with maximum-length sequences. J Audio Engr Soc 37:419–444

91. Ryan C, Furlong D (1995) Effects of headphone placement on headphone equalization for binaural reproduction. 98nd Conv Audio Engr Soc, Paris, France, preprint 4009. Abstr in: J Audio Engr Soc 43:401

92. Sandvad J (1995) Dynamic aspects of auditory virtual environments. 100th Audio Engr Soc Conv, Copenhagen, Denmark, preprint 4226. Abstr in: J Audio Engr Soc 44:644

93. Sandvad J, Hammershøi D (1994) Binaural auralization: comparison of FIR and IIR filter representation of HIRs. 96th Audio Engr Soc Conv, Amsterdam, The Netherlands, preprint 3862. Abstr in: J Audio Engr Soc 42:395

94. Sandvad J, Hammershøi D (1994) What is the most efficient way of representing HTF filters? Proc. NORSIG'94, Ålesund, Norway, 174–178

95. Sapp M, Kleber J (2000) Universal Signal Processing System. ACUSTICA/Acta Acustica 86:185–188

96. Savioja L, Huopaniemi J, Lokki T, Vaananen R (1999) Creating interactive virtual acoustic environments. J Audio Engr Soc 47:675–705

97. Schmitz A, Vorländer M (1990) Messung von Aussenohrstossantworten mit Maximalfolgen-Hadamard-Transformation und deren Anwendung bei Inversionsversuchen (Measurement of external-ear impulse responses with maximum-length-sequence-Hadamard transformation and their application in inversion experiments). Acustica 71:257–268

98. Schobben D, Aarts R (2002) Three-dimensional Headphone Sound Reproduction Based on Active Noise Cancellation. 113th Audio Engr Soc Conv, Los Angeles, CA, preprint 5713. Abstr in: J Audio Engr Soc 50:977

99. Shaw E A G (1971) Acoustic response of external ear with progressive wave source. J Acoust Soc Amer 51:S150 and personal communication

100. Shaw E A G (1973) Acoustic response of external ear replica at various angles of incidence. 86th Ann Meet Acoust Soc Amer, and personal communication

101. Shaw E A G (1982) External ear response and sound localization. In Gatehouse W (ed): Localization of sound: theory and applications 30–41, Amphora, Groton CN

102. Theile G (1986) On the standardization of the frequency response of high-quality studio headphones. J Audio Engr Soc 34:956–969

103. Theile G, Spikofski G (1984) Vergleich zweier Kunstkopfsysteme unter berücksichtung verschiedener Anwendungsbereiche (Comparisons of two dummy head systems with due regard to different fields of applications). Fortschr Akust DAGA'84, Darmstadt, Germany, 223–226

104. Torres R R, Kleiner M, Dalenbäck B-I (2002) Audibility of "diffusion" in room acoustics auralization: An initial investigation. ACUSTICA/acta acustica 86:919–927

105. Wenzel E M (1992) Localization in virtual acoustic displays. Presence 1:80–107

106. Wenzel E M, Arruda M, Kistler D J, Wightman F L (1993) Localization using nonindividualized head-related transfer functions. J Acoust Soc Amer 94:111–123

107. Wenzel E M, Wightman F L, Foster S H (1988) A virtual display for conveying three-dimensional acoustic information. Proc Human Factors Soc 32nd Ann Meet, Anaheim, CA, 86–90

108. Wenzel E M, Wightman F L, Kistler D J (1991) Localization with non-individualized acoustic display cues. ACM Conf Comp-Hum Interact, New Orleans, LA, 351–359

109. West J E, Blauert J, MacLean D J (1992) Teleconferencing system using head-related signals. Appl Acoust 36:327–333

110. Wiener F M, Ross D A (1946) The pressure distribution in the auditory canal in a progressive sound field. J Acoust Soc Amer 18:401–408

111. Wightman F L, Kistler D J (1989) Headphone simulation of free-field listening. I: Stimulus synthesis. J Acoust Soc Amer 85:858–567

112. Wightman F L, Kistler D J (1989) Headphone simulation of free-field listening. II: Psychophysical validation. J Acoust Soc Amer 85:868–878

113. Wightman F L, Kistler D J (1999) Resolution of front-back ambiguity in spatial hearing by listener and source movement. J Acoust Soc Amer 105:2841–2853

114. Wilkens H (1972) Kopfbezügliche Stereophonie – ein Hilfsmittel für Vergleich und Beurteiling vershciedener Raumeindrücke (Head-related stereophonie – an aid for the comparison and critical examination of different room effects). Acustica 26:213–221

115. Xiang N, Blauert J (1993) Binaural Scale Modelling for Auralization and Prediction of Acoustics in Auditoriums. Appl Acoust 38:267–290

116. Xiang N, Genuit K (1996) Characteristic maximum-length sequences for the interleaved sampling method. Acustica 82:905–907

117. Yamaguchi Z, Sushi N (1956) Real ear response of receivers. J Acoust Soc Jpn 12:8–13

10 Hearing-Aid Technology

Inga Holube[1] and Volkmar Hamacher[2]

[1] Institute of Hearing Technology and Audiology, University of Applied Sciences
   Oldenburg–Ostfriesland–Wilhelmshaven, Oldenburg
[2] Siemens Audiologische Technik GmbH, Erlangen

Summary. Within the last 15 years, hearing instruments have strongly improved
due to the application of modern technologies. This chapter provides a look at the
possibilities and restrictions of the technologies used and at the signal-processing
algorithms available today for the most advanced commercial hearing instruments.
Due to ongoing development in semi-conductors the application of even more complex algorithms is expected for the near future. The first part of this chapter focuses
on the different designs and chip technologies seen with today's hearing instrument.
Consequently, signal-processing algorithms are reviewed which are applied in digital
hearing instruments. Finally, we deal with the components of hearing instruments
and with methods for fitting the instruments to the individual's hearing impairment.

10.1 Hearing-Instrument Design

Hearing instruments are available in different sizes and shapes and with different electro-acoustical parameters, e. g., gain, output power. The first available portable hearing instruments had pocket-size. Nowadays, most hearing
instruments as fitted in Europe and North-America are "behind-the-ear",
BTE, or "in-the-ear" devices – see Fig. 10.1. As to in-the-ear devices, there
are the following subgroups, "completely-in-the-canal" instruments, CIC, "in-
the-canal" instruments, ITC, and instruments covering the concha, ITE.
Which of these is chosen in a particular case depends, among other things,
on size and shape of the ear canal of the individual hearing-instrument user.

The interior of a BTE hearing instrument is shown in Fig. 10.2. The hearing instrument is mainly composed from the following components. A microphone to convert the acoustic input signals to electrical signals, a user-
adjustable volume control, an OTM-switch to chose between microphone,
telecoil and the off-position, a battery compartment, a receiver – i.e. the
loudspeaker of a hearing instrument – and the ear hook. The signal processing is implemented on the integrated circuit, IC, or the "hybrid". In ITEs,
commonly hybrids are used. They are composed from ICs and external passive components, e. g., capacitors, mounted on ceramics.

Further crucial selection criteria are the maximum gain required and,
consequently, the battery type. In fact, the larger the size of the hearing
instrument is, the higher the maximum gain and the bigger the usable battery
can be – Tab. 10.1.

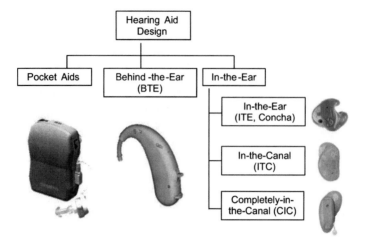

Fig. 10.1. Design variations of hearing instruments

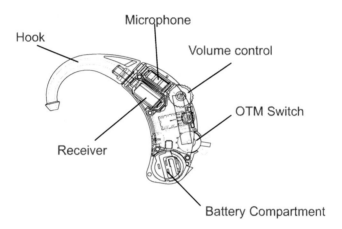

Fig. 10.2. Interior of a BTE hearing instrument

Table 10.1. Maximum gain dependent on battery size

Type	Max. Gain	Battery Type	Capacity
Pocket	95 dB SPL	AA	ca. 1500 mAh
BTE	80 dB SPL	13, 675	260, 575 mAh
ITE	60 dB SPL	13	260 mAh
ITC	45 dB SPL	312	140 mAh
CIC	40 dB SPL	A10, A5	70, 35 mAh

10.2 Chip Technologies

Currently, three general classes of hearing instruments are available. First, until the end of the 1980s, only analog hearing instruments that were adjusted according to the degree of hearing loss with screwdriver-controlled potentiometer trimmers were used. Secondly, IC technologies allowed for the development of analog hearing instruments that could be programmed digitally with dedicated programming devices or with personal computers. Thirdly, in 1996, a novel class of hearing instruments was introduced. These are digitally programmable and offer digital signal processing. Digital hearing instruments have several fundamental differences from analog instruments. Analog devices consist of a microphone, a preamplifier, a means processor such as a tone control or an automatic gain control, AGC, an amplifier, and a receiver – see Fig. 10.3. An acoustic input signal is converted by the microphone to an electronic input signal. This electronic input signal is amplified by the preamplifier and the frequency response is shaped by the tone control. After this shaping, the signal is again amplified and then converted back to an acoustic output signal by the receiver. Both the acoustic and electronic signals in the processing path are analog.

In analog hearing instruments that are digitally programmable, certain parameters of the analog components, such as the tone control or the compression ratio of the AGC, are stored in a memory and can be modified for different hearing instrument users.

In contrast to an analog instrument, a digital instrument has an analog-to-digital converter, a digital signal processor, and an digital-to-analog converter. The amplified electronic signals are converted into digitized signals which are processed by the digital signal processor before they are converted back to analog electronic signals – see Fig. 10.4. In most digital hearing instruments, the output stage and the receiver take charge of performing the digital-to-analog conversion. For a more detailed description of analog and digital signals refer to [11].

Digitally-programmable analog hearing instruments are generally built by use of one IC that contains all electronic parts, e. g., transistors, capacitors, resistors. ICs can be characterized by their structure size, e. g., 1.5 μm, which

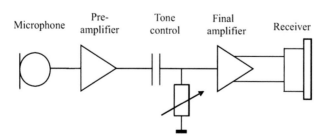

Fig. 10.3. Signal processing in a typical analog hearing instrument

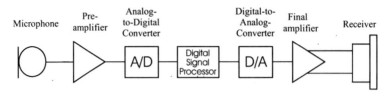

Fig. 10.4. Signal processing in a typical digital hearing instrument

refers to the spacing between different geometric patterns as manufactured with several chemical materials and processes. The structure size determines the necessary area on the chip to implement the different electronic parts. The smaller the structure size, i. e. the closer the components can be placed, the more electronic parts can be integrated on the chip. As well, with a smaller structure size, the chip can be smaller for the same number of electronic parts.

The ICs used in hearing instruments must be as small as possible to produce cosmetically-feasible instruments. In comparison to analog hearing instruments, today's digital hearing instruments can be manufactured with an ever decreasing structure size, currently, e. g., 0.13 µm.

Figure 10.5 shows the reduction in structure size as achieved within the last 14 years. Therefore, digital chips can contain more advanced and more complex signal-processing algorithms than analog chips. In the rest of this chapter, we shall focus on these fully-digitized hearing instruments.

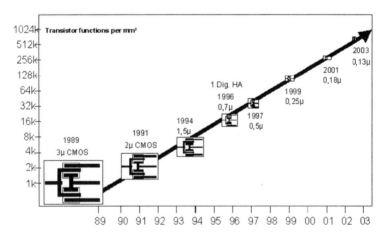

Fig. 10.5. Miniaturization progress in the chip technology

10.3 Advantages of Digital Signal Processing

Digital signal processing has several advantages over analog signal processing, including miniaturization, power consumption, internal noise, reproducibility, stability, programmability, and signal-processing complexity.

- *Miniaturization.* IC technology allows for small-sized digital signal processors containing complex signal processing algorithms. This is especially important for the design of cosmetically appealing hearing instruments. If current signal-processing capabilities as contained in a digital CIC would be implemented using analog technology, the hearing instrument would be the size of a body-worn instrument.
- *Low Power Consumption.* Along with miniaturization comes the advantage of reduced power consumption. Reduced power consumption permits the usage of smaller batteries, which is again advantageous in the design of cosmetically-desirable instruments.
- *Low Internal Noise.* The number of bits, i. e. the word length used in the digital signal processing, controls the internal noise of a digital signal processor. Therefore, the internal noise of a digital hearing instrument is independent of the algorithms – whereas the internal noise of an analog hearing instrument increases with increasing complexity of the signal processing.
- *Reproducibility.* In a digital system, the algorithms always perform the same precise calculations – in contrast to analog technology where the output can vary depending on the exact values of the components in the instrument as a result of, e. g., discrepancies in the production process.
- *Stability.* Digital signal processing is stable because the technology resists external influences. Performance of analog instruments can vary in response to external influences, such as temperature. Instability of digital signal-processing algorithms only results from inadequate calculations.
- *Programmability.* When using digital technology, the signal processing and its parameters can be widely modified. In analog technology, such flexibility could only be achieved by changing components.
- *Complexity.* Digital signal processing allows for the implementation of complex signal-processing algorithms. One example is the analysis of the input signal and appropriate signal modification for noise reduction based on the result of this analysis.

10.4 Technological Restrictions

When using digital signal processing in hearing instruments, some technological restrictions must be taken into consideration.

- While the same microphones are used in analog and digital hearing instruments with a dynamic range of about 90 dB, the dynamic range in a digital hearing instrument is additionally limited by the word length of the signal processing. The word length of an A/D-converter used in digital hearing instruments is usually 16 to 20 bit, but only 12–16 bit are effectively usable. The further bits are switching irregularly and contribute to the internal noise of the hearing instrument. The usable word length corresponds to a dynamic range of about 72 to 96 dB. Nevertheless, the dynamic range of the processed input signals can be increased by signal processing. One possibility is to use a high-level compressor at the input stage, which limits high input levels with a high compression ratio. Another possibility is to use a squelch function which reduces gain for low input levels.

– The calculation accuracy, and therefore the dynamic range, is also limited by the word length of the internal signal processing. Care should be taken that specific calculations, e. g., multiplications of small values, require a higher accuracy than other calculations, such as addition. Therefore, word lengths of 16 to 32 bit are used depending on the calculations, which corresponds to a dynamic range of 96 to 192 dB.

– An additional restriction is the sampling frequency, which determines the frequency bandwidth of the hearing instrument. The sampling frequency has to be larger than twice the signal bandwidth, which therefore has to be limited with a low-pass filter to avoid aliasing effects. Today's commercial hearing instruments use a sampling frequency of about 12 to 32 kHz when processing a bandwidth of about 5 to 12 kHz.

– The dynamic range of the final amplifier can be increased, compared to analog technology, by using digital signal processing. While typical receivers in analog hearing instruments, e. g., Class-D receivers, reach a dynamic range of 70 to 75 dB SPL, the output stage of digital hearing instruments provides a dynamic range of more than 85 dB SPL.

– Digital signal processing can be implemented either on hardwired ICs, using parameters adjustable to the individual hearing loss, or on freely-programmable processors using different software packages for different signal-processing tasks. The first technology is more efficient with respect to required chip area and power consumption.

10.5 Digital Hearing Instruments

Figure 10.6 shows a block diagram of a typical modern digital hearing instrument. It contains up to three microphones, which can be combined to form particular directional characteristics. The input signals are decomposed into different frequency bands, e. g., by means of a filter bank or by Fast-*Fourier*-Transform, FFT. In each frequency band, the input signal is amplified, noise is reduced, speech-signal components are enhanced and the dynamic range is adjusted by compression, all adjusted individually to the needs of hearing-instrument users. Further, feedback that may occur from the hearing instru-

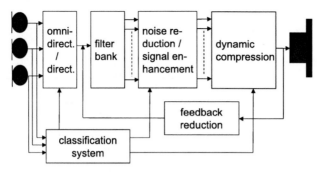

Fig. 10.6. Block diagram of a digital hearing instrument

ment's output back to its input is reduced by an appropriate algorithm. An automatic classifier distinguishes between different listening situations and controls the functionality of the signal-processing algorithms accordingly. The blocks shown in the figure will be described in further detail in the following sections.

10.6 Directional Microphone Systems

Multi-microphone noise-reduction systems are an effective method for improving the intelligibility of speech in noisy situations. They act essentially as spatially-selective noise-reduction systems in that they enhance the hearing instrument's sensitivity to sound from desired directions over sound from other directions. The sound from the other directions is normally considered as noise and is, consequently, deemed as not desirable. Multi-microphone systems are a feasible means towards the general goal of improving the signal-to-noise ratio, SNR.

Besides audibility of the primary signal being essential as such, it is predominatly the SNR, expressed in dB, which determines overall intelligibility. Hence directional microphones are primarily designed to enhance the SNR, and thus will nearly always yield improvements in intelligibility, e. g., [13, 19, 21, 23]).

Omni-directional microphones utilize a single microphone and a single sound inlet and are equally sensitive to sound from all directions. Directional microphones also utilize a single microphone, yet with two sound inlets at different positions and an acoustic damper to introduce an internal time delay. In commercially available products with two omni-directional microphones, the output of one microphone is time-delayed relative to the output of the second microphone, and both signals are subsequently subtracted.

Specific directional patterns, as shown on a polar plot, are determined by the ratio of the internal time delay and the external time delay corresponding to the microphone distance, T_i/T_e, see Fig. 10.7. In addition, the

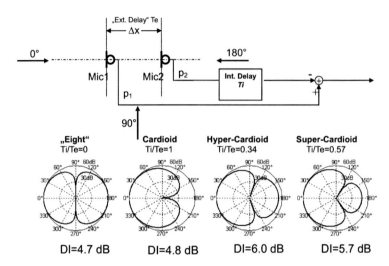

Fig. 10.7. Delay-and-subtract beam former with two microphones. Polar diagrams and directivity index, DI, for different ratios of internal and external delays between the two microphone signals

"directivity index", DI, is shown. DI is defined as the reduction in level in a diffuse sound field when switching between omni-directional and directional characteristic while keeping the sensitivity for frontal sound incidence, $0°$, constant. Theoretically, the maximum DI for these two-microphone systems is 6 dB. However, among other reasons, due to influences of the head and to mismatch between the microphones and reverberation in natural spatial-listening situations, the DI is reduced in many practical applications.

The space available in a hearing instrument for two microphones is very restricted. Therefore, only very limited distances between the two microphone inlet ports can be realized. In addition, a directional microphone has a high-pass frequency response with a slope of 6 dB/oct. On the one hand, the smaller the distance between the microphones, the higher the edge frequency of the high-pass characteristic and therefore, the less low frequencies are available. On the other hand, the further apart the microphones are positioned, the lower are the frequencies of destructive interference which show up as minima in the frequency response – see Fig. 10.8. A compromise between these limitations is given by a distance of about 12 mm between the two inlet ports.

The following restrictions have to be considered when using two-microphone systems:

– As already mentioned, the head has a negative impact on the polar pattern of a two-microphone system, e. g., due to shadowing and diffraction of the sound waves.

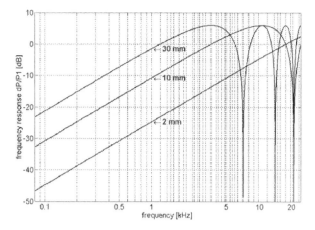

Fig. 10.8. Frequency responses for the frontal direction of a two-microphone system for different distances between the microphones

– Exact matching of the microphones is of paramount impact on the directivity rendered. If, e. g., the sensitivity of the two microphones differs by just 3 dB, the DI at 500 Hz is already reduced from 5.7 to 0.9 dB. Accordingly, at 4 kHz, the DI is reduced from 5.3 to 4.8 dB.
– The two inlet ports to the microphones have to be mounted in the same horizontal plane as where the sound sources are positioned. Otherwise, the effective distance between the two microphones would be reduced.
– Vented ear moulds, i. e. moulds with a narrow canal drilled into them, admit some sound to directly travel into the ear canal, such passing by the hearing instruments. This effect reduces the directivity at low frequencies.
– While measuring polar diagrams, the dynamic compression algorithm within the hearing instrument has to be de-activated. Otherwise the results would show less directivity as the gain for high input levels from the front relative to other directions would be decreased.
– One important advantage of two-microphone systems relative to conventional directional microphones is the possibility to apply matching procedures to the two microphones. This leads to higher robustness, as sensitivity variations will be compensated for.
– A further important advantage can be achieved by controlling the microphone sensitivities according to the acoustical situation encountered. In other words, the directional effect can be made a programmable parameter which can be selected or deleted, depending on the user's needs in different listening situations. Attenuation is maximal for sound sources located in the direction of a minimum of the polar pattern.

Unfortunately, by choosing a particular time delay, the polar pattern of two-microphone systems usually becomes fixed already during the manufacturing or the fitting process. It is especially this restriction, which can be

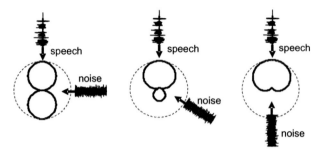

Fig. 10.9. Switching of the polar diagram dependent on the direction of unwanted sounds in an adaptive directional microphone

addressed by using adaptive two-microphone systems. In this application, the pattern is changed depending on the direction of the unwanted sound by adjusting the internal time delay while the sensitivity for the frontal direction is maintained – see Fig. 10.9.

Further improvements can be realized by using a three-microphone system [18]. A respective block diagram is shown in Fig. 10.10. This three-microphone array is used in a commercially available product in an end-fire arrangement with 8 mm microphone spacing. The A/D-conversion is not shown. The output of the three-microphone system is labelled 2nd order in the figure.

Respectively, the polar diagram for this specific implementation is shown in Fig. 10.11. The theoretical maximum of DI is increased to 9 dB. Unfortunately, the sensitivity to microphone mismatch is severely increased and the frequency response shows a high-pass characteristic with a slope of 12 dB/oct, which results in a higher internal noise when using the same amplification

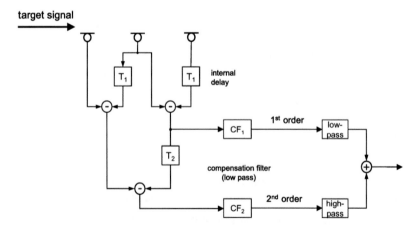

Fig. 10.10. Block diagram of a three-microphone system

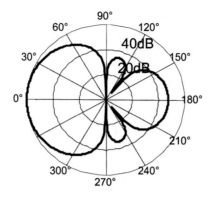

Fig. 10.11. Polar diagram of a three-microphone system

for the low frequencies as compared to the omni-directional microphone or two-microphone cases.

These drawbacks can be partly reduced by a suitable combination of a two-microphone with a three-microphone system – see also Fig. 10.10. Two microphones are enabled for the low-frequency region, resulting in a 6-dB/oct slope of the frequency response and in reduced mismatch sensitivity, whereas three microphones are used in the high-frequency region, such increasing the DI to a theoretical 9 dB. With this mechanism, the articulation-index-weighted directional index, AI-DI, which focuses on the DI at spectral components which are important for speech, can be increased up to 7 dB. Since an increase in AI-DI will result in an improvement of speech intelligibility in noise, it is worthwhile investing engineering effort into multi-microphone systems. Yet unfortunately, the size of the hearing instruments and the drawbacks described above set limitations at this point. Although microphone arrays mounted outside the hearing instrument could potentially increase the AI-DI beyond the value of a three-microphone system, they have not succeeded in the market so far.

10.7 Frequency-Dependent Filtering

Hearing instruments must be able to decompose incoming signals into different frequency bands to compensate for the variety in the frequency configurations seen in hearing impairments. Then, the signal components in the different frequency bands can be processed independently of each other.

The separation of the incoming signals is usually performed by a uniformly- or non-uniformly-spaced filter bank. Typically, non-uniformly-spaced filter banks are realized with time-domain filter banks, whereas uniformly-spaced filter banks are realized with an FFT-algorithm. The filtering can result in quite different amounts of output signals which are determined by the chosen cross-over frequencies between the frequency bands and the steepness of the filter slopes. Commercially available products decompose into up to 20 different bands. The more frequency bands, and the steeper the filter slopes are, the finer the possible control of the signal manipulation in later processing where the frequency response is to be shaped according to the individual hearing losses.

The time delay as present in a digital hearing instrument is determined by far most at this initial stage in the signal processing. It has preferably to be limited to a value of below 10-15 ms. Otherwise mismatches would occur, e. g., between lip-reading and acoustic signals, between the patient's own voice transmitted through bone conduction and air conduction through the hearing aid, and between direct signals travelling through the vent in the ear mould or other leakages of the hearing instrument and the sound processed through the hearing instrument. The latter mismatch can result in a comb-filter effect which causes deep notches and up-to-6 dB peaks in the frequency response due to interference. Figure 10.12 shows, as an example, frequency-dependent time delay-measurements as taken from a commercially-available hearing instrument with a non-uniformly-spaced filter bank. The same measurements for a uniformly-spaced filter bank, presumably realized with FFT-processing, result in an approximately frequency-independent time delay of about 6 ms.

Non-uniformly spaced filter banks in the time domain are particularly popular as they mimic the frequency-domain processing in the human auditory system. Thus, when using an FFT algorithm, the uniformly-spaced frequency bins are typically combined into frequency bands with bandwidths increasing with center frequency.

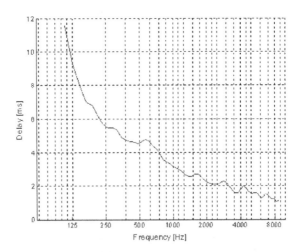

Fig. 10.12. Frequency-dependent time delay of a commercially-available hearing instrument based on a non-uniformly-spaced filter bank, measured with a sinusoidal sweep

10.8 Noise Reduction

Individuals with hearing impairment frequently report that speech is difficult to understand in listening situations where background noise is present. Directional microphone systems are often not sufficient to compensate for the 4 to 10 dB increase in signal-to-noise ratio which hearing impaired persons require [5]. In addition, directional microphone systems are not applicable in smaller ITE devices. Therefore, noise reduction algorithms are the only option to reduce background noises in those cases.

In the past, a variety of algorithms have been developed and evaluated that attempt to improve performance in this difficult listening environment, see, e. g., [4, 7, 8, 20, 22]. These algorithms have enjoyed different degrees of success with regard to reported patients' benefit and satisfaction as well as in terms of commercial advantage.

The introduction of digital hearing instruments has led to the development of advanced signal processing that provides advantages over analog instruments. Several manufacturers have introduced signal-processing algorithms that detect speech and noise independently and process the two types of signals differently. While the underlying algorithms of these systems may vary, the expected result is the same, namely, improved listening comfort and, hopefully, improved speech intelligibility in background noise.

All noise-reduction algorithms have in common that the frequency dependent gain of the hearing instrument is adjusted depending on the listening situation or, more specifically, depending on the amount of speech and noise in the input signal. To distinguish between speech and noise, detailed knowledge about the properties of speech is necessary. For example, speech can be separated from noise by its harmonic structure as present during voiced phonemes. In another algorithm to reduce unwanted noise, called spectral subtraction, the spectral noise parameters are estimated first – see Fig. 10.13. Then, the estimates are used to control a *Wiener* filter that weighs the different frequency bands according to the amount of noise in each of them.

Yet another algorithm uses the typical modulation frequencies present in speech signals, which, in fact, show a maximum in the region of 2 to 8 Hz, see [16]. The modulation frequencies are found in the spectrum of the signal envelope. They originate from the rhythm that speech incorporates, e. g., due to the durations of words, syllables and phonemes. Interestingly, they are independent from the individual voice and of the language. The importance of the speech envelopes has received increasing attention in the speech-perception and speech-recognition literature, e. g., [6, 12]. The prominent region of the modulation frequencies of unwanted background noise is mostly located at frequencies different from those of speech [15].

The algorithm shown in Fig. 10.14 analyzes the envelope of the input signal. If the typical modulation frequencies for speech in the region up to about 12 Hz are not present, the gain in this respective frequency band is reduced. The amount of gain reduction applied depends on the modulation

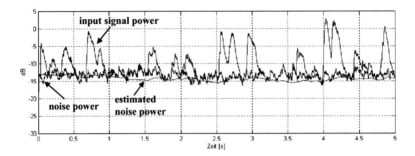

Fig. 10.13. Noise-power estimation used in a spectral-subtraction algorithm. The noise power was estimated from the input signal which contained a mixture of speech and noise

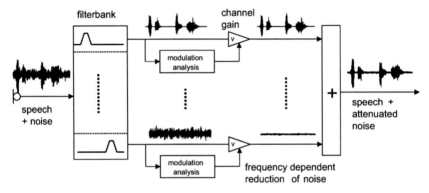

Fig. 10.14. Block diagram of a modulation-frequency-based noise reduction algorithm

frequencies and the modulation depths. As the modulation depth increases, the reduction in gain decreases.

The temporal behaviour of noise-reduction algorithms is determined by their underlying mechanisms. Whereas the spectral-subtraction algorithm is acting comparatively fast, modulation-based filtering needs up to several seconds to analyze the absence of speech and to reduce the channel-specific gain. To avoid unwanted reduction of speech components, gain must recover within 500 ms at most whenever speech is restarted.

All noise-reduction algorithms as described here are based on the assumption that noise is – in at least some respect – different from speech. They have in common that they work best for stationary noises without sudden changes. While it will stay to be a principle problem that hearing instrument cannot readily distinguish between wanted and unwanted sounds, there is still much room for improvements of existing algorithms, and solutions for quiet a variety of different listening situations will presumably be developed.

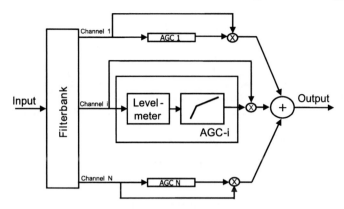

Fig. 10.15. Block diagram of a multi-channel dynamic compression algorithm

10.9 Dynamic Compression

A characteristic of sensorineural hearing loss is the reduced dynamic range, namely, the reduced range between threshold sensitivity and loudness-discomfort level, LDL. A general task of a hearing instrument is to amplify soft sounds until they are audible and to amplify loud sounds such as to be perceived as loud by the individual listener – but not to exceed the listener's LDL. The target amplification is derived from a fitting rationale which may be adapted from published literature, such as DSL (i/o) [3] or NAL-NL1 [2]), or may be proprietary to manufacturers of hearing instruments.

To compensate for the reduced dynamic range, compression algorithms are implemented that automatically adjust the gain for different input levels. Such an adjustment may be based on the amplitudes in individual frequency bands or on the broadband signal level. Today, multi-channel-compression instruments are used – see Fig. 10.15 – since in most listeners the reduced dynamic range varies as a function of frequency. Different systems include different numbers of channels and different dynamic behaviour, i. e. attack and release times. For a summary of the research on multi-channel compression systems see [1, 9, 14]. Commercially available products employ 2 to 20 channels and short as well as long attack and release times. For a more detailed discussion on the advantages and disadvantages of different parameter settings – see [10].

10.10 Feedback Reduction

An important quality issue for hearing instruments is acoustic feedback. The ringing sound that signifies acoustic feedback often bothers hearing-instrument users or their conversation partners. In the feedback condition – see Fig. 10.16 – the output signals from the receiver are fed back into the

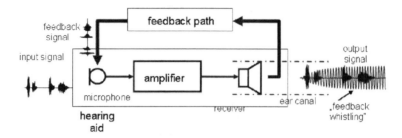

Fig. 10.16. Block diagram of different signal paths in a feedback condition

microphone and amplified again in the signal processor. If the amplification in the hearing instrument outbalances the attenuation in the feedback path and, at the same time, the input signal is in phase with the signal from the feedback path, ringing is induced.

In some instances, feedback can be avoided by reducing the amplification of a certain frequency region while fitting the hearing instrument. Yet, this approach can sometimes have a negative effect on speech intelligibility. Another possible feedback-reduction technique is the usage of narrow-band-reject filters, also called notch-filters. These filters are adaptively effective at frequencies corresponding to the ringing sound. Unfortunately, the hearing instrument has to be in the feedback condition before a notch-filter can be positioned and, further this condition is often fulfilled at several frequencies synchronously. Therefore, the notch-filters need a certain bandwidth in order to reduce adjacent frequencies as well.

Yet another solution uses an adaptive filter – see Fig. 10.17. The respective filter is generated by continuously analyzing the feedback path. The output signal of the adaptive filter is adjusted such as to minimize the feedback components in the input signal to the hearing instrument. While this algorithm

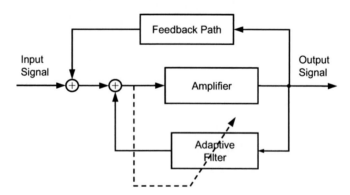

Fig. 10.17. Block diagram of a feedback reduction algorithm

has the potential of having the least impact on speech intelligibility, sound quality might be reduced, especially for sounds like music.

A very common complaint of hearing-instrument users is the unfamiliar sound of the user's own voice in consequence of the occlusion effect [17]. This effect could, for example, be reduced by increasing the diameter of the vents in the ear moulds. Unfortunately, larger vents result in increased feedback. Even with the algorithms available today, compromises have to be accepted in daily practice. Therefore, the performance of feedback-reduction algorithms has to be improved in the future.

10.11 Classification of Listening Situations

Signal processing algorithms, as applied in hearing instruments, are increasing in number as well as in complexity. Further improvements, e. g., for noise reduction, are often limited to certain listening situations and have no or a negative effect in other situations. For example, some noise-reduction algorithms reduce background noises but also reduce the sound quality of music – see Fig. 10.18. Another example along these lines are directional microphone systems. They result in a reduction of lateral sound sources. Unfortunately, desired sound sources, e. g., from the back of a car, are also attenuated due to the same principle whenever the user of the hearing instrument does not face the sound source.

The increased complexity of the hearing instruments results in a demand for decision units which automatically select appropriate algorithms and parameter configurations for specific listening situations. A block diagram for such a classification system is shown in Fig. 10.19. This system extracts representative features for the different listening situations from the input signal. A classification algorithm then compares the extracted features to those of

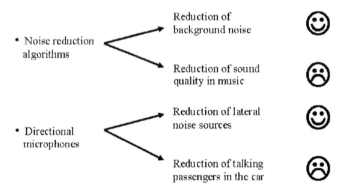

Fig. 10.18. Examples for advantages and disadvantages of different noise reduction schemes

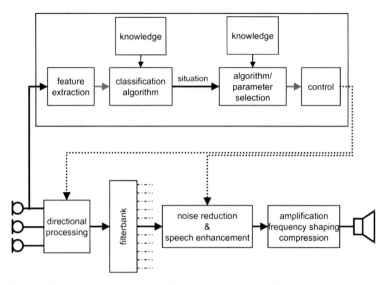

Fig. 10.19. Block diagram of a classification system for different listening situations

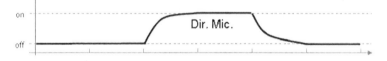

Fig. 10.20. Microphone switching based on the decision of the classification system. Note that noise is present in the middle interval only

listening situations which have been analyzed during a learning period. Based on a selected prototype situation, appropriate signal processing algorithms or parameter settings are transmitted to a control unit which, then, adjusts the hearing instrument accordingly.

It has to be taken into account, however, that sudden switching between different hearing-instrument settings can confuse the users of hearing instruments, especially if they are currently not aware of any changes in the listening environment. This problem can be reduced by a continuous transition between the different settings with long transition times. Such a mechanism also limits false reactions of the hearing instrument to short-time incorrect classification results – see Fig. 10.20 and Fig. 10.21.

As of now, classification systems allow for the differentiation of a very limited set of general listening situations only. Typically, these are speech, speech in noise and noise. In the near future, more complex signal-processing

Noise

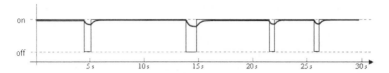

NNNNNSNNNNNNNNNNNNNSSNNNNNNNNNSNNNNSNNNN

Fig. 10.21. Microphone switching of the directional microphone based on the decision of the classification system. Note that noise is always present but that classification is erroneous

algorithms and increasing processing power will allow for the extraction of a growing number of features and, consequently, the separation of further, more specific listening situations. The envisaged improvement of classification systems will, hopefully, result in more optimized algorithms and parameter settings for individual listening situations and, consequently, in improved user benefit.

10.12 Hearing-Instrument Components

The signal-processing algorithms described in this chapter are implemented on the hearing-instrument chip. In addition, other components are necessary, which also contribute to the quality of the entire system. Most important are the transducers. They are as small as possible to allow cosmetically-appealing solutions and are optimized for sensitivity and maximal output power. Nevertheless, their frequency response should be as flat as possible and their distortions should be as small as possible. These requirements are not satisfactorily fulfilled in the receivers used in hearing instruments today – see Fig. 10.22 – thus leaving room for improvement.

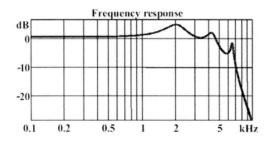

Fig. 10.22. Frequency response of a hearing instrument transducer

10.13 Hearing-Instrument Fitting

During hearing-instrument fitting, the audiologists adjust the devices with respect to the individual hearing losses. An appropriate PC-based fitting program calculates proposals for the fitting parameters based on several statements of the hearing-instrument users. Individual preferences may require deviations from the calculated adjustments. For that purpose, several fitting parameters can be accessed directly or indirectly via a problem-oriented fitting guide. In some cases, sound examples are available to verify the hearing-instrument fitting or to optimize loudness impression and sound quality. However, although signal-processing algorithms have dramatically improved in recent years, common fitting procedures are still focusing on frequency- and level-dependent gain. Audiologist have not yet acquainted themselves sufficiently with the tasks of how to find most suitable settings for parameters such as time constants, strength of noise reduction, and appropriate speech-enhancement procedures – especially as these have to be tailored to the individual hearing-instrument user.

10.14 Summary and Outlook

This contribution includes an overview on properties, algorithms and components of modern hearing instruments as well as on the limitations which they still face. The functionality of hearing instruments has significantly and rapidly increased during the last years due to the progress of technology. Future developments should consider the possibilities of binaural signal processing and the combination with telecommunication systems as well as improvements of the receiver's sound quality and the reduction of the occlusion effect. Further, more sophisticated fitting procedures must be developed and be made available for routine usage in the audiologist's practice.

Acknowledgement

The authors would like to thank *H. Hudde, D. Hammershøi, E. Fischer* and *T. Finkemeyer* for comments on an earlier version of the manuscript.

References

1. Braida L D, Durlach N I, De Gennaro S V, Peterson P M, Bustamante D K (1982) Review of recent research on multi-band amplitude compression for the hearing impaired. In: Studebaker G A and Bess F H (eds) Monographs in contemporary audiology, The Vanderbilt hearing-aid report, Upper Darby PA, 133–140

2. Byrne D, Dillon H, Ching T, Katsch R, Keidser G (2001) NAL-NL1: Procedure for fitting nonlinear hearing aids: characteristics and comparison with other procedure. J Amer Acad Audiol 12:37–51

3. Cornelisse L E, Seewald R C, Jamieson D G (1995) The input/output formula: A theoretical approach to the fitting of personal amplification devices. J Acoust Soc Amer 97:1854–1864

4. Cudahy E, Levitt H (1994) Digital hearing instruments: A historical perspective. In: Sandlin R E (ed) Understanding digitally programmable hearing aids. Allyn and Bacon, Needham Heights MA, 1–13

5. Dillon H (2001) Hearing aids. Boomerang press, Sydney

6. Drullman R, Festen J M, Plomp R (1994) Effect of temporal envelope smearing on speech reception. J Acoust Soc Amer 95:1053–1064

7. Fabry D A (1991) Programmable and automatic noise reduction in existing hearing instruments. In: Studebaker G A, Bess F H (eds) The Vanderbilt hearing-aid report II, York Press, Parkton MD, 65–78

8. Graupe D, Grosspietsch J K, Basseas S P (1987) A single-microphone-based self-adaptive filter of noise from speech and its performance evaluation. J Rehabil Res Developm 24:119–126

9. Hickson L M H (1994) Compression amplification in hearing aids. Amer J Audiol 3:51–65

10. Holube I (2000) Multikanal-Dynamikkompression: Konzepte und Ergebnisse. In: Union der Hörgeräte-Akustiker (ed) Referate des 45. Int Hörgeräte-Akustiker-Kongr, Median-Verlag, Heidelberg, 117–129

11. Holube I, Velde T M (2000) DSP Hearing Instruments. In: Sandlin R E (ed) Textbook of hearing aid amplification. 2nd ed, Singular, San Diego CA, 285–321

12. Houtgast T, Steeneken H J M (1985) A review of the MTF concept in room acoustics and its use for estimating speech intelligibility in auditoria. J Acoust Soc Amer 77:1069–1077

13. Mueller H G, Wesselkamp M (1999) Ten commonly asked questions about directional microphone fittings. Hearing Rev Suppl: High performance hearing solutions: hearing in noise 3:26–30

14. Moore B C J (1995) Perceptual consequences of cochlear damage. Oxford Univ Press, Oxford

15. Ostendorf M, Hohmann V, Kollmeier B (1997) Empirische Klassifizierung verschiedener akustischer Signale und Sprache mittels einer Modulationsfrequenzanalyse. Fortschr Akust DAGA'97, Dtsch Ges Akust, Oldenburg, 608–609

16. Plomp R (1984) Perception of speech as a modulated signal. In: van den Broeche M P R, Cohen A (ed) Proc 10th Int Congr Phonetic Sc Utrecht, Foris Publ, Dordrecht, 29–40

17. Pörschmann C (2000) Influences of bone conduction and air conduction on the sound of one's own voice. ACUSTICA/acta acustica 86:1038–1045

18. Powers T A, Hamacher V (2002) Three-microphone instrument, designed to extend benefits of directionality. Hearing J 55:38–45

19. Ricketts T, Dhar S (1999) Comparison of performance across three directional hearing aids. J Amer Acad Audiol 10:180–189

20. Stein L, Dempesy-Hart D (1984) Listener-assessed intelligibility of a hearing instrument with self-adaptive noise filter. Ear and Hearing 5:199–204

21. Valente M, Fabry D A, Potts L G (1995) Recognition of speech in noise with hearing aids using dual microphones. J Amer Acad Audiol 6:440–449

22. Van Tassel D J, Larsen S Y, Fabry D A (1988) Effects of an adaptive filter hearing instrument on speech recognition in noise by hearing-impaired subjects. Ear and Hearing 9:15–21
23. Wolf R P, Hohn W, Martin R, Power T A (1999) Directional microphone hearing instruments: how and why they work. Hearing Rev Suppl, High Performance Hearing Solutions 3, Hearing in Noise, 14–25

11 Auditory Virtual Environments

Pedro Novo

Institute of Communication Acoustics, Ruhr-University Bochum, Bochum

Summary. A virtual environment aims at creating situations in which humans have perceptions that do not correspond to their physical environment but to a virtual one. The present chapter focuses on auditory virtual environments, i.e. on the auditory component of virtual environments as seen independently from the remaining modalities. Section 11.1 introduces and discusses concepts, such as auditory event and plausibility, required to the understanding of the remaining chapter. Section 11.2 presents the various components of a typical auditory virtual environment along with the techniques and models most often employed by each of them. Section 11.3 presents specific implementations of auditory virtual environments and addresses a selection of current research issues, namely, multi-modality, presence, quality and joint reality. Section 11.4 closes the chapter with a summary of the main conclusions.

11.1 Introduction

A virtual environment aims at creating situations in which humans have perceptions that do not correspond to their physical environment but to a virtual one. To achieve this goal, computer-controlled models generate signals for each modality which are subsequently fed into the appropriate transducers, e.g., screens and loudspeakers, and presented to the subjects. The most important modalities are vision, audition, proprioception, i.e. position, orientation, force, and tactility [10]. The present chapter focuses on auditory virtual environments, AVEs, i.e. on the auditory component of virtual environments as seen independently from the remaining modalities – e.g., [5,10,49,56,87].

The generation of an AVE involves physical and perceptual aspects. In order to avoid ambiguities, it is important to precisely define the terms employed on each case. The terminology proposed in [10], regarding sound events as referring to the physical aspect of the phenomena of hearing and auditory events as referring to what is auditorily perceived, is also adopted here.

As discussed in [10], auditory events are not necessarily caused by sound events. For example, the auditory nerve can be artificially stimulated to provide a stimulus for the formation of an auditory event. On the other hand, there is no direct causal relation between a sound event and an auditory event. For example, blocking the ear canals impedes sound waves from arriving at the ear drums, which prevents the formation of an associated auditory event. Also, a subject will not always hear the same when exposed to the same sound signals. In fact, cross-modal and cognitive effects may influence what is auditorily perceived. With these restrictions in mind and for the sake of conciseness, we will employ in the present chapter expressions such as "auditory event evoked by a sound event" to signify an association, occurring for the majority of subjects, between a specific sound event and a specific auditory event under specific conditions.

A typical AVE system involves sound source, environment and listener models, which can be implemented using three different approaches. One of them, called the *authentic approach*, aims at achieving an authentic reproduction of existing real environments. The objective here is to evoke in the listener the same percepts that would have been evoked in the corresponding auditory real environments. The demand for authentic virtual environments is, however, limited, e. g., for archiving or for A/B comparisons. Another approach, called the *plausible approach*, aims at evoking auditory events which the user perceives as having occurred in a real environment – either known or unknown to him. In this approach only those features which are relevant to each particular application need to be considered. The focus of the present chapter is on plausible AVEs, as a large number of virtual environment applications relevant for communication acoustics, e. g., tele-presence, training, and entertainment, fall within this category. A third approach that could be called the *creational approach* aims at evoking auditory events for which no authenticity or plausibility restraints are imposed. This approach finds application in, e. g., computer games.

The remaining of this chapter is organized as follows. Section 11.2 presents an overview of the main AVE components. Section 11.3 presents a selection of currently available AVE systems and discusses various current research issues. Section 11.4 summarizes the main conclusions.

11.2 Auditory-Virtual-Environment Components

11.2.1 Introduction

As will become clear throughout this section, an AVE system may employ a physical or a perceptual approach. In either case the set of databases and computer programs making up an AVE system may be divided into sound source, environment, signal processing and reproduction modules. Figure 11.1 depicts these modules as well as the main data-flow paths. The signal-processing module processes the sound-source signal together with the filters modelling environmental effects – obtained either through a physical or perceptual approach – and outputs the result in a specific reproduction format.

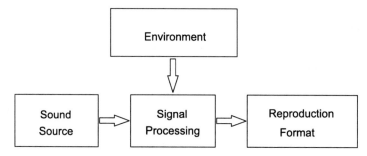

Fig. 11.1. Block diagram of the four main modules and data flow paths of a typical auditory-virtual-environment system.

11.2.2 Sound Sources

Sound-Source-Signal Generation

A sound signal can be obtained through recording, synthesis or a combination of both [87]. In case of recording, it should occur in anechoic conditions, i. e. without environmental reflections, as these would later add to the reflections calculated in the AVE system. The sampling rate and amplitude resolution should be sufficiently high to capture the full audible signal bandwidth and dynamic range. The CD standard, i. e. 16-bit amplitude coding and 44100-Hz sampling rate, which provides a dynamic range of approximately 96 dB and a frequency range of approximately 22 kHz, is adequate for most applications. Further information on practical and theoretical issues involved in sound recording can be found in, e. g., [84].

Although employing recorded sounds has advantages, such as a high fidelity to the original sound, it requires every individual sound to be individually recorded. However, a sound signal can also be synthesized. Among the various methods available for sound synthesis are the physically based methods, which allow the model to be controlled in a physically meaningful way [97]. The synthesis of speech sounds represents a field of its own and a complete chapter in this volume is dedicated to this subject [55].

One of the most popular physically based methods is the digital-waveguide method [96]. Its basic element is a bi-directional delay line, arranged in a regular array and interconnected at the intersections. Each node is connected to its neighbours by unit delays.

To synthesize sound signals similar to those that would have been generated at the listeners' ears by distributed sound sources, e. g. crowd sounds, the following concept may be employed. In a diffuse sound field the sound waves, impinging at the listeners' ears from all directions, can be hypothesized to originate from a large number of sound sources homogeneously distributed around the listener [10]. Also, the correlation coefficient between the sound signals at the entrance of the ear canals of a listener immersed in a diffuse sound field is frequency dependent, assuming negative values in specific frequency ranges [59]. Furthermore, signals presented at the listeners' ears in (partial) phase opposition may evoke two distinct and extended auditory events [10]. This phenomena was, in fact, experienced by listeners immersed in a diffuse sound field as twin extended auditory events located symmetric to the listeners' median plane [71–73]. The precise location of the auditory events is head-size dependent, e. g., a child or an adult head.

Therefore, to evoke in the listener the auditory events corresponding to the sound field generated by distributed sound sources it is sufficient to synthesize a stereo signal that exhibits the desired correlation coefficient, e. g., [59] and to present it over headphones [11, 74]. The signal synthesis can be performed as follows. (i) Filtering the source signal into critical bands – until approximately 1500 Hz [10], (ii) applying within each critical band the correlation algorithm described in [42] and (iii) summing together the resulting signals. Above 1500 Hz a correlation coefficient of approximately zero can be employed.

Sound-Source Directivity

Real sound sources exhibit preferential directions of emission, which provide the listener with valuable information on both the sound-source type, position and orientation, e. g., the orientation of a talker. Directivity models can be implemented with directional filters applied to a point source or by setting-up several simultaneous point sources [33]. They have been employed in a variety of applications, such as in modelling human speech [28, 41], and musical-instrument directivity [48].

11.2.3 Environment Models

Introduction

In a bounded environment the sound waves impinging at the listeners' ears have either been reflected/diffracted at the environment surfaces or have propagated directly from the source to the listener. To model these phys-ical phenomena three main classes of methods are available: wave-based, geometric-based and statistics-based methods, e. g., [54, 61, 77]. As an alter-native, the perceptual approach may be employed. In this case, the objective is to evoke specific auditory events by employing signal-processing algorithms without a connection to any physical model – e. g., [75]

Physical Methods

Wave Methods

Analytical solutions of the wave equation are only available for the simplest geometries, and therefore numerical methods have to be employed to solve most problems of practical interest. Among them, is the boundary-element method, BEM, and the finite-element method, FEM, [49]. In a 3-D domain the BEM requires only the discretization of the 2-D boundaries while the FEM requires a full 3-D discretization of the geometrical calculation domain. The FEM allows an easy determination of the cavity "eigen-frequencies" [77] and is able to accommodate more variation within the computational domain than the BEM. The computational requirements of both FEM and BEM increases very rapidly with frequency, which still limits real-time processing to very low frequencies and small rooms.

The digital-waveguide method can also be used to model the propagation of sound waves within a bounded environment [87]. Still, this method exhibits direction-dependent dispersion of the wave fronts, which can be minimized by employing non-rectangular meshes [24], interpolation methods [86] or frequency warping [88]. Furthermore, the computational requirements for real-time calculation of a three-dimensional domain are beyond current capabilities.

Geometrical-Acoustics Methods

When the sound wavelength is small compared to the global surfaces' dimensions and large compared to their roughness, waves can be approximated by rays. This approximation, known as the geometrical-acoustics approximation, allows a considerably faster calculation than that achieved by wave methods. However, phenomena such as interference and diffraction are not easily taken into account. The most commonly employed geometrical-acoustics methods are the ray-tracing method and the image-source method.

Ray-Tracing Method: In the ray-tracing method, [52, 53], the sound source emits rays that are followed throughout the domain until they either become attenuated below a specified threshold, leave the domain or reach the receiver – the latter being modelled as a detection volume. Specular reflection at the boundaries is the most employed rule. However, diffuse reflections can also be modelled, but at the cost of extra computation, e. g., [50].

Image-Source Method: In the image-source method, e. g., [1, 17], the calculation of the reflection paths from the source to the listener is performed by sequentially mirroring the sound source at the room boundaries. Although reflections of any order can be calculated, it becomes computationally very expensive for late, high-order reflections [100].

Beam-Tracing Method: In the beam-tracing method [29], bundles of rays, i. e. beams, are traced throughout the domain. The beam paths are stored in a beam-tree database which is accessed during run-time. In its current form, this method is able to compute early propagation paths from a static source to a moving receiver at interactive rates for real-time auralization in large architectural environments.

Statistical Methods

Statistical methods, such as statistical energy analysis, SEA, [61], have often been employed for noise-level prediction in coupled systems in which sound transmission by structures plays an important role. These methods have also been employed in other areas and in conjunction with other methods. For example, in [18] a statistical-geometrical method is proposed in which the steady-state sound-pressure distribution and the room impulse response are calculated. This method considers a statistical distribution of sound energy inside an arbitrarily-shaped enclosure as a distribution of energy packets which are radiated from the sound sources and re-radiated diffusely from the enclosure walls. A general equation for the motion of the sound parti-cles is derived, where different transitions are independent and the motion is considered as a *Markov* process. This method has shown to be computationally efficient and current efforts are directed towards achieving real time performance.

Interaction of Sound Waves with Surfaces and the Propagation Medium

When employing physical methods, the interaction between sound waves and the surfaces needs to be explicitly modelled. The result of this interaction depends on the angle of incidence of the sound waves and on the geometric and acoustic properties of the surfaces [19]. Accurately modelling this phenomenon with all its aspects and in real-time is presently not possible and therefore simplifications should be introduced [40]. Another important phenomenon occurs when a surface is interposed between a sound source and a listener. The surface acts as an occluder and produces alterations in the location and timbre of the associated auditory events [27].

In addition, during propagation, a sound wave is partially absorbed by the medium. The inclusion or otherwise of this phenomenon in an AVE system is subject to an amount of dissipation, which mainly depends on the medium temperature, its humidity and on the distance covered by the travelling wave [4, 40].

Perceptually Based Methods

The physically based approach is limited to physically realizable models, which implies that the parameters can not be independently modified. For example, in a given room a change in the absorption coefficients will also change the reverberation time.

An alternative to the physically based approach is the perception-based approach, which, rather than being faithful to a specific physical model, focuses on evoking specific auditory perceptions [47]. Presently, the computational power required by a perception-based approach is significantly lower than that required for a physically based approach. Still, with the continuous increase in the perceptual-based models' complexity, this situation may be reversed in the future.

Models of Reverberation

Real-time calculation of the impulse response of most rooms of practical interest is beyond the capabilities of currently-available processors. To address this issue, the method described in [56] can be adopted. It consists of computing the direct sound and early reflections in detail, but employing a statistical model for the late reflections. These models, usually designated as reverberation models, are employed by physical and perceptual models alike.

For reasons linked to efficiency and naturalness, reverberation models often assume that the sound field is nearly diffuse and the corresponding impulse-response envelope decays exponentially. Furthermore, reverberation models are designed to exhibit a dense pattern of reflections to avoid fluttering, a reverberation time which decreases as a function of frequency to simulate the low-pass-filtering characteristics of air-absorption and materials, and partly incoherent signals at the listeners' ears in order to produce a good spatial impression.

The first successful attempt to produce artificial reverberation employed comb and all-pass filters [89]. Among a variety of models that were subsequently proposed are digital-waveguide [95] and feedback-delay networks [82]. The latter consist of delay lines combined through a matrix in a feedback loop [44, 98]. The procedure described in Sect. 11.2.2, referring to the synthesis of distributed sound sources, has the potential to be employed for the modelling of late reflections, i. e. reverberation modelling.

Further information on reverberation and reverberation models can be found, e. g., in [13, 30, 44, 69].

11.2.4 Reproduction Formats

The human external ear, pinna, head and torso, transforms an incoming sound wave into sound-pressure signals at the two ear drums. The monaural and inter-aural cues resulting from this process, i. e. spectral cues and interaural time and level differences, are employed by the auditory system in the formation of auditory events.

Head-related transfer functions, HRTFs, encode the modifications resulting from the interaction of a sound wave with the listeners' external ear and have been measured for a number of subjects, directions and source distances, e. g., [34, 91, 106], employing a variety of techniques, e. g., [67, 80, 107]. Reproduction formats can make use or not of HRTFs. Those that use HRTFs aim to reproduce sound signals at the entrance to the ear canals which incorporate the linear transformations that would otherwise have been introduced by the external ear. Those reproduction formats not using HRTFs aim at reproducing either the complete sound-wave field or only some aspects of it, e. g., inter-aural level differences, either locally or over an extended area [83].

Reproduction Formats Employing HRTFs

Headphone-Based Systems

In these systems a monophonic time-domain signal is filtered with the left- and right-ear HRTFs and the result is presented via headphones. The inverse headphone transfer function is also employed in order to de-convolve the signal from the headphones' own filtering. Furthermore, if the auditory scenario is not to rotate with the listener's head, the sound field should be updated according to the listeners' head movement. For this purpose a head-tracking system is required, e. g., [10, 63].

Humans have their highest localization accuracy in the horizontal plane particularly in the region right in front of the listener and the required head-tracker accuracy is therefore defined by this feature. The localization blur, i. e. the smallest change in a specific attribute of a sound event that is sufficient to produce a change in the location of the auditory event [10], is usually not lower than $1°$ azimuth, although it is signal dependent. This degree of accuracy can be met by current trackers [92].

HRTFs vary across individuals and optimal results are obtained when the HRTFs are the individual ones of the listener. Employing non-individualized HRTFs, e. g., those from another listener or from an artificial head, may cause a strong degradation of the spatial-hearing capacity in some individuals [35]. Other individuals experience only a small increase in front-back reversals – i. e., when the sound event is located in the front hemisphere and the corresponding auditory event is located in the rear hemisphere, or vice-versa – but the ability to localize in azimuth remains good. However, a significant degradation in the acuity of elevation estimation and a change in the perceived signal colouration is common to most subjects [93, 103, 104].

Employing a head-tracking system also helps to minimize or even eliminate front-back reversals. This is because the changes in the inter-aural ear-input-signal differences, when the head is turned, are opposite for sound waves arriving from the front with regard to those from the rear.

Inside-the-head locatedness, i. e., intracranial positioning of the auditory events, may also occur when employing a headphone-based system [23]. This effect may be alleviated or even eliminated if environmental reflections are present. Tests performed with signals generated by the distributed sound sources algorithm – see Sect. 11.2.2 – resulted in a considerable better externalization of the auditory events as compared to the externalization obtained when employing two de-correlated sound signals [74]. This phenomena has potential for further fundamental research and applications.

Loudspeaker-Based Systems

When a loudspeaker-based system is employed, a monophonic time-domain signal is filtered with the left and right ear HRTFs, and also with the inverse loudspeaker transfer function in order to de-convolve the signal from the loudspeaker's own filtering. In addition the loudspeakers inverse contra-lateral transfer functions may be employed in order to cancel the left-loudspeaker signal at the listener's right ear and vice-versa – so called "transaural systems", [90]. If the listening position and orientation are not to be restricted to a small area and angle, respectively, a head-tracking system is required [31,58]. Furthermore, the listening room should be appropriately damped so as to prevent reflections from disturbing the formation of the desired auditory events.

Reproduction Formats Not Employing HRTFs

Within this class, the most-employed reproduction formats in AVEs systems are "vector-based amplitude panning", VBAP [81], AMBISONICS [62] and "wave-field synthesis", WFS [6]. VBAP allows arbitrary loudspeaker placement over the 3-D space and makes use of a 3-D generalization of the principle employed in standard stereo, namely, amplitude panning. AMBISONICS performs a local sound-field synthesis using spherical harmonics for the decomposition and composition or re-composition of the sound field.

Wave-field synthesis employs 2-D arrays of loudspeakers and has the ability to reproduce the wave field over an entire listening area. Each individual loudspeaker is fed with signals calculated to produce a volume flux proportional to the normal component of the particle velocity of the original sound field at the corresponding position [7,8,14]. This method can also be used with sets of linear-loudspeaker arrays. In this case, the evoked auditory events are limited to one plane, e. g., the horizontal plane. Wave-field synthesis provides a stable virtual-sound-source location behind or in front of the arrays, and allows for reverberation to be reproduced as plane waves from different directions [15,16,22].

Recently, ultra-sound beams have been successfully employed to generate audible sounds by making use of a process of non-linear distortion [79].

11.2.5 Signal Processing

The task of the signal processing module is to process the sound-source signals with filters which model environmental effects as obtained either through a physical or perceptual model, and to output the result in a specified reproduction format. Among a variety of alternatives, this section concentrates on the signal-processing approach as adopted at the Institute of Communication Acoustics at Bochum. For further details of this particular implementation, please refer to Sect. 11.3.2.

In a geometrical-acoustics approximation – see Sect. 11.2.3 – the sound-field model calculates the propagation paths from the sound source to the listener. For each path several auralization parameters are passed on to the signal processing module. These can be arranged in three groups: (a) a delay corresponding to the propagation time, (b) spectral modifications due to one or multiple reflections plus the directional source characteristics and (c) the direction of incidence relative to the listener's head. These three groups of parameters result in three processing steps which form an auralization unit namely, a temporal step, a spectral step and a spatial step [57]. Figure 11.2 illustrates this cascade of processes with a ray that encounters two walls before arriving at the listener. The upper panel of Fig. 11.2 shows the actual propagation path and the lower panel the corresponding block diagram with gains, g, delays, z^{-n}, and reflection filters. Stage 7 of the block diagram is reproduction-format dependent. In this example, an HRTF-based reproduction format is employed – see Sect. 11.2.4,

In order to reflect the changes occurring in a virtual environment, the auralization parameters need to be updated periodically. However, if the time between two computation frames is too short for all parameters to be updated, the most perceptually salient parameters, e. g., the direct-sound incidence direction, should be updated first and the remaining actions can be performed incrementally in the subsequent frames.

Certainly, not all the parameters need to be re-calculated in every frame. A list compiled in [92] describes the events that require a complete re-calculation of the sound field and those events that require only partial

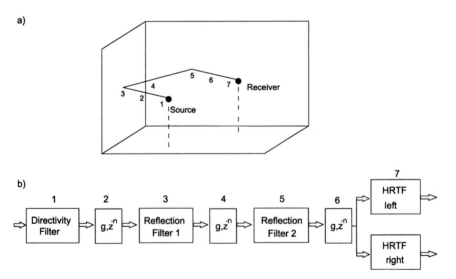

Fig. 11.2. (a) Spatial view and (b) block diagram of a second-order auralization unit – $g \ldots$ gains, $z^{-n} \ldots$ delays – adapted from [99]

re-calculation. Complete re-calculation is required when translational movements of sources or listeners, or modifications of the virtual environment geometry occur. Only a limited re-calculation is necessary when a rotation of sound sources or listeners, changes of directional characteristics of sound sources or in the absorption characteristics of the environment occur. Besides that, in a dynamic situation, i. e. when sound sources and/or receivers move about quickly, the path lengths change over time and the resulting effect, known as *Doppler* shift, should be accounted for [99].

Moreover, the number of filter coefficients available for the auralization can be made variable. For example, the direct sound requires an accurate spatial representation and, therefore, a detailed filter should be employed. Early reflections require accurate modelling of the spectral modifications, but the spatial precision can be less accurate. The later the reflections, the less important are the spatial cues and several reflections may even be bundled together before being sent through the same spatial filter [10].

An AVE system can also be generated by employing a database of pre-calculated or pre-recorded binaural impulse responses. In this case, the impulse response is not calculated in real time, e. g., reflection by reflection, but is previously stored in a database. Therefore, the real-time tasks consist of the retrieval and convolution of the relevant binaural impulse response with an anechoic sound signal. This approach allows for the convolution of longer impulse responses than those possible when the reflections are calculated in real time. The disadvantage is that the creation of the database is very time consuming and the database becomes very large if both the listeners' orientation and position are considered. Further, the method is only feasible with a limited number of sound sources – such as loudspeakers in a control room [38].

The update rate and latency with which the sound field should be re-calculated are related to the two most relevant requirements of an interactive simulation system, namely, "smoothness" and "responsiveness" [3, 56]. A smooth AVE is perceived as being updated without interruption and a responsive AVE is perceived as being updated without delays. From the literature one can find that, to achieve an appropriate smoothness for sound sources having an angular velocity up to $360°/s$, the total system latency should not exceed 58 ms, e. g., [105], further, that update rates below 20 Hz and latency periods of 96 ms and above may significantly degrade both the time and the spatial accuracy of sound-source localization – see [12, 85].

11.3 Current Systems and Research Issues

11.3.1 Introduction

In the previous decade, digital signal processors, DSPs, have been the solution of choice for fast signal-processing hardware. Although computationally efficient, DSP systems require a significant amount of specific programming and, therefore, portability to other systems involves a high effort. However, in recent years, the rapid increase of off-the-shelf microprocessors' processing power has opened up the possibility of implementing AVEs entirely on personal computers, PCs. This carries, among other advantages, lower hardware costs and an immense improvement in the AVEs' portability.

The remainder of this section is organized as follows. Section 11.3.2 presents a selection of currently available AVE implementations and Sect. 11.3.3 points to a selection of current research issues in AVEs.

11.3.2 Systems

Physically Based Systems

The SCATIS system [12,20] is an audio/haptic virtual-reality generator developed during the first half of the nineties. This system was developed within the SCATIS European project, involving Head Acoustics, Aalborg University and the Ruhr-University of Bochum as developers of the audio modality. SCATIS is capable of presenting complex auditory environments to listeners in a plausible way, i.e. such that the listener feels "present" in the environment being modelled. The virtual environment may encompass multiple direct and reflected sounds from spatially non-stationary sources. The experimental paradigm of SCATIS is as follows.

> "A subject is exposed to a virtual space with various – invisible – auditory/tactile objects distributed in it. He/she will localize and identify these virtual objects auditorily, and be able to reach towards them and grasp them individually. Upon tactile contact, contour, texture and thermal attributes of the virtual object will be perceived. The task of the subject is to manually move the objects around, i.e. to rearrange their spatial position and orientation according to an experimental plan. Auditory feedback is given." [12]

Direct sound and early reflections are auralized in real time by distributing the corresponding virtual sources among a cluster of digital signal processors, 80 DSP56002 by Motorola. This is sufficient for auralizing 64 virtual sources in parallel at a sampling rate of 48 kHz while employing 2×80 coefficients per auralization unit. Sets of head-related transfer functions, HRTFs, measured with a resolution of $11.25°$ in azimuth and $22.5°$ in elevation, are employed. Linear interpolation is carried out off-line to enhance the available resolution to below 2 degrees. An average frame rate of 60 Hz and a latency of 60 ms is achieved, which, according to the results reported in Sect. 11.2.5, assures an adequately smooth and responsive AVE. Among other applications, the SCATIS system has been successfully employed in psychoacoustics research [20]. The SCATIS AVE has recently been ported to a PC environment under the new name IKA-SIM [94]. Tests performed with a state-of-the-art PC have indicated that a performance similar to that achieved by the DSP SCATIS system is obtained.

The DIVA system is an audio-visual virtual environment developed at the Helsinki University of Technology [87]. This system, implemented on UNIX workstations, integrates sound synthesis, room-acoustics simulation and spatialized reproduction, combined with synchronized animated motion. DIVA has been applied, e. g., in a virtual concert performance [37] which allows two simultaneous users – a conductor and a listener – to interact with the system. With his/her movements, the conductor controls an orchestra that may contain both real and virtual musicians. At the same time, the listener may freely move about in the concert hall.

To date, several systems based on measured binaural room-impulse-response databases have been developed. Among them are the "binaural room scanning system", BRS [38], targeted at DSP platforms and a PC-based system recently proposed in [68]. The IKA-CONVOLVER is a system developed at the Ruhr-University of Bochum, also for PC platforms. On a state-of-the-art PC, this system is able to perform, in real-time, the convolution of a binaural room impulse response with a duration of approximately 1 s with very low latency, i. e. a few milliseconds, and high update rate, thus guaranteeing a responsive and smooth system. These systems employ head-tracked headphone systems – see Sect. 11.2.4.

Perceptually Based Systems

Among the systems employing a perceptual approach, the SPACIALISATEUR [45–47], is a prominent one. It employs a reduced number of mutually independent perceptive factors, each related to an instrumentally measurable criterion characterizing the transformation undergone by the audio signal. Examples of these pairs of perceptive/instrumental factors are (a) late reverberance vs. late decay time, (b) heaviness and liveness vs. variation of decay time with frequency, (c) brilliance and warmth vs. variation of early energy with frequency and (d) running reverberance vs. early decay time.

Microsoft's DIRECTX is the most widely used standard for real-time multi-media PC applications. Audio environmental effects are taken into account by manipulating filter parameters such as the decay time, ratio of the decay time at high frequencies to the decay time at low frequencies and the delay time of the first reflection relative to the direct path. Another perceptually based system, makes use of standard multi-channel loudspeaker formats [39]. Among other capabilities, it allows for controlling source-distance and room-size perception. Reference to further systems can be found, e. g., in [64].

11.3.3 Current Issues in AVEs

Multi-Modality

Multi-modal virtual environments have the potential of enriching the environment in which the users are immersed. However, the integration of an AVE with other modalities should be carefully performed so as to avoid the occurrence of undesired effects [102]. For example, auditory displays can enhance the utility of a spatial visual display by aurally guiding the visual search. However, even imperceptible delays between sensory information may have an effect on task performance and productivity by causing stress and fatigue [76]. A thorough review of audio-visual interaction is presented in chapter 5 of this volume [51]. The integration of the auditory and haptic modalities has gained increased attention in recent years. For example, research on the temporal factors that affect this integration is reported in [2].

Presence

Since the early nineties, the concept of virtual presence and the concept of tele-presence have received much attention. Perceptual presence can be described as the feeling of being immersed in an environment – be it virtual or real – and tele-presence as the feeling of being at a remote environment. Among the vast literature on the subject, there is a study with adults who have suddenly lost their sense of hearing [32]. These patients report that the environment around them had suddenly become lifeless. The authors propose that the crucial element to create a sense of presence is the auditory background, comprising the incidental sounds made by objects in the environment, rather than the communication and warning signals that typically capture our attention. In another study it is reported that the ability to hear one's own voice contributes to our sense of presence [78].

Quality

The concept of quality associated to sound has been the object of much interest in recent years. In [9] sound quality is defined as the adequacy of a sound in the context of a specific technical goal and/or task. In [75] it is proposed to use the concept of "usability" as associated to quality. Also, it has been proposed a framework for evaluating AVEs, based on the comparison of real-head recordings with physics-based room-acoustic modelling and auralization [60]. In [21] an analysis of the structure of auditory signs employing semiotic theory is presented. Quality evaluation is expected to play an increasingly important role in the future by providing a framework within which AVEs quality will be predictable and therefore amenable to design. See chapters 6, 7 and 9 of this volume in the context of sound-quality evaluation [26, 43, 65].

Joint Reality

The idea of mixing real and virtual environments, "joint reality" or "augmented reality", was probably first realized with assisted reverberation systems. These systems have been installed in concert halls with the objective of enhancing the acoustics by changing, through electro-acoustics means, its reverberation characteristics [101].

A recently published study presents results obtained with in-ear headphones with incorporated microphones [36]. The results indicate that the auralization of the virtual sound sources performed with individual HRTFs can produce a seamless integration of the virtual sources in the real environment. Furthermore, it is reported that, in order to adequately superpose virtual sound sources on real environments, information on the geometric and acoustic properties of the real environment is required. This information can be obtained a priori or may be extracted in real time from the binaural signals recorded at the two ears. The latter approach is considerably more flexible and constitutes a challenging research topic. In fact, recent findings in the field of hearing in perfectly diffuse sound fields, e. g., [71], as well as anticipated progress in the neighboring field of hearing in non-perfectly diffuse sound fields are expected to contribute to advances in e. g., auditory scene analysis- Further implementations of auditory joint-reality systems can be found, e. g., in [25, 70].

11.4 Conclusions

Auditory virtual environments, AVEs, constitute the auditory component of virtual environments and employ models of sound sources, propagation media and listeners. Two approaches to building an AVE system were presented. One of them consists of a physically based approach where the direct sound and early reflections are calculated, based on the environment geometry and acoustic properties. The second approach, known as the perceptual approach, focuses on evoking specific auditory events by employing signal-processing algorithms without a direct connection to a specific physical model. In both approaches a reverberation algorithm may be employed.

Sound-source signals can be recorded or synthesized. Synthesis offers higher flexibility and is considerably less time consuming. Furthermore, physically based synthesis allow the model to be controlled in a physically meaningful way. However, a perceptually based approach can also be employed with advantage. For example, binaural synthesis of the sound field generated at a listeners' ears by a large number of distributed sound sources can be efficiently performed through the use of an algorithm based on the correlation coefficient of diffuse sound fields.

Reproduction formats can be classified according to their use or non-use of HRTFs. HRTF-based reproduction formats explicitly model the result of the interaction of an incoming sound wave with the listeners' pinnae, head and torso. The use of headphones has the advantage of avoiding cross-talk between channels, which occurs when employing loudspeakers. On the other hand, with headphone-based systems, inside-the-head locatedness and front-back reversals of the auditory events may occur. These effects can be minimized or even eliminated by, respectively, introducing reflections and making use of a head-tracking system.

Various formats that do not use HRTFs are available. Among them are the vector-based amplitude panning and AMBISONICS, which allow auditory events to be evoked over all azimuths and elevations but within a relatively small listening area. Wave-field synthesis employs a methodology enabling the reconstruction of the sound-field wavefronts over an extended area. The disadvantage of this format is that it requires a large set-up of loudspeakers, amplifiers and computers.

Smoothness and responsiveness are the two most relevant requirements of an interactive simulation system. The former is related to the update rate and the latter to the latency. Below an update-rate threshold and above a latency threshold the user will perceive, respectively, a discontinuous and a delayed update of the sound field.

There are presently various AVE systems available, but a significant number is still to be found within R&D labs. However, with the rapid increase in microprocessor power, the implementation of AVE systems in off-the-shelf PC systems is becoming more widespread and, consequently, their presence in non-research environments will become more common.

A selection of current research issues in AVEs was highlighted, namely, multi-modality, presence, quality and joined reality. Developments in each of these fields will contribute to make virtual environments more mobile, multi-modal, joint, and hidden, which will eventually render the frontiers between virtual reality and real reality virtually imperceptible.

Acknowlegment

The review presented here reflects, among other issues, work performed at the Institute of Communication Acoustics, Ruhr-University Bochum. I would like to thank Prof. *J. Blauert* and my colleagues *H. Strauss, J. Merimaa* and *A. Silzle*, for many helpful comments and suggestions to the manuscript. I would also like to express my appreciation to the colleagues and friends who have helped to improve the intelligibility of the text.

References

1. Allen J, Berkley D (1979) Image method for efficiently simulating small–room acoustics. J Acoust Soc Amer 65:943–950

2. Altinsoy E (2003) Perceptual aspects of auditory–tactile asynchrony. Proc 10th Int Congr Sound Vibr, Stockholm, 3831–3838
3. Appino P, Lewis J, Koved L, Ling D, Rabenhorst D, Codella C (1992) An architecture for virtual worlds. Presence 1:1–17
4. Bass H, Bauer H (1972) Atmospheric absorption of sound: analytical expressions. J Acoust Soc Amer 52:821–825
5. Begault D (1994) 3-D sound for virtual reality and multimedia. Academic Press Professional, Cambridge, MA
6. Berkhout A (1988) A holographic approach to acoustic control. J Audio Engr Soc 36:977–995
7. Berkhout A, de Vries D, Boone M (1996) Application of wave field synthesis in enclosed spaces: new developments. Proc Forum Acusticum 1969, Antwerpen, Acta Acustica 82(suppl 1):128
8. Blauert J (1974) Vergleich unterschiedlicher Systeme zur originalgetreuen elektroakustischen Übertragung (Comparison of different systems for authentic electro-acoustical transmission), Rundfunktechn Mitt 18:222-227, further: Nachrichtentechn Fachber 48:127-135
9. Blauert J (1994) Product-sound assessment: An enigmatic issue from the point of view of engineering. Proc Internoise 1994, Jokohama, Japan, 2:857–862
10. Blauert J (1996) Spatial Hearing - the psychophysics of human sound localization. Rev edn , The MIT Press, Cambridge MA
11. Blauert J (2004) Private communication
12. Blauert J, Lehnert H, Sahrhage J, Strauss H (2000) An interactive virtual-environment generator for psychoacousic research. I: architecture and implementation. ACUSTICA/acta acustica 86:94–102
13. Blesser B (2001) An interdisciplinary synthesis of reverberation viewpoints. J Audio Engr Soc 49:867–903
14. Boone M, Verheijen N, van Tol P (1995) Spatial sound field reproduction by wave-field synthesis. J Audio Engr Soc 43:1003–1012
15. Boone M, Verheijen E (1997) Qualification of sound generated by wave-field synthesis for audio reproduction. 102nd Audio Engr Soc Conv, Munich. preprint 4457
16. Boone M, de Bruijn W, Horbach U (1999) Virtual surround speakers with wave field synthesis. 106th Audio Engr Soc Conv, Munich. preprint 4928
17. Borish J (1984) Extension of the image model to arbitrary polyhedra. J Acoust Soc Amer 75:1827–1836
18. Coelho L, Alarcão D, Almeida A, Abreu T, Fonseca N (2000) Room-acoustics design by a sound energy transition approach. ACUSTICA/acta acustica 86:903–910
19. Cox T, D'Antonio P (2004) Acoustic absorbers and diffusers: theory, design and application. Spon Press, London
20. Djelani T, Pörschmann C, Sahrhage J, Blauert J (2000) An interactive virtual-environment generator for psychoacoustic research. II: collection of head-related impulse responses and evaluation of auditory localization. ACUSTICA/acta acustica 86:1046–1053
21. Dürrer B, Jekosch U (2000) Structure of auditory signs: a semiotic theory applied to sounds. Proc Internoise 2000, Nice, 4:2201–2204
22. deBruijn W, Piccolo T, Boone M (1998) Sound recording techniques for wave-field synthesis and other multichannel sound systems. 104th Audio Engr Soc Conv, Amsterdam. preprint 4690

23. Durlach N, Rigopulos A, Pang X, Woods W, Kulkarni A, Colburn H, Wenzel E (1992) On the externalization of auditory images. Presence 1:251–257

24. Duyne S, Smith J (1995) The tetrahedral digital waveguide mesh. Proc 1995 IEEE Worksh Appl Signal Processg, Audio Acoustics, New Paltz, NY, 1–4

25. Eckel, G (2001) The LISTEN vision. Preconf Proc ACM SIGGRAPH Euro-graphics Campfire Acoust Rendering Virt Environm, SnowBird UT, 55–58

26. Fastl H (2005) Psychoacoustics and sound quality. Chap 6 this vol

27. Farag H, Blauert J, Alim O (2003) Psychoacoustic investigations on sound–source occlusion. J Audio Engr Soc 51:635–646

28. Flanagan J (1960) Analog measurements of sound radiation from the mouth. J Acous Soc Amer 32:1613–1620

29. Funkhouser T, Tsingos N, Carlbom I, Elko G, Sondhi M, West J, Pingali G, Min P, Ngan A (2004) A beam tracing method for interactive architectural acoustics. J Acoust Soc Amer 115:739–756

30. Gardner W (1998) Reverberation algorithms. In: Kahrs M, Brandenburg K (eds) Applications of digital signal processing algorithms to audio and acoustics, Kluwer, Norwell, MA, 85–131

31. Gardner W (1998) 3-D audio using loudspeakers. Kluwer Academic Publ, Norwell, MA

32. Gilkey R, Weisenberger J (1995) The sense of presence for the suddenly deaf-ened adult. Presence 4:357–363

33. Giron F (1996) Investigations about the directivity of sound sources. Doct diss, Ruhr–Univ of Bochum, Shaker Verlag, Aachen

34. Hammershøi D, Møller H, Sørensen M F, Larsen K A (1992) Head-related transfer functions: measurements on 40 human subjects. Proc 92nd Audio Engr Soc Conv, Vienna. preprint 3289

35. Hammershøi D, Møller, H (2005) Binaural Technique - Basic Methods for Recording, Synthesis and Reproduction. Chap 9 this vol

36. Härmä A, Jakka J, Tikander M, Karjalainen M, Lokki T, Nironen H, Vesa S (2003) Techniques and applications of wearable augmented reality audio. 114th Audio Engr Soc Conv, Amsterdam. preprint 5768

37. Hiipakka J, Hänninen R, Ilmonen T, Napari H, Lokki T, Savioja L, Huopaniemi J, Karjalainen M, Tolonen T, Valimaki V, Valimaki S, Takala T (1997) Virtual orchestra performance Visual. Proc SIGGRAPH'97. ACM, Los Angeles, 81

38. Horbach U, Karamustafaoglu A, Pellegrini R, Mackensen P, Theile G (1999) Design and applications of a data–based auralisation system for surround sound. Proc 106th Audio Engr Soc Conv, Munich. preprint 4976

39. Horbach U, Karamustafaoglu U, Pellegrini R, Corteel E (2000) Implementa-tion of an auralization scheme in a digital-mixing console using perceptual parameters. 108th Audio Engr Soc Conv, Paris. preprint 5099

40. Huopaniemi J, Savioja L, Karjalainen M (1997) Modelling of reflections and air absorption in acoustical spaces -a digital filter design approach. Proc IEEE Worksh Appl Sig Processg Audio and Acoustics, WASPAA'97, New Paltz NY, session 5.2

41. Huopaniemi J, Kettunen K, Rahkonen J (1999) Measurement and modelling techniques for directional sound radiation from the mouth. Proc IEEE Worksh Appl Sig Processg Audio Acoustics, New Paltz NY, 183–186

42. Jeffress L, Robinson D (1962) Formulas for coefficient of interaural correlation for noise. J Acoust Soc Amer 34:1658–1659

43. Jekosch U (2005) Assigning meanings to sound: semiotics in the context of product-sound design. Chap 8 this vol

44. Jot J, Chaigne A (1991) Digital delay networks for designing artificial reverberators. Proc 90th Audio Engr Soc Conv, Paris. preprint 3030

45. Jot J (1992) Ètude et rèalization d'un spatialisateur de sons par modèles physiques et perceptifs, [Study and Implementation of a Spatializer using Physical and Perceptual Models] Doct diss, French Telecom, Paris, E 019

46. Jot J, Warusfel O (1995) A real-time spatial-sound processor for music and virtual–reality applications. Proc Int Computer Music Conf, Banff, 294–295

47. Jot J (1999) Real-time spatial processing of sounds for music, multimedia and interactive human-computer interfaces. Multimedia Systems 7:55–69

48. Karjalainen M, Huopaniemi J, Valimaki V (1995) Direction-dependent physical modeling of musical instruments. Proc 15th Int Congr on Acoustics, ICA'95, Trondheim, 3:451–454

49. Kleiner M, Dalenbäck B I, Svensson P (1993) Auralization - an overview. J Audio Engr Soc 41:861–875

50. Korany N, Blauert J, Alim O (2001) Acoustic simulation of rooms with boundaries of partially specular reflectivity. Appl Acoust 62:875–887

51. Kohlrausch A, van der Par S (2005) Audio-visual interactions in the context of multi-media applications. Chap 5 this vol

52. Krokstad A, Strom S, Sorsdal S (1968) Calculating the acoustical room response by the use of a ray tracing technique. J Sound Vib 8:118–125

53. Kulowski A (1985) Algorithmic representation of the ray-tracing technique. Appl Acoust 18:449–469

54. Kuttruff H (2000) Room acoustics. 4th edn, Spon Press, Taylor and Francis, London

55. Lacroix A (2005) Speech-production: acoustics, models and applications. Chap 13 this vol

56. Lehnert H, Blauert J (1992) Principles of binaural room simulation. Appl Acoust 36:259–291

57. Lehnert H, Richter M (1995) Auditory virtual environments: simplified treatment of reflections. 15th Int Congr Acoust, ICA'95, Trondheim, 3:265–268

58. Lentz T, Behler G (2004) Dynamic cross-talk cancellation for binaural synthesis in virtual reality environments. 117th Audio Engr Soc Conv, San Francisco. preprint 6315

59. Lindevald I, Benade A (1986) Two-ear correlation in the statistical sound fields of rooms. J Acoust Soc Am 80:661-664

60. Lokki T, Järveläinen H (2001) Subjective evaluation of auralization of physics-based room acoustics modeling. 7th Int Conf Audit Displ, ICAD'01, 26–31

61. Lyon R, Dejong R (1995) Theory and applications of statistical energy analysis. Butterworth-Heinemann, Boston

62. Malham D, Myaat L (1995) 3-D sound spatialization using ambisonic techniques. Comp Music J 19:58–70

63. Meyer K, Applewhite H, Biocca F (1992) A survey of position trackers. Presence 1:173–200

64. Miller J, Wenzel E (2002) Recent developments in SLAB: a software-based system for interactive spatial sound synthesis. Proc Int Conf Audit Displ, ICAD'2002, Kyoto, 403–408

65. Möller S (2005) Quality of transmitted speech for humans and machines. Chap 7 this vol

66. Møller H (1992) Fundamentals of binaural technology. Appl Acoust 36:171–218

67. Møller H, Sørensen M, Hammershøi D, Jensen C (1995) Head-related transfer functions of human subjects. J Audio Engr Soc 43:300–321

68. Moldrzyk C, Ahnert W, Feistel S, Lentz T, Weinzierl S (2004) Head-tracked auralization of acoustical simulation. 117th Audio Engr Soc Conv, San Francisco. preprint 6275

69. Moorer J (1979) About this reverberation business. Comp Music J 3:13–28

70. Müller-Tomfelde C (2002) Hybrid sound reproduction in audio-augmented reality. In: Virtual, synthetic and entertainment audio. Proc of the Audio Engr Soc 22th Int Conf, Espoo, 58–63

71. Novo P (2004) Experiments with distributed sound sources in real and virtual environments. Proc Joint Congr CFA/DAGA, Strasbourg, 683–684

72. Novo P (2004) Aspects of hearing in diffuse sound fields. Proc IV Congr Ibero-Amer Acustica, Guimarães, Portugal, paper 68, 131-132

73. Novo P (2004) Binaural signal analysis of diffuse sound fields. J Acoust Soc Am 116:2475

74. Novo P (2004) Computationally efficient and auditorily plausible synthesis of distributed sound sources - HEARCOM Project Report - Ruhr-Univ Bochum, Bochum

75. Pellegrini R (2001) Quality assessment of auditory virtual environments. Proc 7th Int Conf on Audit Displ, (ICAD'2001), 161–168

76. Perrott D, Sadralodabai T, Saberi K, Strybel T (1991) Aurally aided visual search in the central vision field: Effects of visual load and visual enhancement of the target. Human Factors 33:389–400

77. Pietrzyk A (1998) Computer modelling of the sound field in small rooms. In: Audio and Acoustics on Small Spaces. Proc Audio Engr Soc 15th Int Conf Copenhagen, 24–31

78. Pörschmann C (2001) One's own voice in auditory virtual environments. ACUSTICA/acta acustica 87:378–388

79. Pompei F (1999) The use of airborne ultrasonics for generating audible sound beams. J Audio Engr Soc 47:726–731

80. Pralong D, Carlile S (1994) Measuring the human head-related transfer functions: construction and calibration of a miniature in-ear recording system. J Acoust Soc Amer 95:3435–3444

81. Pulkki V (1997) Virtual sound source positioning using vector based amplitude panning. J Audio Engr Soc 45:456–466

82. Rochesso D, Smith J (1997) Circulant and elliptic feedback–delay networks for artificial reverberation. IEEE Speech Audio Processg Mag, SAP, 5:51–63

83. Rumsey F (2001) Spatial audio. Focal Press, Oxford

84. Rumsey F, McCormick T (2002) Sound and recording 4th edn, Focal Press, Oxford

85. Sandvad J (1996) Dynamic aspects of auditory virtual environments. Proc 100th Audio Engr Soc Conv. preprint 4226

86. Savioja L, Välimäki V (1997) Improved discrete-time modelling of multi-dimensional wave propagation using the interpolated digital waveguide mesh. IEEE Int Conf Acoust Speech Sig Processg, ICASSP'97, Munich, 1:459–462

87. Savioja L, Huopaniemi J, Lokki T, Väänänen R (1999) Creating interactive virtual acoustic environments. J Audio Engr Soc 47:675–705

88. Savioja L, Välimäki V (2000) Reducing the dispersion error in the digital waveguide mesh using interpolation and frequency–warping techniques. IEEE Trans Speech Audio Processg, (SAP), 8:184–194

89. Schroeder M (1962) Natural sounding artificial reverberation. J Audio Engr Soc 10:219–223

90. Schroeder M, Atal B (1963) Computer simulation of sound transmission in rooms. IEEE Conv Rec, 7:150-155

91. Schroeter J, Pösselt C, Opitz M, Divenyi P, Blauert J (1986) Generation of binaural signals for research and home entertainment. Proc 12th Int Congr Acoust, Toronto, 1:B1–6

92. Shinn-Cunningham B, Lehnert H, Kramer G, Wenzel E, Durlach N (1995) Auditory displays. In: Gilkey R, Anderson T (eds) Binaural and spatial hearing in real and virtual environments, 611–664, Lawrence Erlbaum, Mahwah NJ

93. Silzle A (2003) Quality of head-related transfer functions - some pratical remarks. Proc 1st ISCA Tutorial Res Worksh Auditory Quality of Sytems, Essen, 101–108

94. Silzle A, Novo P, Strauss H (2004) IKA-Sim: a system to generate auditory virtual environments. 116th Audio Engr Soc Conv, Berlin. preprint 6016

95. Smith J (1985) A new approach to digital reverberation using closed waveguide networks. Proc 1985 Int Comp Music Conf, Vancouver. 47–53

96. Smith J (1992) Physical modelling using digital waveguides. Comput Music J 16:74–91

97. Smith J (1996) Physical modelling synthesis update. Comput Music J 20:44–56

98. Stautner J, Puckette M (1982) Designing multi-channel reverberators. Comp Music J 6:52–65

99. Strauss H (1998) Implementing *Doppler* shifts for virtual auditory environments. 104th Conv Audio Engnr Soc, Amsterdam. preprint 4687

100. Vorländer M (1989) Simulation of transient and steady-state sound propagation in rooms using a new combined ray-tracing/image-source algorithm. J Acoust Soc Amer 86:172–178

101. Warusfel O, Blauert J, Wessel D (2002) Synopsys of revereberation enhancement systems. Proc Forum Acusticum 2002, Seville, paper RBA–02–009-IP

102. Welch R, Warren D (1986) Intersensory interactions In: Boff K, Kaufman L, Thomas J (eds) Handbook of perception and human performance, Vol 1: Sensory processes and perception. John Wiley, New York

103. Wenzel E (1992) Localization in virtual acoustic displays. Presence 1:80–107

104. Wenzel E, Arruda M, Kistler D, Wightman F (1993) Localization using non-individualized head-related transfer functions. J Acoust Soc Amer 94:111–123

105. Wenzel E (1997) Analysis of the role of update rate and system latency in interactive virtual acoustic environments. 103rd Audio Engenr Soc Conv, New York. preprint 4633

106. Wightman F (1989) Headphone simulation of free–field listening I : Stimulus synthesis. J Acoust Soc Amer 85:858–867

107. Xiang N, Blauert J (1993) Binaural scale modelling for auralization and prediction of acoustics in auditoria. Appl Acoust 38:267–290

12 The Evolution of Digital Audio Technology

John N. Mourjopoulos

Audiogroup, Wire Communications Laboratory, Electrical & Computer
Engineering Department, University of Patras, Patras

Summary. Digital audio technology is allowing music and speech to be easily and
readily accessible to most people, since these signals are treated as digital media
and, hence, are significant components of the information-age revolution. From its
commercial introduction via the audio Compact Disc, CD, approximately 20 years
ago, this technology had a meteoric evolution which have seen the introduction
of numerous methods, techniques, systems and formats and has allowed the users
to benefit from reductions in the size of digital audio equipment and its cost. An
overview of these developments is presented here with a critical assessment of their
significance, along with a reference to many important publications and events. It is
shown that this technology is mainly rooted on three constituent evolutionary components, namely, (a) digital electronics and computer technology, (b) DSP theory
and techniques, (c) auditory modelling. Based on the analysis of these components,
some conclusions are drawn, which allow the prediction of future trends concerning
the evolution of this technology.

12.1 Introduction

12.1.1 Overview

From its commercial introduction via the "Compact Disc", CD, more than
20 years ago [88], digital audio technology is now reaching a maturing stage,
allowing audio to be an integral component of the information-age revolution. Today, the great majority of pre-recorded sounds reaching a listener are
derived from digital sources, digital audio technology being part of the everyday experience for audio professionals and music listeners alike. With this
technology, storage, manipulation, transmission and reproduction of audio
signals can be achieved with extreme speed, accuracy and flexibility – due to
the binary representation of sampled acoustic events. The users also benefit from improvements in system performance, especially compared to earlier
analogue audio technologies, as well as a continuous reduction in component
size and cost which is characteristic of digital electronics.

Today, two divergent trends are emerging in this technology's application.
The first is towards improved quality for sound reproduction and a transition
from stereo to multi-channel delivery, as it is manifested with the introduction
of the audio "Digital Versatile Disc", "DVD-Audio" [36], and "Super Audio

CD", SACD [72], formats. The second trend is towards reduced audio bit-rate, utilizing data compression, e. g., ISO/MPEG or propriety standards [20, 73], so that acceptable-quality sounds can be accessed easily via computer and communication networks or readily reproduced from miniaturized or limited-storage-capacity devices. As a result of these past developments, audio has become an integral component of audio-visual home entertainment systems, communication devices and multi-media presentations. Furthermore, digital audio technology is progressively substituting most devices and tasks traditionally implemented via analogue electronics and components. In this sense, digital audio technology may be considered to represent a progression of practices and techniques known during many past decades in electro–acoustics and audio engineering. However, it will be argued here that digital audio technology offers a significantly expanded scope for evolution and introduces novel modes of application in audio engineering. Leaving aside such clear dependence on past analogue audio practices, digital audio technology developments will be examined from the perspective of three constituent evolutionary components, which have provided the necessary technological infrastructure, namely,

1. digital electronics and computer technology,
2. DSP theory and techniques, and
3. auditory modelling.

For each of these components, significant past events will be traced and future trends will be analyzed. Nevertheless, it will be also shown that this technology has not only given consumers and professionals alike access to high-quality audio, but has also helped to highlight many problems, some related to human auditory perception and others to signal-processing methodologies. Many of these problems remain still unresolved so that there are significant theoretical and practical issues to be tackled by future scientific research as well as further audio-product development.

12.1.2 Systems, Formats and Market Trends

The numerous digital audio devices and systems [79] may be grouped into the following classes.

– General-purpose home systems, stereo, multi-channel and audiovisual, e. g., CD-players, DVD-Video players, decoders,
– dedicated high-resolution home-audio systems, e. g., DVD-Audio, SACD,
– portable and wireless devices, e. g., *mp3*-players, Minidisk, "Digital Audio Broadcasting", DAB, decoders, automotive audio systems,
– computer audio devices and software, e. g., sound-cards, software editors, processors, virtual instruments,
– professional-audio systems, e. g., mixing consoles, recorders, processors, codecs, measuring devices.

As with the "Digital-Audio CD", CD-DA, the great majority of these devices is employing PCM for coding the binary audio data – for the principles of PCM see [43], this volume – standardized according to the *Red-Book* specifications, for $N = 16$ bit sample resolution, $f_s = 44100$ Hz sampling frequency – this is a 22.6-μs sampling period per channel – and $c = 2$ channels (stereo) [21, 79]. The audio-data bit-rate is then evaluated according to

$$R = f_s \cdot N \cdot c. \tag{12.1}$$

Hence for CD-DA, the resulting rate is 1.4112 Mbit/s. From this initial format, numerous other standards and formats have emerged over the years, as is illustrated in Fig. 12.1, due mainly to the following reasons,

– adaptation to computer, network and multi-media requirements, e. g., audio-file formats such as .WAV and .AIFF [79],
– improvement of audio quality, increase of storage capacity and channel delivery, e. g., 6-channel reproduction, sampling frequency up to 192 kHz and sample resolution up to 24 bit for DVD-Audio [36],
– reduction of the overall audio bit-rate for applications where either channel bandwidth or storage space are constrained, e. g., MPEG-1, MPEG-2 and propriety standards such as Dolby Digital, DTS, ATRAC, RealAudio, WMA, APT-X [20, 73],
– introduction of differential 1-bit encoding as opposed to traditional multi-bit PCM, termed Direct-Stream Digital, DSD, for both consumer and professional applications, for the SACD format and related devices [72].

Currently, digital audio delivered from CD-DA optical-disc players represents the most successful application with approximately 800 Million players and approximately 20 Billion CD-DA discs sold worldwide since this format's introduction. For the past 10 consecutive years, disc sales have increased, reaching 2.5 Billion discs in 2000, although, since about 2001, sales have declined [46]. Since 1999, the DVD-Video format which allows access to multi-channel audio and digital video at home – being an audio-visual format – is proving to be the fastest-growing consumer-electronic product in history, reaching 10 Million DVD players worldwide [16]. Between 2000 and 2001, the number of homes in Western Europe with DVD-Video players has risen fourfold to 5.4 Million, whereas DVD-disc sales have also increased by approximately 150%. Furthermore, the dedicated audio high-resolution formats of DVD-Audio and SACD are also emerging, having improved with respect to CD-DA fidelity and allowing multi-channel reproduction. However, early sale figures for these formats are far lower than record industry predictions.

At the multi-media end of the market, reliance on limited-bandwidth digital audio delivery, e. g., through the Internet, as well as limited computer-processing and storage resources has resulted in the widespread use of lower-quality, compressed audio formats. The *on-line* music delivery market – often based on compressed-audio streaming techniques [106] – is growing at a very

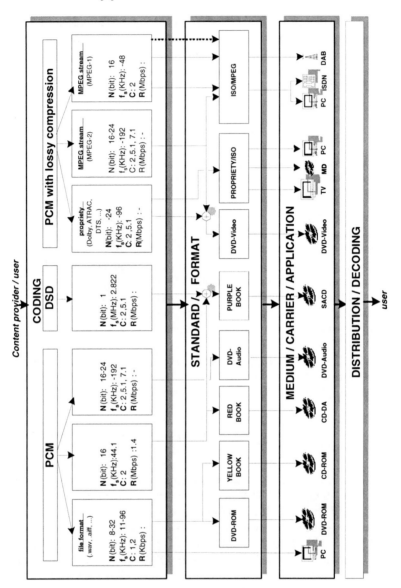

Fig. 12.1. Most prominent audio-coding methods, formats and related applications

fast pace, with an estimated total size of more than 700 Million EUR by 2005. As an example, the 1998 annual report by one of the Internet audio providers already claimed 54 Million users since the format's introduction in 1995, with a growth of 175,000 player downloads per day [16]. This example highlights a significant transformation in the audio and related industries. Traditionally, consumer audio-electronic manufacturers were focusing on technologies and standards, while recording industry was responsible for content and retailer trade was responsible for distributing these hardware and software products. It is now becoming evident that most of the above market segments can be integrated into service industries delivering technologies and content to networked home-audio devices [78]. Currently, such networked systems are PC-based although, in the near future, they will be based on networked audio-visual recording/reproduction devices, i.e. "digital hub/set-top" [101].

Similarly, the current PC-based audio device market, including sound cards, self-powered speakers, etc., is an ever-expanding area with an approx. 25% annual-growth rate, with total sales already covering a significant portion of the audio and Hi-Fi- product market. A significant sector of this market is covered by software-based audio products, such as editors, processors, virtual instruments. Portable digital audio devices, e.g., *mp3*-players, are becoming very popular with young people, with a projected worldwide market size of 20 Million units by 2005.

Comparable detailed figures are not readily available for the professional audio end of the market. However, it is evident that a significant portion of the studio and broadcasting audio components and systems, e.g., mixing consoles, recorders and processors, have been progressively implemented as all-digital devices. Nevertheless, the market segment for installations of public-address systems, PA, and audio-systems still appears to rely on analogue devices to a large extent.

12.2 Technology Evolution

12.2.1 A Brief History

The theoretical roots of digital audio technology may be traced on early information theory works by *Shannon* [95], *Rice* [84], *Nyquist* and the principle of Pulse-Code Modulation, PCM, as devised by *Reeves* in 1938 [83]. Nevertheless, before the sixties computers and electronics did not allow the practical implementation of such systems. Early applications of digital methods into audio were for computer-music synthesis, e.g., *M. Mathews* [85], and the pioneering work on acoustic modelling by *Schroeder, Logan* and *Atal* [89,91,92], which have preceded by nearly a decade the commercial appearance of real-time digital audio devices [14]. Other significant early developments were for the implementation of PCM audio recording/playback prototypes [42,50,69,87], audio DSP [15], hard-disc editing [40,49] and compact-disc prototypes, later to be standardized as the CD-DA format [21,88].

In perspective, the seventies may be considered as the formative period of this technology, since most theoretical principles were introduced at that time, most of them relating to DSP methods – as will be discussed in more detail in Sect. 12.2.2.

However, practical limitations in electronics, computer-processing power and storage technologies, restricted these developments to remain at a research level. The technology evolved at fast pace during the eighties, mainly due to the emergence of real-time-DSP chips [18, 63] and optical-disc storage media, so that the first products appeared in the consumer and professional audio market. Many new techniques were realized, and many known analogue audio devices were re-developed in the digital domain. This transition of existing audio products and services into the digital domain has required significant development effort and cost, so that a disruptive effect was generated within the audio industry [16, 60]. This effect was somehow balanced-out by the emergence of audio-data-compression methods, based on perceptual audio coding [19, 51]. These developments led in the nineties to the introduction of the various ISO/MPEG standards [20, 73], as well as many novel digital audio products for transmission, broadcasting and storing compressed audio. Within these standards, in the dawn of the new century, have emerged the evolutionary technologies of high-resolution multi-channel digital audio [36,72]. An overview of these past events has also been included in [59]. Table 12.1 provides a brief list of these developments.

12.2.2 Evolutionary Mechanisms

Assessing the above events, it is possible to trace certain dominant underlying evolutionary mechanisms which have enabled these novel methods to be realized by providing the necessary infrastructure, which it is envisaged to also drive future developments.

Digital Signal Processing

Since the mid-sixties, DSP has emerged as a powerful tool for analysis, manipulation and understanding of signal properties. Historically, audio and acoustics represented one of the first areas for the introduction of novel DSP techniques, possibly due to the existing tradition in analogue audio-signal processing, due to the relatively low requirements for computer processing and also due to the direct sensory, i.e. auditory, evaluation of processing results in listening tests. The *IEEE's Signal-Processing Society* was formed out of the Institute of Radio Engineers' *IRE Professional Group on Audio*, formed in 1948. In the mid-sixties, this group evolved into *IEEE's Audio and Electroacoustics Signal-Processing Group*, leading to the past decade's

Table 12.1. Significant developments in digital audio technology

Date	Technique/Application	Name/Application
1961	Digital artificial reverb	*Schroeder & Logan* [89,92]
1960s	Computer music	various, e. g., *Mathews* [85]
1967	PCM-prototype system	NHK [42]
1969	PCM Audio recorder	Nippon Columbia
1968	Binaural technology	*Blauert* [10,12,13]
1971	Digital delay line	*Blesser & Lee* [14]
1973	Time-delay spectroscopy	*Heyser* [44]
1975	Digital music synthesis	*Chowning* [22]
1975	Audio-DSP emulation	*Blesser* et al. [15]
1977	Prototypes of CD & DAD	Philips, Sony, e. g., see [88]
1977	Digital L/S measurement	*Berman & Fincham* [9]
1978-9	PCM / U-matic recorder	Sony
1978	32-ch digital multi-track	3M
1978	Hard-disk recording	*Stockham* [49]
1979	Digital mixing console	*McNally* (BBC)
1981	CD-DA standard	industry e. g., see [21,88]
1983	Class-D amplification	*Attwood*
1985	Digital mixing console	Neve
1987-9	Perceptual audio coding	*Brandenburg, Johnston* [19,51]
1991	Dolby AC-3 coding	Dolby
1993	MPEG-1 Audio standard	ISO, see, e. g., [20,73]
1994	MPEG-2, BC, standard	ISO, see, e. g., [20,73]
1995	DVD-Video standard	industry
1997	MPEG-2, AAC, standard	ISO, see, e. g., [20,73]
1997	SACD proposal	Sony, see, e. g., [72]
1999	DVD-Audio standard	industry, see, e. g., [36]

impressive proliferation of research and application activities in DSP, many outside the domain of audio and acoustic signals [27,52]. The evolution of audio DSP has been highly influenced by the pioneering work of *Schroeder* on digital-filter modelling for acoustic systems [89,91,92], together with the theoretical analysis of *Heyser* [44]. Later, DSP advances from the better-funded field of speech technology have also been adapted to wide-band audio applications. Lately, the most successful area in audio DSP has been the perceptual-coding methods – see [20,43,73] this volume – which, for the past decades, have been the basis for many standards and applications. Table 2 presents a brief overview of these developments.

The latest advances in audio DSP tend to follow and re-enforce trends observed in most other DSP fields, which extend and depart from earlier basic principles [5]. For example, in many audio DSP applications, single-rate has given way to multi-rate filter-bank processing, time-invariant to adaptive processing and linear to non-linear signal processing. This last category of

Table 12.2. Some DSP methods relevant to audio technology

Period	Theory/Method	Application
1950-60	Power spectrum estimation [76, 84]	Audio analysis, spectrograms
	Fast *Fourier* Transform, FFT [24, 25]	Most applications
	Digital filter models for room acoustics [89, 91, 92]	Digital reverb, virtual acoustics, etc.
	Time-frequency analysis for audio/acoustics [44]	Audio analysis, measurement
	Homomorphic filtering [74, 97]	Audio signal enhancement
1970	Digital audio theory [17]	ADC, DAC and system design
	Linear prediction [65], Short-time FFT [3], de-convolution [96, 97]	Coding, music synthesis, active noise cancellation, ANC transducer control, etc.
	FIR, IIR filter design [77]	Digital audio systems
	Digital measurement and system analysis [9, 54, 81, 105]	Audio analysis, measurement
	Modelling audio and acoustic systems [4, 33, 70]	ANC, echo cancellation, transducer/acoustic control
	FM music synthesis [22]	Digital music synthesizers
	Channel-error correction [88]	Optical-disc systems
1980	Adaptive filtering,LMS [102]	ANC, transducer / acoustic control, coding, etc.
	Oversampling methods [2]	ADC, DAC and system design
	Workstation DSP [40, 49, 67, 86, 93]	Software-based systems
	multi-rate processing [28]	Codecs and other systems
	Audio/acoustic de-convolution [23, 70]	Transducer/acoustic control
	Measurement/analysis [62, 90]	Audio analysis, measurement
	Dither theory [98]	ADC, DAC and system design
	Perceptual coding [19, 51]	Audio-data compression and standards, e. g., MPEG-1
1990-2000	Non-linear compensation [56]	Transducer control
	Advanced audio coding and evaluation [6, 59, 73]	Audio data compression and standards, e. g., MPEG-2, -4, ...
	Arrays, beam-forming [41, 57, 66]	Transducer, acoustic control
	ANC & echo cancellation [71]	Industrial, communication and hearing-aid devices
	Physical modelling [85], Structured Audio [100] and automatic audio transcription [1, 64]	Music synthesis and technology, standards, e. g., MPEG-4

more recent DSP methods, which includes *Volterra* systems [82], higher-order spectra, neural networks and related techniques, together with the emerging statistical signal processing methods [37], hold great promise for future audio research and applications. Significantly, audio DSP was always attempting to integrate models of human perception, which are largely based on non-linear,

i. e. signal-dependent, mechanisms either in the warped-frequency-scale domain [58], or based on time-varying input signal-depended thresholds, as is the case with masking as being exploited by the perceptual coders [19,43,51]. Hence, as will be further discussed below, the scope of audio DSP has been greatly expanded and enriched via the fusion with such inherently non-linear and multi-dimensional models of auditory perception.

Digital Electronics, Computer and Networking Technologies

Since the early days, the evolution of digital audio technology has been closely linked to the digital electronics, microprocessor, computer, storage media and related technologies. Sound is a real-time signal. Consequently, for the implementation of any commercially viable product, a real-time response and local storage, typically 10 Mbyte/min, are prerequisites, something that it is not always necessary for research applications. Practically any digital audio system will realize complicated mathematical operations on digital real-time audio signals, with minimum input-to-output latency equal to one sample period – i. e. 22.6 μs for one-channel 44.1-kHz audio. Many of the established audio DSP algorithms operate on such sample-by-sample processing basis, though many operations require manipulation of frequency domain data, which are derived in a block-processing basis, an approach which can also offer speed advantages for specific applications – at the expense of latency.

Processing Power

Although for the implementation of audio DSP a variety of alternative hardware options may be adopted, ranging from the use of off-the-shelf microprocessors, field-programmable gate arrays, FPGAs, to custom-integrated circuits, ASICs, a specific class of optimized microprocessors, termed "DSP chips" has proved to be the most popular approach [32, 53]. The architecture of these DSP chips evolved together with DSP algorithm requirements, a trend that has started in 1982, when Texas Instruments introduced the TMS32010 processor with specialized hardware to enable it to compute a multiplication in a single clock cycle [63]. Currently, improved DSP-chip architectures are available, based on enhanced DSP architectures – e. g., parallel execution units and "very-long instruction words", VLIW, [32,99,101]. Nevertheless, the emergence of desktop audio, i. e. PC-based audio applications, has led to the utilization of general-purpose-computer CPUs, as opposed to the earlier trend of using dedicated accelerator-DSP cards attached to PCs. This trend has been supported by the ever-increasing clock frequency of such CPUs – e. g., in the region of GHz, as opposed to approx. 250 MHz for the latest DSPs – and their later adaptation to signal processing tasks. Currently such general-purpose processors, GPPs, can easily challenge DSP chips for audio-processing speed, allowing numerous real-time-audio software products

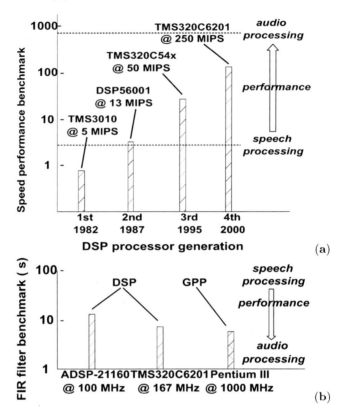

Fig. 12.2. Real-time processor audio performance – adapted from [45]. (a) Speed of different DSP processor generations, (b) DSP and GPP for FIR-filter task

to emerge for the PC environment. In order to evaluate processor power for a specific audio-processing task, application-specific benchmarks have evolved that assess each system over a variety of relevant algorithms and tasks [32,45]. By application of such criteria to DSP chips of different generations, it is becoming evident that only after the mid-nineties processing power became sufficient to accomplish most real-time audio-related tasks – see Fig. 12.2.

By observing similar results for current DSPs and GPPs, it is also evident that since the mid-nineties GPPs became faster, so that real-time desktop audio – e. g., effect processors, editors, synthesizers – could be efficiently implemented in software without special hardware other than a sound-card. Hence, for approximately the last decade, digital audio products can be realized either as stand-alone dedicated devices or as software attached to the common desktop-computer platforms. Let us consider typical examples for audio-related processing tasks and examine the evolutionary trends, attempting to extrapolate past and current data to predictions up to the end of the current decade. As was also noted in [68], processing power is increasing by

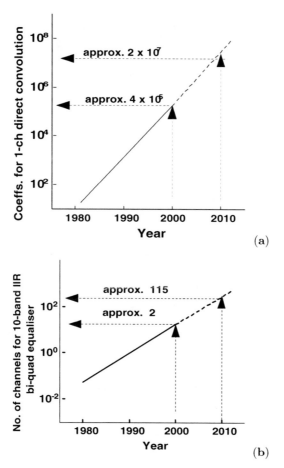

Fig. 12.3. Real-time audio filtering trends – adapted from [45,68]. (**a**) length for block-based FIR filter, (**b**) channels for sample-based 10-band IIR filters

approximately 2.25 times every two years, so that this trend conforms to the *Moore*'s Law of digital electronics [30,68]. Figure 12.3 describes such trends for typical sample-based or block-based audio-equalization applications – for 44.1 kHz audio. These figures indicate that by 2010, or even earlier, power will suffice for all currently existing audio applications, though low-latency sample-based processing is just now becoming sufficient for demanding applications.

Storage Capacity

Following the methodology set by the examples in the previous section, it can be predicted that by 2010, or earlier, more than 500 Gbyte capacity will be readily available for most hard-disc-related audio storage/retrieval tasks,

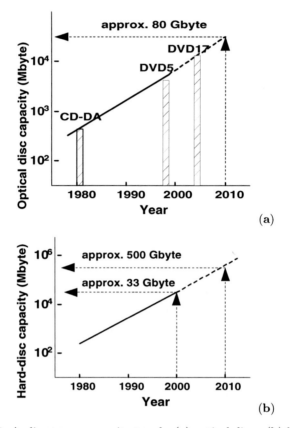

Fig. 12.4. Audio storage capacity trends. (a) optical discs, (b) hard discs

space for approx. 500 CD albums – see Fig. 12.4. For optical-disc media, future evolution will be strongly linked to standards, but at least 80 Gbyte – approx. four times more if blue laser technology is adopted – will be available by 2010 the latest for read-only and possibly recordable optical-disc media. In any case, the above predictions indicate that storage availability will soon exceed most current audio requirements. The evolution of solid-state memory – flash memory, EPROMs, etc. – as future alternative to optical discs indicates that available capacity is already becoming sufficient for storing a single CD-DA album in uncompressed form. Hence, such memories can be employed for lower-quality audio applications, based on down-mixed or compressed audio.

Audio Networking

It is now becoming accepted that audio can be delivered to the user through the Internet in compressed format – e. g., *mp3*, *RealAudio*, [29,106]. Although networking is dependent on many local and non-technological factors, the trends indicate that Internet speed is not growing at the required rate – see

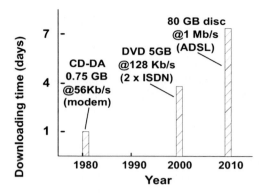

Fig. 12.5. Internet audio delivery speed trends, for different source media

Fig. 12.5 – and will be only sufficient for lower-quality, compressed and/or down-mixed audio.

Considering now the case of local distribution at home, the current wireless *Bluetooth* platform, nominally at 1 Mbit/s and up to 10 m distances between devices, is not sufficient for uncompressed audio transmission [35] – see Fig. 12.6. Nevertheless, WLAN IEEE 802.11g is now sufficient for transmitting CD-quality and multi-channel audio – from DVD-Video – and the extension of this protocol's features, IEEE 802.11e, will allow multi-channel high-resolution audio transmission, provided the cost of these links will become reasonable.

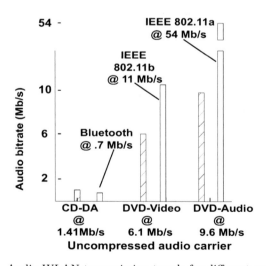

Fig. 12.6. Audio WLAN transmission trends for different source media

Analysis and Modelling of Human Perception

From its introduction, audio engineering has attempted to implement systems conforming to the properties of human hearing and perception [48,107]. Perhaps the most significant difference of digital audio to other consumer electronic technologies, is this unique mixture of engineering practices with objective and subjective perceptual evaluation, at best employing fully developed models of hearing mechanisms – e. g., [6]. Such DSP-based perceptual models offer significant advantages such as optimization to human receiver and reduction of complexity and order. Typical examples of this synergy can be considered to be the increasing significance of data compression and binaural technologies. In both cases, two different peripheral auditory mechanisms, i. e. masking and spatial hearing, well defined through earlier psycho-acoustic research [12,13,107], have been adapted into practical systems. The evolution of real-time perceptual-compression codecs, initially on dedicated DSP chips and more recently in software, has made possible numerous standards and applications to emerge, so that sound can be efficiently transmitted through networks and stored in miniaturized devices.

Similarly, binaural technology is becoming widely employed in many audio applications ranging from architectural-acoustics evaluation to 3-D audio rendering in Internet and multi-media presentations [7,11–13,26,55,57]. It is envisaged that many other audio applications such as source-position-sensing devices, sound-quality-evaluation tools, cocktail-party processors, special microphones for acoustically adverse conditions, etc., will soon emerge from further utilization of binaural technology [11].

However, digital audio technology is also enabling further research on many open questions concerning human auditory perception. In this sense, the previously described engineering exploitation of peripheral perceptual functions appears to be only the beginning for future models which will utilize the largely unexplored field of cognitive psychology [12] – see also [13], this volume. In order to describe these aspects of human performance, the cognitive models will have to remove parts of the known redundancy of the physical principles that govern current audio-engineering practices. This will be especially beneficial to the most power-intensive and theoretically challenging applications such as sound field modelling and control.

12.3 Emerging Technologies

It is always possible that in the following years audio technologies might emerge that could have an unpredictable and disruptive effect in the industry [16]. Nevertheless, it is most likely that future novel technologies would be the result of the interaction of emerging methods – e. g., advanced audio compression, virtual reality, artificial intelligence, expert and agent systems [1], auto-configurable algorithms, automatic and aesthetic-driven audio transcription, advanced interfaces – with promising future consumer electronic applications [60], such as software defined product, portable, low-power devices, home networking, broadband wireless residential multi-media, distributed, remote and peer-to-peer computing, context addressable storage. In these developments, it is also likely that MPEG audio will continue to play a significant role [59]. In the immediate future, some of the emerging and promising areas, likely to become products, are briefly discussed in the following sections.

12.3.1 The All-Digital Audio Chain

In the near future, audio peripherals such as power amplifiers, loudspeakers, microphones could be implemented digitally and linked via wireless means. All-digital power amplification introduces such integrated solutions from the source up to the loudspeaker, while providing high power efficiency – on the order of 90% – small size and low heat dissipation. Up to now, high-power digital amplifiers are implemented via a chain of subsystems, performing audio coding, e. g., oversampling/distortion compensation, 1-bit conversion, e. g., PWM, class-D amplification and output filtering [38]. Single-chip solutions may soon appear, but are currently available only for low-power applications.

Direct acoustic transduction of digital audio data remains a challenging research area. Some alternative solutions are emerging, based on arrays of small 1-bit-actuator elements or binary multiple-voice-coil loudspeakers [34, 41, 47]. Nevertheless, efficient, high-power, low-frequency direct digital audio transduction from small-sized radiating elements still remains an open research field. For microphones, the integration of the transduction element with the analog-to-digital circuit, ADC, has led into the development of tiny microphones built within silicon integrated circuits, ICs [75, 94]. The possibility of further on-chip integration with wireless devices – see also Sect. 3.3 – will allow small-sized, low-cost, interactive acoustic sensing, useful for many applications.

12.3.2 Sound-Field Control

Optimal acoustic reproduction, especially in reverberant enclosures, introduces many theoretical, perceptual and engineering challenges. Wave-field synthesis, i. e. accurate reconstruction of an original sound field via a large number of transducers [8], appears as an alternative to current-generation multi-channel digital audio systems, especially given the potential future availability of all-digital controllable, self-powered speakers. Loudspeaker systems with digitally controlled directivity may further allow accurate sound-beam steering even at low frequencies, hence optimizing audio reproduction into rooms or public spaces [41, 80].

Digital filters, for quite some time now [23, 31, 104], have been employed for single-channel or multi-channel equalization and cross-over design for optimization of loudspeaker responses and the room sound field, often combined with transaural processing for 3-D sound-image rendering. However, such methods can only partially compensate for room reverberation, and their perceptual robustness cannot be easily guaranteed. These methods will benefit from advances in acoustic monitoring elements, e. g., on-chip wireless-microphone arrays. Yet, further research is clearly required into the perception of complex sound fields by the human listeners and evolution of relevant DSP algorithms.

12.3.3 Portable, Networked and Wireless Devices

Digital audio may soon be transmitted wirelessly within networked devices in the home and office environments. *Bluetooth* allows interactive compressed audio transmission to distances up to 10 m [35] and the recent WLAN protocols, IEEE 802.11 b and 802.11 g, will allow increased audio bandwidth for multi-channel applications – see Fig. 12.6. These developments, together with the reduced size and power consumption requirements for solid-state storage media and low-power all-digital amplifiers, allows for small, battery-operated portable sound systems to emerge, possibly adapted into "wearable-audio" form [39, 103].

12.4 Conclusions

The previous sections have shown that past evolution was initiated mainly at industry and university research centers and institutions in Japan, Europe and the United States of America, namely, Philips, Bell Labs., NHK, Sony, BBC, Fraunhofer IIS, among others. For the next decade, highly flexible audio delivery will be possible, either via traditional optical-disc carriers, or via the Internet and wireless local networks. However, technical restrictions will still impose limitations on the data rate achieved by these systems, so that audio quality will be appropriately graded. If DVD-Audio and SACD succeed in the marketplace, they will offer high-fidelity multi-channel audio, potentially taking over from CD-DA as dedicated digital audio formats. DVD-Video and compressed multi-channel digital audio formats will clearly dominate audio-visual applications, being integral components of future multi-media, Internet and WLAN media-delivery systems. Desktop audio applications will also further proliferate into the consumer and professional markets, allowing flexible software-based audio storage, manipulation, reproduction and communication with other personal and home-audio devices.

Such developments shall lead to increased use of wireless portable audio systems, which will provide head-related audio devices with compressed-quality stereo, multi-channel or 3-D audio utilizing binaural technology. Such portable systems will wirelessly communicate with desktop computers or disc-players. Room-related audio systems will be based on loudspeakers or loud-speaker arrays. It is reasonable to predict that the reduction in size and cost of digital electronics will allow wireless, self-powered operation and control of each loudspeaker unit, so that adaptive and equalized operation may be achieved, either for room-optimized multi-channel reproduction or listener-specific trans-aural 3-D-audio rendering. Larger scale public audio systems and installations may also benefit from such developments, especially for accurate full-bandwidth directivity control and steering.

It has been shown that, for the next decade, processing power and storage capacity will suffice for the above applications. Computer and communication networks as well as wireless protocols will also partially meet many of the audio-related bandwidth requirements, especially for audio coded via existing or future MPEG and propriety standards. These developments will increase the amount of audio data stored, accessed, delivered and manipulated by the consumer and the audio professional, so that apart from logistical problems, robust and audio-transparent data security must be introduced into such systems. Nevertheless, many perception-related aspects of audio engineering will have to be resolved for this technology to convincingly meet and exploit many of the previously defined tasks.

References

1. Aarts R, Dekkers R (1999) A real-time speech-music discriminator. J Audio Engr Soc 47:720–725
2. Adams R (1986) Design and implementation of an audio 18-bit ADC using oversampling techniques. J Audio Engr Soc 34:153–156
3. Allen J B (1977) Short term spectral analysis, synthesis and modification by Discrete Fourier Transform. IEEE Trans Audio Speech Sig Processg, ASSP 25:235-238
4. Allen J B, Berkley D (1979) Image method for efficiently simulating small-room acoustics. J Acoust Soc Amer 66:943–950
5. Atlas L, Duhamel P (1999) Recent developments in the core of digital signal processing. IEEE Sig Processg Mag 16:16-31
6. Beerends J, Stemerdink J (1992) A perceptual audio quality measure based on a psychoacoustic sound representation. J Audio Engr Soc 40:963–978
7. Begault D (1991) Challenges to the successful implementation of 3-D sound. J Audio Engr Soc 39:864–870
8. Berkhout A J (1988) A holographic approach to acoustic control. J Audio Engr Soc 36:977–995
9. Berman J, Fincham L (1977) The application of digital techniques to the measurement of loudspeakers. J Audio Engr Soc 26:370–384
10. Blauert J (1968) A contribution to the persistence of directional hearing in the horizontal plane (in German). Acustica 20 :200–206
11. Blauert J (1999) Binaural auditory models. Proc 18th DANAVOX Symp, Scant-icon, Kolding

12. Blauert J (1997) Spatial hearing: the psychophysics of human sound localization. MIT Press, Cambridge MA
13. Blauert J (2005) Analysis and synthesis of auditory scenes, Chap 1 this vol
14. Blesser B, Lee F (1971) An audio delay system using digital technology. J Audio Engr Soc 19:393–397
15. Blesser B, et al. (1975) A real-time digital computer for simulating audio systems. J Audio Engr Soc 23:698–707
16. Blesser B, Pilkington D (2000) Global paradigm shifts in the audio industry. J Audio Engr Soc 48:861–872
17. Blesser B (1978) Digitization of audio: a comprehensive examination of theory, implementation and current practice. J Audio Engr Soc 26:739–771
18. Bloom P J (1985) High-quality digital audio in the entertainment industry: an overview of achievements and challenges. IEEE Audio Speech Sig Processg Mag, ASSP 2:2–25
19. Brandenburg K (1988) High quality sound coding at 2.5 bits/sample. 84th Audio Engr Soc Conv preprint 2582, Paris
20. Brandenburg K, Bossi M (1997) Overview of MPEG audio: current and future standards for low-bit-rate audio coding. J Audio Engr Soc 45:4–21
21. Carasso M G, Peek J, Sinjou J (1982) The Compact Disc digital audio system. Philips Techn Rev 40:149–180
22. Chowning J (1973) The synthesis of complex audio spectra by means of frequency modulation. J Audio Engr Soc 21:526–534
23. Clarkson P M, Mourjopoulos J N, Hammond J K (1985) Spectral, phase and transient equalization of audio systems. J Audio Engr Soc 33:127–132
24. Cooley J (1992) How the FFT gained acceptance. IEEE Sig Processg Mag 9:10–13
25. Cooley J, Tukey J (1965) An algorithm for the machine calculation of complex Fourier series. Math of Comp 19:297–301
26. Cooper D, Bauck J (1989) Prospects for transaural recording. J Audio Engr Soc 37:3–19
27. Cox R (2002) The ghost of ICASSP past. IEEE Sig Proc Mag 19:7–9
28. Crochiere R, Rabiner L (1985) multi-rate digital signal processing. Prentice-Hall, Englewood Cliffs NJ
29. Dietz M, Popp H, Brandenburg K, Friedrich R (1996) Audio compression for network transmission. J Audio Engr Soc 44:58–72
30. Eldering C, Sylla M, Eisenach J (1999) Is there a Moore's law for bandwidth? IEEE Comm Mag 37:117–121
31. Elliot S, Nelson P (1989) Multi-point equalization in a room using adaptive digital filters. J Audio Engr Soc 37:899-907
32. Eyre J, Bier J (1999) DSPs court the consumer. IEEE Spectrum 36:47–53
33. Flanagan J L, Lummis R (1970) Signal processing to reduce multi-path distortion in small rooms. J Acoust Soc Amer 47:1475–1481
34. Flanagan J L (1980) Direct digital-to-analog conversion of acoustic signals. Bell Syst Techn J 59:1693–1719
35. Floros A, Koutroubas M, Tatlas N A, Mourjopoulos J N (2002) A study of wireless compressed digital-audio transmission. 112th Audio Engr Soc Conv, preprint 5516, Munich
36. Fuchigami N, et al. (2000) DVD-Audio specifications. J Audio Engr Soc 48:1228–1240

37. Godsill S, Rayner P (1998) Digital audio restoration: a statistical model based approach. Springer, Berlin Heidelberg New York
38. Goldberg J M, Sandler M B (1991) Noise shaping and PWM for all-digital power amplifier. J Audio Engr Soc 39:449–460
39. Gough P, Eves D (2000) Wearable electronics. Philips Res Password 3:5-9
40. Griffiths M, Bloom PJ (1982) A flexible digital sound-editing program for minicomputer system. J Audio Engr Soc 30:127–134
41. Hawksford M (2001) Smart directional and diffuse digital loudspeaker arrays. 110th Audio Engr Soc Conv, preprint 5362, Amsterdam
42. Hayashi R (1969) Stereo recorder. NHK Tech Rep 12:12–17
43. Heute U (2003) Speech and audio coding: aiming at high quality and low data rates. Chap 14, this vol
44. Heyser R (1971) Determination of loudspeaker arrival times, part I. J Audio Engr Soc 19:734–743
45. http://www.BDTI.com (2001) Berkley design technologies benchmark results 2000. Accessed February 2002
46. http://europa.eu.int/comm/eurostat/ (2002) Statistics on audiovisual services. Accessed Febr. 2002
47. Huang Y, Busbridge S C, Fryer P A (2000) Interactions in a multiple-voice-coil digital loudspeaker. J Audio Engr Soc 48(6):542–552
48. Hunt V (1954) Electroacoustics. Amer Inst Physics, Acoust Soc of Amer Publications, Sewickley PA
49. Ingebretsen R, Stockham TG (1984) Random-access editing of digital audio. J Audio Engr Soc 32:114–122
50. Iwamura H, et al (1973) Pulse-code-modulation recording system. J Audio Engr Soc 21(9):535–541
51. Johnston J (1988) Transform coding of audio signals using perceptual criteria. IEEE J Selected Areas Comm 6:314–323
52. Kahrs M (1997) The past, present and future of audio signal processing. IEEE Sig Processg Mag 14:30–57
53. Kahrs M, Brandenburg K (ed) (1998) Applications of digital signal processing to audio and acoustics. Chap 5, Kluwer, Norwell, MA
54. Kates J (1977) Digital analysis of loudspeaker performance. Proc IEEE 65:377–380
55. Kleiner M, Dalenbäck B I, Svensson P (1993) Auralization – an overview. J Audio Engr Soc 41:861–875
56. Klippel W (1992) The mirror filter – a new basis for reducing nonlinear distortion and equalizing response in woofer systems. J Audio Engr Soc 40:675–691
57. Kyriakakis C, Tsakalides P, Holman T (1999) Surrounded by sound. IEEE Sig Processg Mag 16:55–66
58. Laine U, Karjalainen M, Altosaar T (1994) Warped linear prediction (WLP) in speech and audio processing. Proc IEEE Int Conf Audio Speech Sig Processg, ICASSP 94:349–352
59. Lindsay A, Herre J (2001) MPEG-7 and MPEG-7 Audio – an overview. J Audio Engr Soc 49:589–594
60. Lipoff S (2002) Consumer electronics technology megatrends. IEEE Consumer Electron Soc Newsletter Feb 2002:11–14.
61. Lipshitz S (1998) Dawn of the digital age. J Audio Engr Soc 46:37–42

62. Lipshitz S, Scott T, Vanderkooy J (1985) Increasing the audio measurement capability of FFT analyzers by microcomputer postprocessing. J Audio Engr Soc 33:626–648
63. Magar S, Caudel E, Leigh A (1982) A microcomputer with digital signal processing capability. Proc IEEE Solid-State Circ Conf, 32–35
64. Maher R (1990) Evaluation of a method for separating digitized duet signals. J Audio Engr Soc 38:956–979
65. Makhoul J (1975) Linear prediction: a tutorial review. Proc IEEE 63:561–580.
66. Marro C, Mahieux Y, Simner K (1998) Analysis of noise reduction and dereverberation techniques based on microphone arrays and postfiltering. IEEE Trans Speech Audio Processg, SAP 6:240–259
67. Moorer J A (1982) The audio signal processor: the next step in digital audio. Digital audio. Collected papers AES Premiere Conf, Rye NY, 205–215
68. Moorer J A (2000) Audio in the new millennium. J Audio Engr Soc 48:490–498
69. Myers J, Feinburg A (1972) High-quality professional recording using new digital techniques J Audio Engr Soc 20:622–628
70. Neely S, Allen J B (1979) Invertibility of a room impulse response. J Acoust Soc Amer 66:165–169
71. Nelson P, Hammond J K, Elliott S (1990) Active control of stationary random sound fields. J Acoust Soc Amer 87:963–975
72. Nishio A, et al. (1996) Direct stream digital audio system. Audio Engr Soc 100th Conv, preprint 4163, Copenhagen
73. Noll P (1997) MPEG digital audio coding. IEEE Sig Processg Mag 14:59–81
74. Oppenheim A V, Schafer R (1968) Nonlinear filtering of multiplied and convolved signals. Proc IEEE 56:1264–1291
75. Ouellette J (2000) Echoes. J Acoust Soc Amer Newsletter 10:2-3
76. Papoulis A (1962) The Fourier integral and its applications. McGraw Hill, New York NY
77. Peled A, Liu B (1974) A new hardware realization of digital filters. IEEE Trans Acoust Speech Sig Processg, ASSP 22:456–462
78. Perry T (2001) Service takes over in the networked world. IEEE Spectrum 38:102–104
79. Pohlmann K (1995) Principles of digital audio. Mc-Graw Hill, New York NY
80. Pompei J F (1999) The use of airborne ultrasonics for generating audible sound beams. J Audio Engr Soc 47:726–731
81. Preis D (1976) Linear distortion. J Audio Engr Soc 24:346–367
82. Reed M, Hawksford M (1996) Practical modeling of nonlinear audio systems using the Volterra series. 100th Audio Engr Soc Conv, preprint 4264, Copenhagen
83. Reeves A (1938) Electrical signal system. French patent 852,183. British patent 535,860. US patent 272,070
84. Rice S (1944) Mathematical analysis of random noise. Bell Syst Techn J 23:383–332
85. Roads C (1996) The computer music tutorial. MIT Press, Cambridge, MA
86. Sakamoto N, Yamaguchi S, Kurahashi A, Kogure (1981) Digital equalization and mixing circuit design. Audio Engr Soc 70th Conv. preprint 1809, New York
87. Sato N (1973) PCM recorder. J Audio Engr Soc 21:542–558
88. Schouhamer Immink K A (1998) The compact disc story. J Audio Engr Soc 46:458–465

89. Schroeder M R (1961) Improved quasi-stereophony and "colorless" artificial reverberation. J Acoust Soc Amer 33:1061–1064
90. Schroeder M R (1981) Modulation transfer functions: definition and measurement. Acustica 49:179–182
91. Schroeder M R, Atal B S (1963) Computer simulation of sound transmission in rooms. IEEE Int Conv Report Part 7:150-153
92. Schroeder M R, Logan B F (1961) Colorless artificial reverberation. J Audio Engr Soc 9:192–197
93. Sekiguchi K, Ishizaka K, Matsudaira T, Nakajima N (1983) A new approach to high-speed digital signal processing based on microprogramming. J Audio Engr Soc 31:517-522
94. Sessler G (1993) New acoustic sensors. 94th Audio Engr Soc Conv, preprint 3525, Berlin
95. Shannon C E (1948) A mathematical theory of communications. Bell Sys Techn J 27: 379-423, 623-656
96. Silverman H, Pearson A (1973) On deconvolution using the DFT. IEEE Trans Audio Electr, AU 2:112–118
97. Stockham T, Cannon T, Ingebretsen R (1975) Blind deconvolution through digital signal processing. Proc IEEE 63:678–692
98. Vanderkooy J, Lipshitz S (1987) Dither in digital Audio. J Audio Engr Soc 35:966–975
99. various (2002) DSP technology in industry. IEEE Sig Processg Mag 19:10-78
100. Vercoe B, Gardner W, Scheirer E (1998) Structured audio: creation, transmission and rendering of parametric sound representations. Proc IEEE 86:922–936
101. Wallich P (2002) Digital hubbub. IEEE Spectrum 39:26–33
102. Widrow B, Stearns S (1985) Adaptive signal processing. Prentice-Hall, Englewood Cliffs NJ
103. Wijngaarden S, Agterhuis E, Steeneken H (2000) Development of the wireless communication ear plug. J Audio Engr Soc 48:553–558
104. Wilson R, Adams G, Scott J (1989) Application of digital filters to loudspeaker crossover networks. J Audio Engr Soc 37(6):346-367
105. Wyber R (1974) The application of digital processing to acoustic testing. IEEE Trans Acoust Speech Sig Processg, ASSP 33:66-72
106. Xu A, Woszczyk W, Settel Z, Pennycool B, Rowe R, Galanter P, Bary J, Martin G, Corey J, Cooperstock J (2000) Real-time streaming of multi-channel audio data over the Internet. J Audio Engr Soc 48:627-641
107. Zwicker E, Zwicker U (1991) Audio engineering and psychoacoustics: matching signals to the final receiver. J Audio Engr Soc 39:115-126

13 Speech Production – Acoustics, Models, and Applications

Arild Lacroix

Institute of Applied Physics, Goethe-University Frankfurt, Frankfurt am Main

Summary. The mechanism of speech production is explained for the different speech sounds. The role of the most relevant articulators is discussed as well as, to a certain extent, sound excitation. The main cavities of the speech-production system are modelled by concatenation of short homogeneous acoustic tubes, interconnected by appropriate adaptors. Based on the concept of forward- and backward-travelling waves, signal-flow graphs are given, describing pressure, flow or power waves. Losses can be considered either by lumped impedances or distributed along the tube by properly designed filters. The estimation of a lossless un-branched tube system can easily be achieved by linear prediction from the speech signal. If losses are included and the tube system is branched, optimization algorithms can be applied for the parameter estimation. The termination at the glottis is assumed to be either fixed or time dependent according to the glottal opening. The mouth opening is described by a simple frequency-dependent termination. Applications in the fields of speech synthesis, source coding and recognition are briefly discussed.

13.1 Introduction

The production of speech is the result of a complex feedback process in which, besides acoustical processing, information processing in the peripheral auditory system and in the central nervous system and, consequently, perception are involved. In Fig. 13.1 this feedback process is shown schematically.

In this contribution only the forward path of the feedback system will be considered. This path includes sound excitation, acoustical wave propagation within the speech-production system, and radiation from mouth and/or nostrils [8, 10]. Sometimes the sound excitation is considered as part of the speech production system. In this contribution the sound excitation is treated separately from the processing in the articulation channel.

13.2 Mechanism of Speech Production

The organs which are contributing to the speech production process can be explained in brief with the aid of a mid-sagittal section of the human head as shown in Fig. 13.2. The organs which are active during articulation are called articulators, namely, tongue, lower jaw, lips, and *velum*.

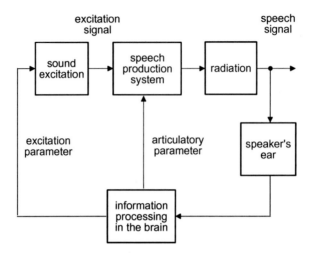

Fig. 13.1. Feedback structure of the speech production process

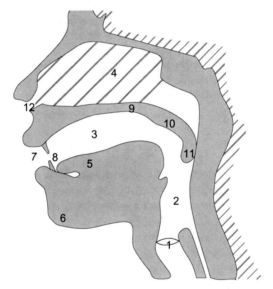

Fig. 13.2. Mid-sagittal section of the human head. Speech organs: 1 vocal cords, 2 pharynx, 3 mouth cavity, 4 nasal cavity with septum, *hatched*, 5 tongue, 6 lower jaw, 7 lips, 8 teeth, 9 palate, 10 velum, 11 uvula, 12 nostrils

Voiced speech sounds like vowels and vowel-like consonants are generated by the vibrating vocal cords, which are driven by a constant air flow from the lungs. The resulting air flow is intermitted due to the fact that the opening between the vocal cords – also known as *glottis* – varies in area between zero and a certain value which mainly depends on the speech sound, the intensity, and the fundamental frequency. The result is an acoustic pulse train of particular shape which can also be characterized by a wide-band spectral function with a decay for increasing frequencies of approximately 12 dB/octave. This kind of excitation is called phonation.

The pulse-like acoustic wave propagates through a complex system of acoustical cavities consisting of *pharynx*, mouth cavity and nasal cavity The nasal cavity is coupled at the velum for certain speech sounds like nasals and nasalized vowels. If the velum is in the upper position, i. e. closed, the nasal cavity is decoupled from the speech production system. In this way, non-nasalized vowels are produced by the excitation of pharynx and mouth cavity alone. For nasal sounds and nasalized vowels the velum is in the lower position, i. e. open. For nasal sounds the mouth cavity is closed at the lips, speech sound [m], at the alveolar position, [n], or at the velar position, [ŋ]. Hence, the sound is transmitted through pharynx, open velum, and nasal cavity while the mouth cavity is still coupled at the velum as a closed acoustical resonator. For nasalized vowels all cavities are used for sound transmission and the sound is radiated from mouth and nostrils simultaneously. It should be mentioned that the nasal cavity is coupled at the velum also for non-nasals and non-nasalized vowels in certain situations due to co-articulation [2].

Most of the consonants, like voiceless fricatives, are not excited by a pulse train but by a wide-band noise signal which results from turbulent air flow generated by constrictions at certain *uvular, velar, palatal, alveolar, dental* or *labial* positions. The corresponding speech sounds are designated according to the place of constrictions, which is named after the point of articulation. For voiced fricatives, like [z], additionally phonation is active yielding speech sounds which are partly voiced, partly unvoiced. But also fully voiced consonants like the glide [j] exist. The lateral [l] is characterized by a mouth cavity separated by the tongue into two ducts. For the vibrants tongue–r, [r], or uvular–r, [R], the sound intensity is modulated by the vibrating tip of the tongue or by the vibrating uvula.

Plosives are generated by a complete closure at the palate, namely, speech sounds [g] and [k], at the alveolar, [d] and [t], or at the labial position, [b] and [p] – with the velum being closed in all these cases. As a consequence, a constant pressure is build up which results in the production of the plosives after the closure is released. Plosives are preceded by a short pause which is necessary to produce the constant pressure. Plosives may be accompanied by aspiration. Whispered speech is excited by turbulent noise, generated by a constriction at the vocal cords.

A complete list of the speech sounds which is widely adopted is known as the IPA-alphabet [14]. For the vowels a systematic arrangement can be depicted, based on the tongue-hump position and the degree of opening between palate and tongue hump or tongue height. An equivalent system exists for consonants where primarily the points of articulation, namely, the places of constriction, are used as discriminative feature.

13.3 The Acoustics of Speech Production

The acoustic wave propagation within the speech production system can be described by the wave equation, which is a partial differential equation. If plane waves can be assumed, which is valid for the range of audible frequencies under consideration, *Webster*'s horn equation [8, 10, 51] is adequate. Due to articulation the coefficients in this equation are time dependent. According to the variation of the glottal area during phonation the source impedance is also time variant – which contributes further to the property of time-variance of the partial differential equation.

The situation is even more complicated as the actual specific kind of excitation and the position where excitation takes place also vary in time, namely, in accordance with the speech sound which is about to be generated. During intervals of glottal opening, interaction is possible between the speech production system and the sub-glottal tract – which consist of *trachea* and lungs. Losses occur due to viscous flow at the boundaries and due to thermal conduction through the walls. The vibrating cheeks or lips also cause losses, especially at low frequencies.

The treatment of wave propagation is considerably simplified when piecewise-uniform tubes and time invariance are assumed, the latter at least during short time intervals.

13.4 Speech-Production Models Based on Discrete-Time Acoustic Tubes

In this section the concept of wave quantities is used, where incident waves are denoted by A and reflected waves by B. There exist three definitions which are related to different physical quantities, namely, pressure waves, flow waves, and power waves. The dimension of the power waves is actually equal to the square root of power.

In Fig. 13.3 a signal-flow graph is shown which enables the representation of two lossless uniform acoustic tubes of length l_0 in terms of pressure waves. If the sampling period, T, of a discrete-time implementation is adapted to the propagation time, $\tau_0 = l_0/c = T/2$, with the sound velocity, c, in air, then we are able to describe the behaviour of the acoustic tube completely in discrete time.

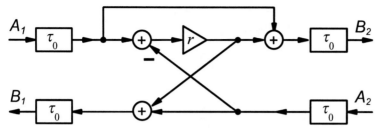

Fig. 13.3. Signal-flow graph describing pressure waves for a junction of two lossless uniform tubes of same length

This is important, e. g., for simulations on computers and for realizations with the aid of signal-processor components. The non-zero reflection coefficient, r, represents the signal flow for pressure waves in the case of cross-sectional-area variations of adjacent acoustical tubes. Fig. 13.4 depicts the situation for the cross-sectional areas, S_1 and S_2, of adjacent acoustical tubes.

The effective reflection coefficient, r, is a simple function of the areas, S_1 and S_2, and of the acoustical tube impedances, Z_1 and Z_2, respectively. Depending on the ratio of areas the resulting reflection coefficient varies between plus and minus one. For $S_1 = S_2$ the reflection coefficient has the value of zero. It is well known that for the open tube end, $S_2 \to \infty$, the reflection coefficient is dependent on frequency [25, 47]. If for an open tube end a constant, non-frequency-dependent, reflection coefficient is assumed, according to *Lord Rayleigh* a geometrical correction of the effective tube length is necessary towards a longer tube [40].

A model of the speech-production system which consists of a concatenation of uniform acoustic tubes is given in Fig. 13.5. Between nasal cavity, pharynx and mouth cavity a connection is necessary for the speech sounds where the velum is in the lower position. In contrast to pharynx and mouth cavities the structure of the nasal cavity is much more complicated in the following way.

- The nasal cavity is divided by the septum into two more or less symmetric branches,
- three pairs of para-nasal cavities are connected to the nasal cavity via thin channels,
- the topology of the nasal cavity is complicated in the region of the *conchae* and the *ethmoid* bone.

Therefore the nasal cavity must certainly be modelled by more complex structures than the ones required for a single cavity.

In Fig. 13.3 the lossless tube elements are connected by a two-port adaptor in terms of the wave-digital-filter concept [9]. As already mentioned, a pressure-wave adaptor has been used in Fig. 13.3. More signal-flow graphs for the two-port adaptor are shown in Table 1 for flow waves, pressure waves,

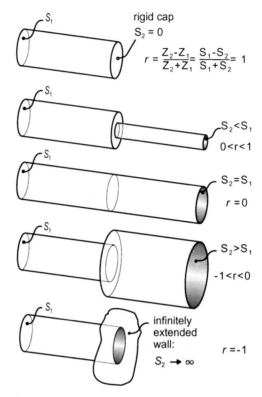

Fig. 13.4. Junction between two uniform acoustic tubes

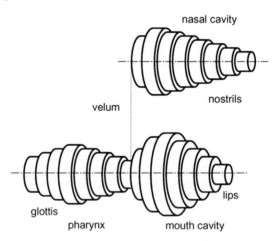

Fig. 13.5. Tube model of speech production. At the *dotted line*, a connection exists if the nasal cavity is coupled to pharynx and mouth cavity with the velum in the lower position

Table 13.1. Different realizations of two-port adaptors: (b) pressure wave adaptor, (c) flow wave adaptor, (e) power wave adaptor. The adaptors (a) and (d) have the same transfer behaviour as the other adaptors, except for a constant gain which depends on the reflection coefficient. In the *center* and *right columns* the associated scattering and scattering transfer matrices are given – from [23]

	S	T
a)	$\begin{pmatrix} r & 1-r^2 \\ 1 & -r \end{pmatrix}$	$\begin{pmatrix} 1 & r \\ r & 1 \end{pmatrix}$
b)	$\begin{pmatrix} r & 1-r \\ 1+r & -r \end{pmatrix}$	$\dfrac{1}{1+r}\begin{pmatrix} 1 & r \\ r & 1 \end{pmatrix}$
c)	$\begin{pmatrix} r & 1+r \\ 1-r & -r \end{pmatrix}$	$\dfrac{1}{1-r}\begin{pmatrix} 1 & r \\ r & 1 \end{pmatrix}$
d)	$\begin{pmatrix} r & 1 \\ 1-r^2 & -r \end{pmatrix}$	$\dfrac{1}{1-r^2}\begin{pmatrix} 1 & r \\ r & 1 \end{pmatrix}$
e)	$\begin{pmatrix} r & \sqrt{1-r^2} \\ \sqrt{1-r^2} & -r \end{pmatrix}$	$\dfrac{1}{\sqrt{1-r^2}}\begin{pmatrix} 1 & r \\ r & 1 \end{pmatrix}$

Table 13.2. Discrete time equivalents for the uniform acoustic tube including losses. (a) same propagation delay in both directions of $\tau_0 = T/2$ with sampling period T, (b) propagation delay forward equal to $2\tau_0 = T$, (c) propagation delay backward equal to $2\tau_0 = T$. In the *center* and *right columns* the associated scattering and scattering transfer matrices are given – from [23]

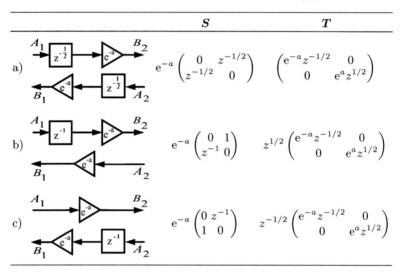

	S	T
a)	$e^{-a}\begin{pmatrix} 0 & z^{-1/2} \\ z^{-1/2} & 0 \end{pmatrix}$	$\begin{pmatrix} e^{-a}z^{-1/2} & 0 \\ 0 & e^{a}z^{1/2} \end{pmatrix}$
b)	$e^{-a}\begin{pmatrix} 0 & 1 \\ z^{-1} & 0 \end{pmatrix}$	$z^{1/2}\begin{pmatrix} e^{-a}z^{-1/2} & 0 \\ 0 & e^{a}z^{1/2} \end{pmatrix}$
c)	$e^{-a}\begin{pmatrix} 0 & z^{-1} \\ 1 & 0 \end{pmatrix}$	$z^{-1/2}\begin{pmatrix} e^{-a}z^{-1/2} & 0 \\ 0 & e^{a}z^{1/2} \end{pmatrix}$

power waves, and for additional adaptors for which no direct physical equivalent can be given [23]. In [20] a direct realization of the scattering matrix for pressure waves leading to a four-multiplier adaptor has already been derived. Three equivalent discrete-time elements are given in Table 2 for the tube elements, which differ in group delay for both directions of propagation.

The fractional-delay elements can be avoided by one of the other tube elements. If the propagation delay should be equal to the physical delay then the tube elements with one delay in forward or backward direction can be used alternately. At the position of the velum a three-port adaptor is necessary for proper distribution of the sound waves between the three main cavities [22].

With the aid of two-port adaptors, three-port adaptors and tube elements as basic components, it is possible to model concatenated and branched systems of acoustic tubes like the one in Fig. 13.5. For voiced speech sounds the tube model is excited at the glottis by an almost periodic pulse train of appropriate shape, with the pulse period being chosen according to the fundamental frequency, which is an important quantity for the correct intonation. A noise-like signal is used as excitation at the glottis position for whispered speech. For consonants the excitation is located at the point of articulation. In these cases, a noise-like signal is fed into the system at this point. For a correct modelling the excitation signal has to be transmitted in both directions, forward and backward.

With the use of suitable adaptors it is possible to model the time-variant behaviour of the speech-production system due to articulation [33,50]. However, it turns out in experiments with speech signals that conventional adaptors may yield good results as well [7, 33]. During phonation the glottis opening varies almost periodically with time. By using a time-variant glottis impedance this effect can be included into the model [34,41].

Losses during wave propagation cause different dependencies of the attenuation with respect to frequency [10].

- Friction caused by viscous flow and thermal conduction through the walls lead both to an attenuation proportional to the square root of frequency,
- wall vibrations result in an attenuation which is primarily effective at low frequencies.

Additional losses are caused by the terminations at glottis, mouth, and nostrils. Frequency-independent losses of the low-loss type can be modelled by attenuation coefficients in both branches of the tube elements in Table 2. Losses can also be introduced into the system by lumped impedances [19,26]. Frequency-dependent losses according to the above mentioned loss types can be introduced by properly designed filters in both branches of each tube element [46]. The characteristic of the radiation can be included by a relatively simple filter at the output of the system [24].

The different types of adaptors allow for precise physical modelling of either pressure, flow or power waves [21]. Nevertheless, although each of the two-port adaptors may be used, different attenuation factors occur during the wave propagation along the concatenated tubes.

Recently an interactive software package, SPEAK, has been developed in our laboratory, which allows for representation, modification and simultaneous analysis of the cross-sections of a tube model as well as its transfer functions in the z-domain, its frequency responses, and its impulse responses [39]. In Fig. 13.6 a screen-plot is given as an example.

13.5 Parameter Estimation

For an un-branched concatenation of acoustic tubes of the same length the estimation of the reflection coefficients is possible on the basis of linear prediction, which goes back to a famous publication of the Frenchman *Prony* [36]. Later, linear prediction approaches have been developed in a way which enables the direct calculation of predictor and reflection coefficients [1,3,15,29,30,32]. The reflection coefficients in which we are primarily interested can be calculated in two ways, as follows.

- After computation of the autocorrelation function of the speech signal a set of linear equations with *Toeplitz* structure is solved with the aid of the *Levinson–Durbin* algorithm [32].

– Inverse filtering by a concatenation of non-recursive two-ports which operate inversely with respect to the assumed tube model [3, 15, 30, 32]. The calculation of the reflection coefficients is performed from stage to stage of the inverse filter. For this purpose there are essentially two formulas in use which have been proposed by *Burg* [3] and by *Itakura* and *Saito* [15]. Although being of different origin, the results obtained from these methods differ only marginally – at least for speech signals.

When analyzing the transfer function of un-branched tube models, it turns out that the modelling results represent all-pole filters which are characterized by a number of resonances, according to the order of the transfer function. However, no zeros exist which could contribute to the frequency response. Nevertheless, if the parameters of these all-pole models are estimated properly, satisfactory results are achieved for natural speech signals, especially for vowels. This means that, at least for vowels, the all-pole filter models the acoustics of the vocal tract adequately for many applications. Yet, if one looks at the vowel spectra in more detail it turns out that apparently even for vowels the all-pole model has its deficiencies.

In Fig. 13.7 a the magnitude response of an all-pole model of order 30 is compared to the DFT-spectrum of the vowel [i:]. The spectral deviations of

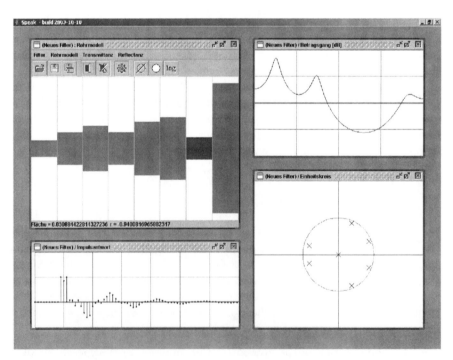

Fig. 13.6. Cross-section of a tube model, magnitude response, impulse response, and z-domain plot, interactively analyzed with the software SPEAK

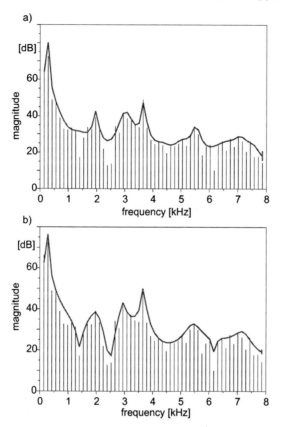

Fig. 13.7. Magnitude response, *solid line*, of (**a**) an all-pole model of order 30, (**b**) a pole-zero model with 20 poles and 10 zeros in comparison with the DFT spectrum of the vowel [i:] – from [44]

the all-pole model, which still exist even for the high order of 30, are apparent. In contrast to the all-pole model, the performance of a pole-zero model with 20 poles and 10 zeros is considerably better – compare Fig. 13.7 b. The pole-zero model has been achieved with the aid of an estimation algorithm described in [44]. The deviations in Fig. 13.7 a may be associated either with the excitation or with partial nasalization.

In Fig. 13.8 a cross-sectional areas are shown which have been derived from the reflection-coefficient estimation of an un-branched tube model. The model is time variable due to a variable glottis coefficient. Additionally the tube termination at the lips is dependent on frequency and differs for individual speech sounds [41, 43]. Prior to the estimation procedure the influence of excitation and radiation has been separated from the speech signal by inverse filtering with up to three real poles being removed. For comparison cross-sectional areas as obtained from NMR-data [49] are shown in Fig. 13.8 b which

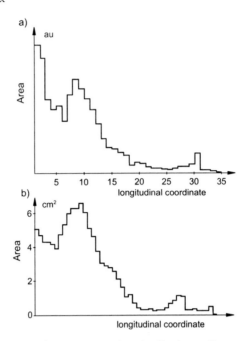

Fig. 13.8. Cross-sectional areas versus longitudinal coordinate, mouth *left*, glottis *right*, (**a**) of a 33-stage tube model [from 43] analyzed from the vowel [a:], (**b**) from NMR-data for the same vowel, yet uttered by a different person [49]

represent the same vowel [a:] as in Fig. 13.8 a, but articulated by a different person. The overall shape of both area functions is similar, indicating that the separation of source and radiation as well as the remaining estimation procedure was successful. Similarly good results have been achieved for other speech sounds under investigation [41].

If the tube estimation is applied to the production of speech with the nasal tract being involved, the situation is much more complicated. It is still possible to determine the parameters, e. g., poles and zeros of a general pole-zero model, as already demonstrated for the example in Fig. 13.7 b. However, the relation between these parameters and the parameters of branched tube models is ambiguous [45]. Good results have been achieved, e. g., for nasal sounds under the assumption that the length of the mouth cavity is known [27].

Inverse filtering can be used if the signals from mouth and nostrils are available separately. An algorithm is then required for the separation of mixed sources, e. g., [18] which is difficult in this case due to the high degree of correlation between the two sources. In our laboratory we have the possibility of recording the signals from mouth and nostrils separately. With this information we can apply inverse filtering in order to yield estimates of mouth cavity and nasal cavity [28]. In [42] an optimization procedure has been developed to estimate the parameters of the entire tube model.

The nasal cavity has been investigated to quite some extent. As has been mentioned above, its topology including three pairs of para-nasal cavities is complicated. In Fig. 13.9, the nasal cavity is shown in a 3D-view in order to give an impression of its complex structure.

The 3-D data for this figure have been obtained from CT recordings at the university hospital in Frankfurt. Based on these data, wave propagation has been analyzed with the aid of a finite-differences method, yielding the impulse responses [38]. In Fig. 13.10, DFT spectra of the impulse responses of both branches of the nasal cavity are compared to magnitude responses of tube models, including one pair of para-nasal cavities [37]. The tube parameters have been estimated from the impulse responses. If more para-nasal cavities are coupled, an even better spectral approximation can be achieved.

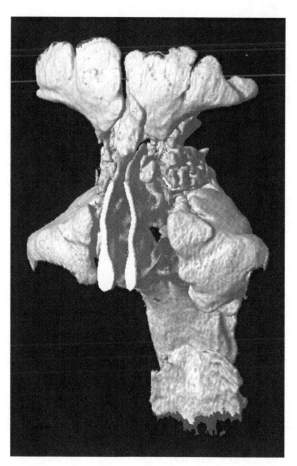

Fig. 13.9. 3-D view of the nasal cavity between nose, *white areas in front*, and velum, *bottom*, including three pairs of para-nasal cavities, i.e. frontal sinus, *top*, maxillary sinus, *left and right*, sphenoidal sinus, *not visible* – from [38]

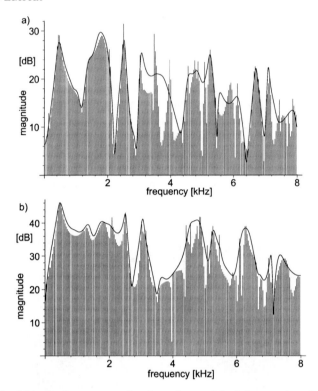

Fig. 13.10. Magnitude response for the tube model of the nasal cavity including one pair of para-nasal cavities, *solid line* (**a**) right branch, (**b**) left branch, in comparison with DFT-spectra obtained from evaluation of the wave equation with the aid of finite differences – from [37]

13.6 Applications

Acoustic tube models have been applied until now primarily in the areas of *speech synthesis* and *speech coding* as follows.

In speech synthesis there exist numerous investigations on the basis of tube models. Part of them has led to complete synthesis systems [4–6, 11, 16, 17, 35, 48]. The close relation between the tube parameters, i. e. the cross-sectional areas, and articulatory parameters, e. g., tongue position, turns out to be advantageous for this purpose. Age, namely, adult vs. childish voice, and sex, male vs. female voice, of the intended speaker can be controlled easily by the total length of the tube system and by proper choice of the fundamental frequency. Prosodic parameters, such as accents and focuses, can be introduced like in natural speech although the extraction of these parameters from text still remains a major problem to be solved in the near future. Certainly, the naturalness of the synthetic speech being achieved depends on successful extraction of prosodic and other related parameters.

In speech coding, tube models are well established at medium to low bit-rates. One reason for this is the insensitivity of the transfer characteristic with respect to quantization of the model parameters – e. g., reflection coefficients. Also, there exists a simple test of stability, which can easily be applied after quantization of the reflection coefficients or related parameters. Well known examples for coding systems based on tube models are the GSM-coding standard [12] and the LPC-vocoder [31] – see also [13].

Applications of tube models in speech recognition are conceivable. A prerequisite, however, would be that reliable separation of different phoneme classes were possible, e. g., from the structure and shape of the tube model. Yet, until now, little work has been done along these lines of thinking, except some simple vocal-tract-length normalization. Further, different speakers could be identified if the tube model were able to represent speaker-specific features.

Acknowledgement

The support of *K. Schnell* and *F. Ranostaj* during preparation of the manuscript for this chapter is gratefully acknowledged. *U. Heute* and *J. Mourjopolous* contributed comments which considerably improved an earlier version.

References

1. Atal B S, Hanauer L S (1971) Speech analysis and synthesis by linear prediction. J Acoust Soc Amer 50:637–655.
2. Bettinelli M, Schnell K, Lacroix A (2002) Separate Messung und Analyse von Mund- und Nasensignalen bei natürlicher Sprache. Proc. 13th Conf. Elektr. Sprachsig. Verarb, ESSV, Dresden 237–244.
3. Burg J (1968) A new analysis technique for time series data. NATO Advanc Study Inst Sig Processg, Enschede.
4. Carré R, Chennoukh S, Mrayati M (1992) Vowel-consonant-vowel transitions: analysis, modeling, and synthesis. Proc. Int Conf Speech Language Processg, ICSLP, Banff
5. Dorffner G, Kommenda M, Kubin G (1985) Graphon – the Vienna speech synthesis system for arbitrary German text. Proc Int Conf Acoust Speech Sig Processg, ICASSP, Tampa, 744–747.
6. Eichler M, Lacroix A (1996) Ein Experimentalsystem zur Sprachsynthese mit einem zeitdiskreten Rohrmodell. Fortschr Akust DAGA'96, 508–509, Dtsch. Ges. Akust, Oldenburg
7. Eichler M, Lacroix A (1996) Schallausbreitung in zeitvariablen Rohrsystemen. Fortschr Akust DAGA'96, 506–507, Dtsch Ges Akust, Oldenburg
8. Fant G (1970) Acoustic theory of speech production. 2nd ed, Mouton, The Hague–Paris
9. Fettweis A (1971) Digital filter structures related to classical filter networks. Archiv Elektronik Übertragungstechnik, AEÜ 25:79–89.

10. Flanagan J L (1972) Speech analysis, synthesis, and perception. 2nd ed, Springer, Berlin–Heidelberg–New York

11. Heike G, Philipp J (1984) Artikulatorische Sprachsynthese: Das Programmsystem LISA. Fortschr Akust DAGA'84, 833–836, Dtsch. Physik Ges,Bad Honneff

12. Heute U, ed (1988) Special issue on medium rate speech coding for digital mobile telephony (10 contributions), Speech Comm 7:111–245.

13. Heute U (2005) Speech and audio coding aiming at high quality and low data rates. Chap 14 this vol

14. Int Phonetic Ass (1989) Report on the 1989 Kiel Convention. J Int Phonetic Ass. 19:67–80. See also: IPA chart, rev 1993, J. Int. Phonetic Ass 23:center page

15. Itakura F, Saito S (1971) Digital filtering techniques for speech analysis and synthesis. Proc. 7th Int Congr Acoust, ICA, 261–265, Budapest

16. Itakura F, Saito S, Koike T, Sawabe M, Nishikawa, M (1972) An audio response unit based on partial autocorrelation. IEEE Trans Communications COM-20:792–797.

17. Jonsson A, Hedelin P (1983) A Swedish text–to–speech system based on an area function model. Proc. Int Conf Acoust Speech Sig Processg, ICASSP, 1340-1343, Boston

18. Jutten C, Herault J (1991) Blind separation of sources, Part I: An adaptive algorithm based on neuromimetic architecture. Sig Processg 24:1–10.

19. Karal E C (1953) The analogous acoustical impedance for discontinuities and constrictions of circular cross section. J Acoust Soc Amer 25:327–334.

20. Kelly J L, Lochbaum C C (1962) Speech synthesis. Proc. 4th Int Conf Acoust, ICA, G42:1–4, Copenhagen

21. Kubin G (1985) Wave digital filters: voltage, current or power waves. Proc Int Conf Acoust Speech Signal Processg, ICASSP, 69–72, Tampa

22. Lacroix A (1978) Source coding of speech signals by improved modeling of the voice source. Proc. ITG Conf Inform Syst Th Dig Comm 103–108, VDE-Verlag, Berlin

23. Lacroix A (1996) Digitale Filter. 4th ed, Oldenbourg, München–Wien

24. Laine U K (1982) Modelling of lip radiation impedance in the z-domain. Proc. Int Conf Acoust Speech Sig Processg, ICASSP, 1992–1995, Paris

25. Levine H, Schwinger J (1948) On the radiation of sound from an unflanged circular pipe. Phys Rev 73:383–406.

26. Liljencrants J (1985) Speech synthesis with a reflection-type line analog. Doct diss, Royal Inst of Technology, Stockholm

27. Liu M, Lacroix A (1996) Improved vocal tract model for the analysis of nasal speech sounds. Proc Int Conf Acoust Speech Sig Processg, ICASSP, II:801–804, Atlanta

28. Liu M, Lacroix A (1998) Analysis of acoustic models of the vocal tract including the nasal cavity. Proc 45th Int Kolloq Ilmenau, I:433–438, Ilmenau

29. Makhoul J (1975) Linear prediction: a tutorial review. Proc IEEE 63:561–580.

30. Makhoul J (1977) Stable and efficient lattice methods for linear prediction. IEEE Trans Acoust Speech Sig Processg, ASSP–25:423–428.

31. Markel J D, Gray AH (1974) Fixed-point truncation arithmetic implementation of a linear prediction autocorrelation Vocoder. IEEE Trans Acoust Speech Sig Processg, ASSP–22:273–282.

32. Markel J D, Gray A H (1980) Linear prediction of speech. 2nd print, Springer, Berlin–Heidelberg–New York

33. Meyer P, Strube H W (1984) Calculations on the time-varying vocal tract. Speech Comm 3:109–122.
34. Meyer P (1984) Ein selbstschwingendes Glottismodell für Echtzeit–Sprachsynthese. Fortschr. Akust, DAGA'84, 829–832, Dtsch Physik Ges, Bad Honnef
35. Meyer P, Strube H W, Wilhelms R (1984) Anpassung eines stilisierten Vokaltraktmodelles an stationäre Sprachlaute. Fortschr. Akust, DAGA'84, 825–828, Dtsch Physik Ges, Bad Honnef
36. Prony R (1795) Essai experimental et analytique. J école polytechn on bulletin du travail fait a cette école, Deuxieme Cahier an IV:24–76, Paris
37. Ranostaj F, Schnell K, Lacroix A (1999) Modellierung des Nasaltrakts. Proc. 10th Conf. Elektron Sprachsig Verarb, ESSV, 58–63, Görlitz
38. Ranostaj F, Lacroix A (2000) Bestimmung des Übertragungsverhaltens des Nasaltraktes aus computertomographischen Daten. Proc ITG/Konvens Conf Speech Comm, 131–134, Ilmenau
39. Ranostaj F, Lacroix A (2003) Ein Experimentalsystem zur Sprachakustik und Sprachproduktion. Proc. 14. Conf. Elektron Sprachsig Verab, ESSV, 280–285, Karlsruhe
40. Rayleigh J W S (1896) The theory of sound. Vol. II §312, 2nd ed, reprint (1945), Dover, New York
41. Schnell K, Lacroix A (1999) Parameter estimation for models with time dependent glottis impedance. Proc 2nd Conf Europ Sig Processg Ass, ECMCS'99, CD-ROM, Krakow
42. Schnell K, Lacroix A (1999) Parameter estimation from speech signals for tube models. Proc. joint ASA/EAA Conf, Forum Acusticum 1999 Berlin, CD-ROM, Berlin
43. Schnell K, Lacroix A (2000) Realization of a vowel–plosive–vowel transition by a tube model. Proc. Europ Sign Proc Conf, EUSIPCO, 757–760, Tampere
44. Schnell K, Lacroix A (2001) Pole-zero estimation from speech signals by an iterative procedure. Proc Int Conf Acoust Speech Sig Proc, ICASSP, CD-ROM, Salt Lake City
45. Schnell K, Lacroix A (2002) Analysis of vowels and nasalized vowels based on branched tube models. Proc. Eur Sig Processg Conf, EUSIPCO, III:65-68, Toulouse
46. Schnell K, Lacroix A (2003) Analysis of lossy vocal tract models for speech production. Proc EuroSpeech'03, 2369-2372, Geneva
47. Schönbach B (1990) Schallausbreitung in gekoppelten Rohrsystemen. VDI–Fortschr Ber Reihe 7, Nr 176, VDI–Publ, Düsseldorf
48. Sondhi M M (1983) An improved vocal tract model. Proc 11th Int Conf Acoust, ICA, 167–170, Paris
49. Story B H, Titze I R, Hoffmann E A (1996) Vocal tract functions from magnetic resonance imaging. J Acoust Soc Amer 100:537–554
50. Strube H W (1982) Time-varying wave digital filters for modeling analog systems. IEEE Trans Acoust Speech Sig Processg, ASSP–30:864–868.
51. Ungeheuer G (1962) Elemente einer akustischen Theorie der Vokalartikulation. Springer, Berlin–Göttingen–Heidelberg

14 Speech and Audio Coding –
Aiming at High Quality and Low Data Rates

Ulrich Heute

Institute for Circuit and System Theory, Faculty of Engineering,
Christian-Albrecht University, Kiel

Summary. The historic "coding gap" between high-rate coding of narrow- and wide-band speech on the one hand, and low-rate coding of narrow-band speech on the other hand, has been bridged more and more during the past 15 years. The GSM coder of 1990 was a very important milestone in this process, as it has prompted increasing research towards better quality and higher compression. These particular research efforts, together with other relevant activities worldwide, have helped closing the gap. In the following, the concepts behind this progress will be explained. A special focus will be put on the basis of this success, namely, on the fact that, finally, a break-through could be achieved for narrow-band speech, allowing for good quality at medium-to-low rates. For wide-band speech this holds true at medium rates. The same concepts as applied to speech coding were also applied to music, yet, ending up with some noticeable conceptual differences. While for speech time-domain approaches prevail, frequency-domain coding turned out to be more successful for audio. A characteristic, there, is extensive exploitation of psycho-acoustical phenomena.

14.1 Introduction and Fundamentals

The topic "speech and audio coding" basically addresses efficient digital representations of discrete-time acoustic signals. Some fundamentals of the algorithms as relevant in this context are explained in the following section.

14.1.1 Signal Digitization – A Very Brief Summary

Bandwidth and Time-Discretization

The continuous signal $x_0(t)$ is sampled at equally spaced instances, $t_k = kT_s$, with $k \in \{\cdots -1, 0, 1, 2 \ldots\}$ and $f_s = 1/T_s$ denoting the sampling frequency. Considering speech and music as band-limited signals with an upper frequency limit of f_c, termed "low-pass signals", the sampling theorem tells us that

$$f_s \geq 2f_c \qquad (14.1)$$

is needed. In the telephone system, for decades a "cut-off frequency" of

$$f_c < 4 \text{ kHz} \ldots \text{ e.\,g., } 3.4 \text{ kHz} \tag{14.2}$$

has been used. A larger value, namely,

$$f_c < 8 \text{ kHz} \ldots \text{ e.\,g., } 7.0 \text{ kHz}, \tag{14.3}$$

defines so-called "wide-band speech" which provides for more naturalness of the speech sounds but rather does not improve intelligibility.

For music a limit of about 7 kHz is still inadequate although $f_c < 5$ kHz is still commonly used in AM radio. Yet, for good quality, considerably higher values would be more reasonable, such as

$$f_c < 22 \text{ kHz} \ldots \text{ e.\,g., } 16 \text{ kHz}. \tag{14.4}$$

Quantization

In the further course of digitization, the samples in time, $x_0(kT_s) \doteq x(k)$, are then discretized in amplitudes as well, i.e. each sample is represented by w binary digits. The "unit" for the number of binary digits has a dimension of 1 and is called bit. This step is carried out by a quantizer, sorting $x(k)$ into the ith of 2^w possible quantization intervals of the size Q_i and describing it by its w-bit binary index, i. Globally, the absolute sizes, Q_i, are proportional to the amplitude range, D, covered by the possible values, $x(k)$. Usually, D is chosen according to

$$D = \lambda \cdot \sigma_x, \tag{14.5}$$

with $\sigma_x^2 = E\{x^2(k)\}$ denoting the signal variance, with zero-mean signals being assumed. For speech signals, $\lambda \geq 4$ is an appropriate constant avoiding too frequent range overflows. One additional bit doubles the number of intervals and, thus, cum grano salis, halves the absolute sizes, Q_i. Vice-versa, halving $\sigma_x \sim D$ allows for dropping one bit without coarser quantization.

In the simplest case, namely, the linear quantizer, also called linear analogue-to-digital converter, ADC, or linear pulse-code modulation, PCM, we have a constant size of all intervals, that is

$$Q_i \equiv Q = D \cdot 2^{-w}. \tag{14.6}$$

All signal values, $x(k)$, within an interval indexed by $i = i(k)$, are then embodied by one identical representative, $[x(k)]_Q$. This quantization causes a deviation of the quantized values, $[x(k)]_Q$, from $x(k)$. The difference is usually modelled as a random error, particularly, in the linear-PCM case, as an additive, white, uniformly-distributed quantization noise of zero mean and a variance of

$$\sigma_Q^2 = Q^2/12. \tag{14.7}$$

14.1.2 Coding Efficiency

Coding efficiency of an audio or speech coder means that the bit-rate, f_B, as needed to represent the consecutive samples in time according to

$$f_B = w \cdot f_s \tag{14.8}$$

is smaller than that of linear PCM – preferably by a substantial amount.

To achieve this goal, so-called "lossless" compression techniques exist. They reduce the bit-rate by removing redundancy such that it can be completely recovered at the receiver. An example is *Huffman* coding, where w code-words are mapped to variable-word-length indices [24]. For instance, in speech-PCM, indices pointing to the small amplitudes, which are comparatively frequent, would use few bits, indices pointing to the rare large amplitudes would use many bits, optimized in such a way that the average bit-rate is reduced. Such approaches are helpful, but delicate in case of a noisy transmission. Furthermore, for the very low bit-rates that speech and audio coders aim at, "lossy" compression seems to be unavoidable. In other words, the decoded – continuous – signal will differ from its original.

However, this unavoidable difference should be as small as possible in the end. In fact, it should not become audible or, at least, should not cause any annoying increase of noise or any further restriction in bandwidth. As has, hopefully, become clear from our reasoning above, these goals cannot be achieved by preserving the PCM sequence. Obviously, manipulations upon the signals in the time and/or spectral domains will be necessary.

Of course, the goal of achieving high efficiency also means that the necessary computational effort as well as the processing delay should be minimal. The contradictory character of these requirements is obvious.

14.1.3 Speech Production – Another Brief Review

Speech sounds are produced by pressing an air stream from the lungs through a tube of locally varying cross sections. Thereby, the a-priori laminar flow is modified in such a way as to generate audible spectral components – see [34] for models of this process.

The latter effect may occur at the vocal folds in the larynx. These folds are, then, forced to open abruptly under air pressure with a period of T_p. Comparable effects may also happen at various possible locations within the tube. There, e.g., turbulences can be caused by constrictions, or pulse-like air "explosions" may be created by suddenly opening a complete closure. The three resulting acoustic-signal types have a – more or less– constant spectral envelope. This means that they can be considered to be – more or less – "white" signals, white noise in the turbulence case, single "white" impulses in the plosive case, and a "white periodic" impulse sequence with – more or less – equal-size spectral lines at multiples of the fundamental frequency, $f_p = 1/T_p$.

The type of the signal produced carries some information already. Periodic components occur when vowels like [a] or (partially-) voiced, "v", speech sounds are uttered, noise-like signals are typical for fricatives like [f] or other (partially-) unvoiced, "uv", speech sounds, and single impulses appear with plosive, "p", speech sounds like [t].

The details, however, which define a particular speech sound as a phoneme, are due to subsequent processing of the three signal types in the so-called vocal tract. The vocal tract is the tube which extends from the origin of the air stream, the glottis, to the mouth and nose openings. The vocal tract is continuously modified by the speaking individual, such that its cross-section/location function varies in time. This process causes varying reflections within the tube and at its terminals. Consequently, the vocal tract acts as a filter which permanently modifies the acoustical signals in such a way that sounds which are recognizable as speech sounds can be produced – among other sounds. For example, for speech sounds which are classified as phonemes, characteristic relative maxima, so-called formants, and relative minima, so-called anti-formants, of the spectral envelope are formed.

This process of speech production can be modelled in a much simplified manner by a set of three generators which excite a linear filter – appropriate amplitude factors, $\sigma_{v,uv,p}$, given, see, e. g., [44,55]. This linear filter can then be described by a running, piece-wise time-invariant transfer function $H^{-1}(z)$ in the z domain or $H^{-1}(e^{j\Omega})$ in the frequency domain, respectively.

Further, a – fixed – pre-filter may be inserted, which would then take care of the fact that the excitation of the vocal tract is not exactly white but slightly decaying with increasing frequency. This so-called glottal pre-filter is often simply modelled as a first-order low pass, termed "de-emphasis", the effect of which can easily be inverted by a corresponding "pre-emphasis" high-pass filter – see Fig. 14.1.

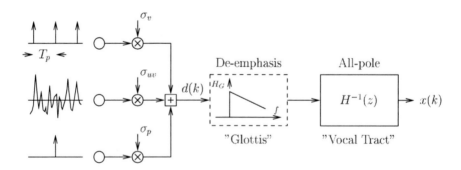

Fig. 14.1. Speech-production model with three generators which excite the vocal tract. The plot includes a "de-emphasis" pre-filter.

If the tube itself is modelled as a simple sequence of n constant-length cylindrical segments with piece-wise constant diameters, it is found that a denominator-only or all-pole transfer function

$$H^{-1}(z) = z^n / \sum_{\nu=0}^{n} \beta_\nu \cdot z^\nu = 1 / \sum_{\nu=0}^{n} \beta_{n-\nu} \cdot z^{-\nu} \qquad (14.9)$$

sufficiently describes the spectral envelope. There are simple relations of the cross sections, A_i, to the reflection factors, k_i, in the tube, the coefficients, β_ν, and the equivalent complex poles $z_{\infty\nu}$ of $H^{-1}(z)$.

As is always the case, filter poles have to do with frequency-response maxima, denoting resonances of the physical system. Especially the pole angles, $\Omega_{\infty\nu}$, in

$$z_{\infty\nu} = |z_{\infty\nu}| \cdot e^{j\Omega_{\infty\nu}}, \nu \in \{1, 2, ..., n\}, \qquad (14.10)$$

are, cum grano salis, equivalents of the formant frequencies, F_ν, normalized to the sampling frequency as follows,

$$\Omega_{\infty\nu} = 2\pi \cdot F_\nu / f_s. \qquad (14.11)$$

This simplified model of speech production is indeed the basis of many successful speech-coding approaches. More details and refinements are addressed in [34]. Unfortunately, no model along these lines of thinking is yet available for audible signals in general, often termed *audio signals* in the field. Music is a characteristic example of audio signals. As to music, the following has to be considered. Since, in essence, any combination of sounds may be defined to be music by artists and, hopefully, will be accepted as such by the listeners, audio coders usually do not rely on a production model, but rather exploit features of the receiver side. Nevertheless, knowledge of this kind is certainly useful for the design of speech coders as well.

14.1.4 Sound Perception – Again a Brief Summary

When a sound reaches a listener, it is first filtered when hitting head and body and passing on to the outer and middle ears. This causes a spectral shaping and, this is a prominent effect, a band-limitation of the relevant frequencies to an upper value as given in (14.4) as well as to a lower boundary of, say, $f_L \geq 20$ Hz.

However, not all signals within these limits are perceivable for the auditory system of listeners. A minimum level is needed to this end, depending on frequency. Components below this (absolute) hearing threshold, which is termed the sensation level of 0 phon, cannot be heard even in quiet surroundings. The actual dependence of the threshold as a function of frequency is, to a high extent, given by filtering due to the outer and middle ears [19].

In the inner ear, a decomposition of the incoming signals into spectral components occurs [23]. An acoustic signal of given frequency causes a travelling wave to proceed into the spiral-shaped cochlea. The wave amplitude gradually grows and reaches a relatively sharp maximum at a frequency-specific resonance point on the basilar membrane. This membrane divides the cochlea into a lower and an upper duct. The upper one is again divided electro-chemically by a further membrane, thus forming a kind of electric "battery". The basilar membrane carries ionic sensors to sense its movements, termed hair cells. These hair cells are particularly susceptible to deflections at the points of maximum excitation. In this way, each segment of the membrane is directly "responsible" for a certain frequency band. Yet, these bands are not separated ideally by rectangular frequency-selection filters, but rather show approximately triangularly-shaped, strongly-overlapping frequency responses. The bandwidths of these cochlea filters increase with increasing center frequency.

In a much simplified model, e. g., [14, 55, 56], this part of the hearing process may be viewed as a set of parallel band-pass filters, i. e. a filter bank with overlapping frequency channels, termed "critical bands", as sketched in Fig. 14.2. Along an equi-spaced Θ-axis, termed *Bark* scale – which assigns numerals to the channels – the center frequencies, f_Θ, and the bandwidths, Δf_Θ, of these channels vary according to

$$\Theta/\text{Bark} = 13 \cdot \arctan(0.76 f/\text{kHz}) + 3.5 \cdot \arctan[(f/7.5\,\text{kHz})^2], \qquad (14.12)$$

$$\text{with } \Delta f_\Theta = 100\,\text{Hz} \quad \text{for } f \le 500\,\text{Hz},$$
$$\text{and } \Delta f_\Theta = f/5 \quad \text{for } f > 500\,\text{Hz}.$$

Depending on the specific application and the related absolute bandwidth, some 24 channels may cover the full audio-frequency band. When appropriate, less or more channels may be defined accordingly.

After this frequency decomposition, the components are further processed in non-linear ways. First, the excitation in the cochlea is not directly described by signal power. As the wave amplitude grows slowly up to its maximum and decays fast, but is still progressiv thereafter, a broader segment of the membrane is excited than just the "best-frequency" region. Further, signal components which give rise to an excitation being smaller than the one which already exists on the basilar membrane, may not have any effect at all. This phenomenon is called "masking" in the field. Masking may occur synchronously in time, but also components which occur shortly before or after the "masker", may be deleted in this way. These effects are called pre- and post-masking. Pre-masking, though, is a faint effect which is normally not dealt with in modelling.

Secondly, the excitation pattern as resulting in the end does not directly describe the – perceived – loudness. Rather, a transformation from the excitation, $E(\Theta)$, in a cochlea segment, Θ, to the specific loudness, $N'(\Theta)$, is carried out according to

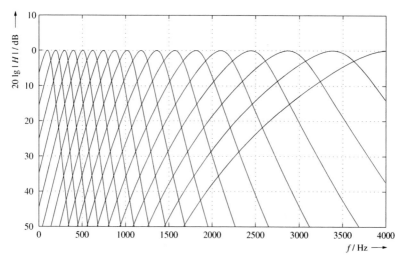

Fig. 14.2. Model of the critical-band filter bank with 18 channels up to 4 kHz

$$N'(\Theta) \sim E(\Theta)^{0.23}. \tag{14.13}$$

The total loudness is derived from integrating $N'(\Theta)$ over Θ. For details, the reader is referred to [56] and [14].

14.1.5 Speech Coding and the "Coding Gap"

In *classical* speech-coding literature, a distinction between "high-rate" and "low-rate" coding schemes is made.

In the seventies, digital speech transmission and/or storage became feasible at telephone quality, sometimes also called "toll quality", referring to the quality of speech with a bandwidth covering 300 Hz to 3400 Hz and, following (14.2), a sampling rate of $f_s = 8$ kHz, and with a signal-to-noise ratio of

$$SNR \doteq 10\lg[S/N] = 20\lg[\sigma_x/\sigma_n] = 35...38\,\text{dB}. \tag{14.14}$$

Here, $N = \sigma_n^2 = \sigma_Q^2$, as in (14.7), would hold for a linear PCM. More generally, $N = \sigma_n^2$ describes the mean-square value of the difference between the input speech and its reconstruction after some intermediate operations, i.e. the variance of the output-signal error. Various simple schemes from logarithmic PCM at a bit-rate of $f_B = 64$ kbit/s, ITU G.711 [26], down to ADPCM at $f_B = 32$ kbit/s, ITU G.726 [28], were applied first.

However, really low rates for the same narrow-band speech, such as $f_B = 2.4$ kbit/s or below, were only possible in such a way that one arrived at speech which sounded quite "synthetic" indeed, namely, by using a technology called VOCODER. The system known as LPC-10 [53] is an example of this technology, where no waveform reproduction is aimed at and, therefore, no SNR description can appropriately be used. Nevertheless, the resulting, intelligible speech was sufficient for a number of successful applications – mainly military ones.

Between these two groups of approaches, high-rate and low-rate, a "gap" had appeared. This gap gave rise to an increase in laboratory research, mainly in the seventies and beginning eighties [21]. As a result, the gap could be filled up progressively by laboratory demonstrators as well as commercial applications, which became available in the nineties.

14.1.6 Higher Bandwidth – Wide-Band Speech and Audio Coding

Following (14.3), wide-band speech comprises frequencies from 50 Hz to 7000 Hz, with a sampling rate of $f_s = 16$ kHz, and offers a more natural, and consequently, higher quality. Hence, early coding attempts for these signals started to work at "high rates". For example, a split-band ADPCM codes the bands below and above 4 kHz separately and needs 64 kbit/s, ITU G.722 [27, 41].

The same system, as standardized in 1986, was at that time also considered to be appropriate for music. Hence it could be seen as an early audio-coding system. But in the beginning eighties the CD appeared on the market, featuring linear PCM for music, using $w = 16$ and $f_s = 44.1$ kHz without any data reduction at all, thus consuming the very high rate of 705.6 kbit/s per mono channel – *plus* error correction bits! – but with a much better quality, known as "CD quality". Therefore, in the later eighties, research began to strive for reduced rates at this quality, though, in a first step, at considerably higher rates than those used for speech signals. Audio coding, thus, became a research field of its own.

14.2 Speech Coding

14.2.1 The Origin of the Gap

Possible Approaches to Lower Rates

Within the high-rate class of coding, a considerable rate reduction was achieved. A simple linear A/D converter needs a word length $w = 11...12$, resulting in $f_B = w \cdot f_s = 88...96$ kbit. Yet, log PCM only needs 64 kbit/s, and with ADPCM the rate goes down to 32 kbit/s, preserving roughly the same SNR quality. Three essential concepts were followed to achieve the bit-rate reduction, namely, non-linear quantization, adaptivity, and, last but not least, linear prediction, LP.

The first step, resulting in logarithmic quantizers, makes the quantization intervals $Q(x)$ proportional to the signal amplitude $|x|$. As any quantizer it utilizes the fact that the human ear is unable to perceive arbitrarily small

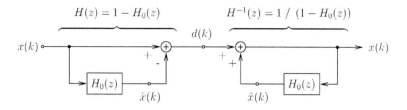

Fig. 14.3. Typical structure of a predictive system

differences, but here the observation is exploited that its sensitivity to absolute differences decreases with increasing signal size. So, for large signals, large errors can be allowed, i. e. perceptually irrelevant signal details are discarded.

As far as adaptive quantization is concerned, adaptivity follows the same idea, but in addition, lets Q follow the "instantaneously" or "syllabically" measured signal size.

Linear Prediction

The idea of adaptation to changing signal characteristics is also used in the third concept named above, namely linear prediction. Due to its relevance we shall devote a full paragraph to this algorithm as follows. A non-recursive filter with a transfer function

$$H_0(z) = \sum_{i=1}^{n} a_i \cdot z^{-i} \tag{14.15}$$

computes a linear combination of past signal values as an estimate, $\hat{x}(k)$, of the next sample according to

$$\hat{x}(k) = \sum_{i=1}^{n} a_i \cdot x(k-i), \tag{14.16}$$

which is subtracted from the actual sample, $x(k)$. The difference or residual signal, $d(k) = x(k) - \hat{x}(k)$, results as the output of a prediction-error filter

$$H(z) = 1 - H_0(z). \tag{14.17}$$

Excitation of $1/H(z)$ by $d(k)$ in the receiver regenerates $x(k)$ – see Fig. 14.3.

Rate Reduction

The minimization of the residual variance, σ_d^2, leads to the optimal predictor-coefficient vector **a**, found from the correlation matrix, **R**, and correlation vector, $\mathbf{r}_0$. Due to this minimization, σ_d is smaller than σ_x, and, following the conclusions from (14.5), $d(k)$ can be quantized with less bits than $x(k)$, even considerably less if both predictor and quantizer are adapted with time. This is no heavy algorithmic load, since the solution

$$\mathbf{a} = \mathbf{R}^{-1} \cdot \mathbf{r}_0 \qquad (14.18)$$

can be computed very efficiently, e. g., using the *Levinson–Durbin* or *Schur–Cohn* recursions.

If the SNR quality is to be preserved, however, it is found that the word-length cannot be less than $w = 4$. Together with $f_s = 8\,\mathrm{kHz}$, this limits the rate reduction to the above-mentioned value of 32 kbit/s.

The Gap

However, large parts of the signal information reside in the predictor and quantizer parameters, either adapted backwards in both transmitter and receiver, in adaptive differential pulse-code modulation, ADPCM [28], or transmitted as additional side-information with $f_B \approx 2\,\mathrm{kbit/s}$, in adaptive predictive coding, APC or Block-ADPCM.

This equals the low-data rate of the LPC-VOCODER mentioned above [53]. This device synthesizes speech from – mainly LP – parameters, yet, resulting at "synthetic quality" only, since the large data volume of the residual signal is replaced by a few descriptive parameters. In fact, in line with the speech-production model of Sec. 14.1.3, a filter with the transfer function $1/H(z)$ is used to model the vocal tract. This is possible, since it has the same all-pole form as found in (14.9). We just have to identify the coefficients in (14.9) and (14.16) as

$$\beta_{n-i} = -a_i, \ i \in \{1, ..., n\}, \ , \ \text{and} \ \beta_n = 1. \qquad (14.19)$$

Beyond that, little further information is needed to control the possible excitation signals, namely, their type, size, and periodicity.

14.2.2 Approaches to Close the Gap

RELP Coding

If fully parametric representations reduce the quality and residual rates cannot be reduced further by employing smaller values of w, then only decreasing the sampling frequency, f_s, is left as a means to reduce f_B, according to (14.8).

This is, however, impossible for speech, since its band-width is fixed according to (14.2), which also also fixes f_s due to the sampling theorem – see (14.1). Yet, such a reduction of the sampling frequency is well possible for the residual which has been whitened by the – necessarily – de-correlating filter, $H(z)$. Since $H(z)$ exploits the auto-correlation of speech to reduce σ_d^2, the correlation as remaining in $d(k)$ must be smaller. Ideally, a pulse-like auto-correlation should remain. This means that the residual power spectrum would be constant, i. e. the residuum became white. This whitening operation can be interpreted as an inversion of the vocal-tract filtering by $H(z)$. The transfer function of the acoustic tube was termed $H^{-1}(z)$ for this reason.

After reduction of the bandwidth of the – more or less – white signal, $d(k)$, to a limit of $f_c/r \dots \quad r > 1$, the remaining – more or less – constant base-band spectrum requires a sampling rate of just f_s/r. In the receiver, a spectrally widened signal can then be applied to $1/H(z)$, which leads to the concept of base-band residual-excited LP systems, BB-RELP. The spectral widening can, for example, be achieved through a simple spectral repetition via up-sampling, i. e. by inserting $(r-1)$ zeros between each two samples.

However, some unnatural, metallic-sounding distortions still remain, even after more sophisticated widening operations. These artifacts are due to inflexible down-sampling, which, in particular, does not take sufficient care of the spectral fine structure of $d(k)$, namely of higher harmonics of the fundamental frequency, $f_p = 1/T_p$, which is often also called "pitch frequency".

Further Redundancy Reduction: LTP

Equally spaced spectral lines are due to periodicity. They carry redundancy like the spectral shape, the latter being more or less removed by $H(z)$. Consequently, they can be removed in a similar way by a further predictor. This so-called pitch predictor uses longer-term correlations, namely, not of distances of n/f_s – e. g., $n \approx 10$ in Block-ADPCM/APC, but rather of $m_p/f_s \approx 1/f_p$, with $m_p \in \{32, \dots, 160\}$ for normal voices. This long time range gives rise to the notation of long-term prediction, LTP – see Fig. 14.4.

The LTP parameter, β_p, may be found by means of the same principle, see Sec. 14.2.1, as is used for determining the LPC coefficients, a_i, in (14.16). If m_p is known, e. g., from a pitch-detection algorithm [20], the variance, σ_e^2, of the second-difference signal, $e(k)$, is again minimized with respect to β_p, in order to render the amplitude range of the quantizer as small as possible. However, the search for m_p can also be included into this minimization, yielding the best delay value in terms of the smallest variance σ_e^2. Following this line of operations, m_p turns out not to be a "good" estimator of pitch, yet, to be most suitable in terms of the criterion σ_e^2.

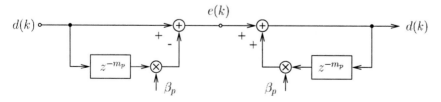

Fig. 14.4. Long-term predictor, 1-tap LTP, and its inversion

As f_p/f_s is generally a positive real number, integer values for m_p are probably sub-optimal. An improvement is achievable by either using more than one coefficient for neighbouring delays, e. g., $\{m_{p-1}, m_p, m_{p+1}\}$ in a "3-tap LTP", or by interpolating $d(k)$ and $e(k)$ and choosing β_p and m_p on a finer time grid, i. e. "fractional-delay LTP".

Flexibilities in Down-Sampling: RPE and MPE

Down- and up-sampling of the base-band signal results, for voiced sounds, in spectral lines which are harmonic, i. e. occur at multiples of f_p within the base band, but then shifted according to f_s/r, irrespective of f_p.

Instead of taking every r^{th} sample of the residual, $d(k)$, in a naive manner, one may think of dissolving the mis-fit of sampling and pitch frequencies by, e. g., a decimation switching between various factors, r, with a reasonable mean value, $\bar{r}$. However, this is much less successful than a simple switch between the possible r phases of the down-sampling grid, according to some optimality criterion – see Fig. 14.5. What remains is a decimated residual, applied in the receiver as a regular-pulse excitation, RPE, of $1/H(z)$. As usually, an inverse LTP has to precede $1/H(z)$ if an LTP is included in the transmitter.

Beyond that, optimized amplitudes may be applied instead of just samples, yielding a generalized RPE. Furthermore, the N/r remaining values, after a block of N data has been down-sampled, could as well be placed at any optimal positions rather than on a grid. This idea leads to the concept of multi-pulse excitation, MPE. Both RPE and MPE can be further simplified by allowing only simple amplitudes, like $\{-1, 0, 1\}$, with a common gain factor per block or for suitably chosen sub-blocks, thus separating shape and gain of the excitation.

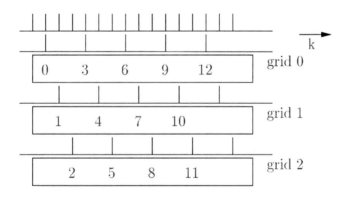

Fig. 14.5. The RPE principle with $r = 3$ possible sub-sampling phases

Vector Quantization

After quantization, a reduced set of samples with a limited number of permitted values remains. Hence, only a particular set of possible excitation blocks is available at the receiver.

An alternative, more direct way to such excitations applies a table or "code-book", with, say, $L = 2^w$ decimated or non-decimated length-N vectors from which the best one is chosen and coded by w bit per N samples. For instance, this would mean for $L = 1024 = 2^{10}$ and $N = 40$ that $w = 10/40 = 1/4$ bit/sample were used in such a scheme, termed vector-quantization, VQ. The same principle can also be applied to parameters like the predictor-coefficient vector, **a**, or any equivalent, transformed coefficient set, such as the so-called reflection-coefficient vector, **k**, the log-area-ratio vector, **lar** [48], or the line-spectral-frequencies' vector, **lsf** [25] – the latter being calculated from a polynomial transformation of $H(z)$.

14.2.3 Closing the Gap

A first important practical realization helping to bridge the coding gap was the GSM full-rate codec , FR – see Fig. 14.6, standardized at the end of the eighties by ETSI and being applied world-wide since 1992 [10]. The GSM-FR codec is based on those of the above ideas which were realizable with the hardware available at that time, and it yielded the best quality as achievable by these means.

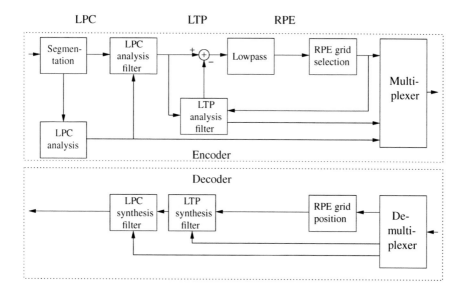

Fig. 14.6. GSM-FR system – RPE-LTP-RELP, with $r = 3$

A simple power measurement is used to define the "optimal" RPE-grid phase, and the low-pass filter does not strictly band-limit the LTP output. Rather, considerable aliasing is admitted, thus allowing for quite smooth spectral widening in the receiver, without introducing deep spectral gaps. Yet, a measurable spectral notch at $f = f_\mathrm{s}/r$ remains, which is typical for GSM-coded speech.

The above mentioned *best-possible quality*, however, can still not be measured by means of the *SNR*. Instead, for quality evaluations, test listeners had to supply judgement on coder-processed test-speech samples, and mean-opinion scores, MOS, had to be calculated from these judgements. Due to the large load of such auditory tests, new instrumental measures of perceived quality, based on psychoacoustics, became a current field of research, as addressed in [38].[1]

Generally, the limited quality of GSM-FR coding as well as the tremendous growth of the number of subscribers prompted a wave of research towards improvements. Firstly, these improvements could be based on an enormous increase in computational power. Indeed, following *Moore*'s law, six years after the GSM definitions, a four-fold algorithmic complexity could be allowed for already in the mid nineties. In this way, the originally-planned mobile car phones could now easily be replaced by hand-held devices – called *handies* in the German language! Another prerequisite for this development was that the batteries became much smaller and more powerful. Nowadays, even a sixteen-fold complexity is envisaged as compared to the state of the art at the time when the original GSM codec had been defined.

Secondly, the enhancements were based on some of the tools mentioned above. However, thirdly and above all, they were made possible by two *decisive* ideas that we shall describe now.

14.2.4 The Basis of the Break-Through

The VQ principle [16] turned out to be a very powerful concept. It has been applied to both the receiver-excitation signal and the parameters. Especially, LSF-VQ achieves very good quality at only 20...24 bit per 20-ms frame, i.e. at a rate as low as 1...1.2 kbit/s.

Improved code-book designs via efficient training, especially for the residual signals, $d(k)$ or $e(k)$, enhance quality. By means of simple entries, e.g., 0 and ±1 only, the computational load was reduced and realizability was achieved. The simple vectors descriptive for the shape could be combined with separate gains and, further, be transformed efficiently into more complex varieties by means of linear combinations.

Above all, however, a better optimization criterion became a decisive factor for improvement. Instead of minimizing the MSE of the residual approx-

---

[1] Usage of the terms subjective for auditory and objective for instrumental is discouraged.

imation, the distortion of the receiver-output signal – being synthesized in the transmitter! – was reduced to a minimum. Furthermore, the analysis-by-synthesis technique, as used for this purpose, applies perceptual weighting in a way as rendered by a crude perception model before the MSE is evaluated. Finally, LTP is included into its closed loop, what can be viewed as a second, pitch-adaptive code-book. Again, not an optimum pitch estimate is aimed at, but the most suitable period according to the optimization criterion applied.

Code-Book Design

Vector quantizers are designed as an N-dimensional generalisation of the classical, scalar "optimum quantizer" [37]. This so-called "*Lloyd–Max* quantizer" uses quantization-interval boundaries and representatives along the x-axis such that the mean-square error, MSE, namely σ_Q^2, is minimized. This leads to a set of partially non-linear equations with the 1-dimensional probability-density function of the signal as an input information. For a VQ, optimum sets of N-dimensional volumes, "*Voronoi* cells", and their N-dimensional representatives are to be found from equivalent expressions, needing the N-dimensional probability-density function as an entry.

Some work has been performed towards tractable probability-density functions for the de-correlated residual signal – remember, an $N \approx 40$ is normal! – successfully applying the theory of spherically-invariant random processes, SIRPs [5], particularly with the aim of finding theoretical limits for the rate reduction.

For practical VQ designs, an iterative procedure with relatively low complexity became very popular [36]. It uses a large number of signal samples rather than a probability-density-function information. In fact, it starts with only one cell and its "center" as a representative vector and, then, continues with splitting the existing cells and thereby doubling their number with every step. The result is not necessarily optimal, but still good enough. Also, and this is even more important, this approach is not restricted to the MSE criterion.

Simplified Code-Books

Originally, the code words may consist of N complete samples. Then L variants can be indexed by $log_2 L$ bit without any simplification of the contents indeed, e. g., neither down-sampling nor any coarse quantization is needed. For the sake of simplifying the further processing, however, both steps are useful. The RPE and MPE concepts yield code words with only N/r non-zero elements. Further, the values may be restricted, e. g., to the set $\{-1, 0, +1\}$, and a sufficient variety may still be allowed for by means of matrix multiplications, "algebraic code-books", and/or an additional set of factors, again separating shape and gain of the signal blocks,"shape-gain VQ". Compare the RPE/MPE part of Sec. 14.2.2 in this context. Similarly, the use of more than one, but smaller code-books with orthogonal entries and an appropriate weighted summation, "vector-sum coding", VS, reduces the computational load during the design phase as well as during application.

Analysis-by-Synthesis

The underlying idea is (i) simple as many good ideas, (ii) more than 40 years old [2, 18], also [48], and (iii) quite universal in its general formulation. Consider, for example, a piece of a speech signal being given in any useful parametric description. It is then to be represented by some – more compact or more informative – model with certain degrees of freedom. Consequently, the free parameters are assigned values such that, with regard to the *initial form of the description*, a best match is found.

For speech coding, this actually means that the derivation of the inverse filter, $H(z)$ – plus the pitch predictor, if applicable – and the scalar or vector quantization of the residual signal cannot be regarded as independent steps. Rather, the best combination of all quantized filter parameters and quantized excitation sequences has to be searched for, being optimized with respect to the reconstructed signal. In theory, the optimum should even hold for the full-length signal. In practice, however, finite-length blocks still are considered only, though possibly with some inclusion of the past behaviour. The (LPC-) filter analysis is mostly performed as usual for a given input-signal block, may be after pre-emphasis filtering. However, *all* possible excitations are indeed applied to the full synthesis process, including delayed versions of preceding excitations, i.e., pitch-periodic components – if available. The term *all* is usually not interpreted too stringently though, since a restriction to sub-sets and/or certain combinations may be necessary for the sake of realizability.

Perceptual Weighting

The above mentioned *best-match approach* has to refer to a valid criterion, of course. MSE would not be appropriate for this purpose, as it maximizes the SNR, which does not reflect listeners' impressions with low-rate coders. To solve the problem, the difference between the synthesized and the original speech signal is spectrally weighted prior to the mean-square calculation and minimization. The weighting is carried out by sending the difference signal through a filter with a transfer function of

$$W(z) = H(z)/H(z/\gamma), \ \gamma \approx 0.9. \qquad (14.20)$$

Here $H(z)$ is the "inverse-vocal-tract" filter of (14.17), resulting from LPC analysis as discussed in Sec. 14.2.1. With $\gamma < 1$, $H(z/\gamma)$ corresponds to a frequency response which is a "damped" version of $H(e^{j\Omega})$. Thus, $|W(e^{j\Omega})|$ varies around a value of 1.0 with minima near the formant frequencies, F_ν, and maxima between them. Thereby, distortions are attenuated when being positioned close to formants, i.e. relatively strong spectral components, and emphasized otherwise. In this way, masking effects are coarsely modelled.

14.2.5 CELP Coders

Analysis-by-synthesis as well as perceptual-weighting techniques can be applied in various systems, including LP-based RPE and MPE coders. However, the most successful combination so far comprises analysis-by-synthesis, perceptual weighting by $W(z)$, see (14.20), vector or code-book excitation, and LP. Coders which use this combination of features are termed CELP systems. Their variants can as well be understood based on the explanations above.

Figure 14.7 shows a typical codec, namely, one of the more recent standards of ITU [30]. It uses the already mentioned approach of a simple-entry

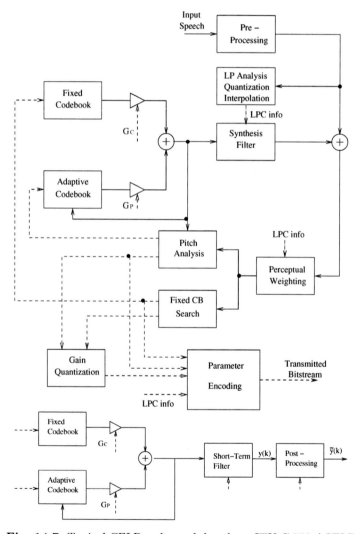

Fig. 14.7. Typical CELP coder and decoder – ITU G.729 ACELP

code-book with matrix manipulation, giving rise to its denotation as algebraic CELP, ACELP. This is a very popular type among the many CELP systems in use nowadays.

Some of them are listed in Table 1, see also [22,49,55]. The two last systems adapt to changing transmission channels and/or user requirements, with eight and nine different rates, respectively. The latter is to be applied also in the enhanced GSM system, EDGE, and in UMTS. The notation eXCELP refers to an extension of the CELP principles by combining it with other techniques and, above all, by varying sub-code-books and even parts of the algorithm depending on a signal-segment classification [50].

Table 14.1. Listing of some CELP codecs

ITU G.729:	Algebraic CELP coder, ACELP; $f_B = 8$ kbit/s, high quality; high complexity, 18 Mops [30]
GSM-HR:	ETSI half-rate coder: vector-sum excitation, VSELP; $f_B = 5.6$ kbit/s, quality $\leq$ GSM-FR; high complexity, 20 Mops [11]
GSM-EFR:	ETSI enhanced full-rate coder: ACELP similar to [30]; $f_B = 12.2$ kbit/s, enhanced quality and error protection compared to GSM-FR; high complexity, 18 Mops [12]
ITU proposal:	ITU coder for high quality at low rates: "eXCELP"; $f_B = 4$ kbit/s; very high complexity – not yet fixed [51]
ITU G.728:	ITU codec with low delay (2 ms, "LD-CELP"): high-order prediction ($n = 50!$), backward adaptation, no LTP; very short codewords ($N = 5$), $f_B = 16$ kbit/s, high quality; high complexity, 20 Mops [29]
MPEG-4 CELP:	MPEG-4 system with scalable data rate: ACELP with MPE/RPE variants; $f_B = 4...24$ kbit/s, variable quality; high to very high complexity (simple narrow-band coding: ≈ 20 Mops, enhanced narrow-band version: ≈ 40 Mops, enhanced wide-band coding: $\approx 50...70...$Mops [17]
GSM-AMR:	ETSI adaptive multi-rate system: ACELP; variable rates $f_B = 4.75...12.2$ kbit/s, variable quality; up to high complexity, 17 Mops [13]
GSM WB-AMR:	ETSI wide-band AMR system: ACELP; low to high rates $f_B = 6.6...23.85$ kbit/s, up to very high quality; up to very high complexity, 35 Mops [54]

The quality of these systems has to be compared to that of the common ISDN telephone and that of the wide-spread GSM-FR codec, i.e. to a relatively *high quality* and to a somewhat *reduced quality*. A good CELP system may indeed yield high quality, much better than RPE-LTP-RELP. But, of course, also the complexities have to be taken into account. While the ITU standards G.711, PCM, and G.726, ADPCM, need almost no computational power, ETSI's GSM needs some 3.5 Mops already, while a good CELP coder may need 5 to 10 times as many, or even more. 3.5 Mops present no problem for a simple digital signal processor (DSP). More advanced DSPs can handle 20 Mops easily, and cost start to really increase at 100 Mops – although hardware performance is still increasing fast.

14.2.6 Frequency-Domain Coding

In the beginning of the GSM definition phase, the majority of the proposals was of a completely different kind [21]. They aimed at medium rates by means of – continuous – sub-band coders, SBC, or – block-oriented – transform coders, TC. In fact, both kinds can be viewed as systems with analysis-synthesis filter banks, either explicitly built with parallel band-pass filters or being realized by filter-bank equivalent DFTs or DCTs. The true differences between them are the numbers M of frequency channels or components, namely, $M = \ldots 8 \ldots 16 \ldots$ in SBC and $M = \ldots 128 \ldots$ in TC. The corresponding down-sampling of the narrow-band signals by $r = M$ created the misleading classification, as for large values of M, the block-processing point-of-view becomes very obvious. Actually however, there is no principal difference in this respect, since in a continuously-operating FIR-filter bank the down-sampling is performed by calculating an output value only at every r^{th} clock interval anyhow.

Figure 14.8 displays a typical adaptive TC system, ATC, with adaptive quantization and bit allocation following the spectral envelope. This system has been realized in [15] with $f_B = 14$ kbit/s, with a quality close to GSM-FR at a similar computational load. Figure 14.9 shows an SBC system with lower quality, but also lower complexity [35].

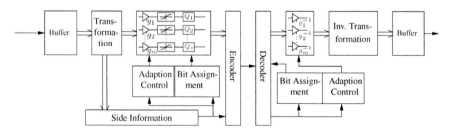

Fig. 14.8. Adaptive transform codec, ATC

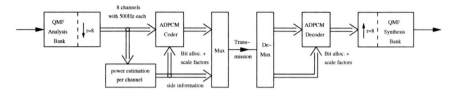

Fig. 14.9. Sub-band codec, SBC, with adaptive bit-allocation and ADPCM

Of course, some of the ideas as applied in CELP systems could be transferred to frequency-domain coders. VQ is readily used to describe spectral envelopes in the bit-allocation procedure – be it by means of a vector of averaged spectral amplitudes or by means of an LPC-derived coefficient vector. The above-mentioned, very efficient LSF-VQ can, for example, be applied directly. Also, cepstral coefficients were proposed for being vector-quantized. Furthermore, the perceptual-weighting filter can be built into the bit-allocation procedure: In the fundamental paper [40], the number of allocated bits per frequency component followed the average spectral envelope directly, yet, one could also make it follow a "damped" contour, thereby taking care of – partial – masking effects.

Many such ideas were examined. Nevertheless, frequency-domain coders turned out to be much less successful than the predictive systems as discussed above. Consequently, they disappeared more or less from the speech-coding scene since about 1990.

Other Ways to Close the Gap

Describing speech segments as a finite sum of sinusoids with suitably varying amplitudes, frequencies, and phases has been investigated intensively in the late eighties, e. g., [52].

Amplitudes are either to be transmitted as single values or to be described by an envelope which, in turn, can be represented by, e. g., LPC parameters. Besides the frequency components, phase angles have to be determined. This is usually done on the basis of simplified models of their time-varying behaviour, e. g., by low-order polynomial functions. A trade-off between quality, computational load and bit-rate can be achieved by selecting more or less refined descriptions, especially for frequencies and phases.

Since, in addition to what has been said above already, the number of components naturally varies from signal block to signal block, a scheme called sinusoidal-modelling, SM, defines, a priori, a system with varying bit stream – unless special measures are taken. To account for this effect, a simplified version of SM employs a frequency/pitch grid in such a way that, after fixing f_p and resorting to an envelope description, the number of parameters per block becomes constant. This harmonic-coding scheme, HC, can be augmented by substituting sinusoids by narrow-band noise in unvoiced sections. From here, the way is not far to hybrid time-frequency codecs with mixed or multi-band excitation: Rather than using sinusoidal *or* noise excitations, *both* kinds of excitation signals are applied, but differently in different frequency bands, see, e. g., MELP, MBE, IMBE, EMBE, which aim at rates near 4 kbit/s and below [55].

SM and HC, in their original forms, are sometimes regarded as frequency-domain approaches. Together with their "relatives", they have not really come to the fore recently.

14.2.7 Wide-Band-Speech Coding

As mentioned in Sec. 14.1.6, wide-band speech may be seen as being close to audio signals regarding its characteristics and requirements. In Sec. 14.3, it will be shown that frequency-domain techniques are being applied success-fully to audio coding. In this context quite some work has also been performed towards coding of wide-band speech, especially when combined with music, in the same ATC or SBC system. At present however, CELP systems with an enlarged prediction order of $n > 10$, dominate the scene. A variant is included in the list of Sec. 14.2.5, as a part of the ETSI WB-AMR standard.

There is one aspect in this coder [3] which is worth mentioning here and for which some pre-standard demonstrators are available [9, 47]. Namely, the "split-band" principle of [27] is enjoying a revival. However, instead of separating spectral components below and above 4 kHz as in past systems, an asymmetrical splitting takes place now. The components at 0...6 kHz – or 6.4 kHz in [3] – are directly ACELP-coded, the remaining band is replaced by noise, or by aliased parts of the lower band similar to BB-RELP – see Sec. 14.2.2). The noise is then filtered by $H^{-1}(z)$ and only a gain factor needs to be transmitted. This term is quite sensitive though, but its word-length can be reduced by linear prediction again. The idea behind this has already been used in an earlier approach to wide-band speech coding, though without *any* upper-band information: In [6, 7] the components above 4 kHz were not transmitted, but substituted in the receiver as above, with a spectral weighting and gain-term estimated from the lower-band data, using CELP techniques.

14.3 Audio Coding – An Even Shorter Overview

This section can be short because the important principles have all been discussed above already. Strategies like adaptive quantization and adaptive bit-allocation, LP, spectral decomposition, VQ, and use of production or per-ception models are also applicable to audio signals.

Nevertheless, there are important differences between successful speech and audio codecs as of today. In fact, for music, frequency-domain coding prevails, whereby the coarse perceptual model of the CELP (or RPE/MPE) weighting filter is replaced by much more detailed models of the hearing

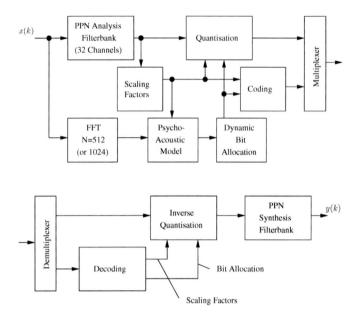

Fig. 14.10. Audio codec of MPEG – MPEG-1, layers I+II

process. These models may include, among other things, outer and middle-ear frequency responses, a critical-band or *Bark*-scale separation on the basilar membrane, non-linearities of the loudness perception, and time- as well as frequency-masking properties of the auditory system – though not in every detail in each coder.

Figure 14.10 shows the block diagram of the MPEG-1/Layers-I+II system, see, e. g., [43]. Its similarity with the ATC and SBC, shown in Figs. 14.8, 14.9, is obvious, yet, with the exception of the inherent psycho-acoustic model. From all given spectral components individual masking contributions are estimated and, from these, a total masking curve is derived. Then the number of quantizer bits is allocated in such a way that the quantization noise is masked. In the most popular variant, the MPEG-1/Layer-3 system – widely know as *mp3* – the 32 sub-bands are subdivided into 18 narrower components by a modified DCT, MDCT. *Huffman* coding is then added, and window-length switching reduces any "pre-echoes". Pre-echoes are a phenomenon which is typical for frequency-domain coding as follows. A sudden acoustical event, like a drum or a bell sound, corresponds to a broad-band short-time spectrum. However, the quantization of this running spectrum causes noise spread all over the reconstructed time segment. Even if the sound attack to be dealt with takes place in the middle or at the end of a longer signal block, it evokes a noise eruption which may already be audible at the beginning of the segment, thus blurring the later part of the event. This effect is reduced by shortening the segment window following detection of an attack. The correspondingly poorer spectral resolution is no disadvantage, because of the broad-band signal. *mp3* achieves CD quality at $f_s = 44.1$ kHz and $f_B = 192$ kbit/s.

The way towards *mp3* began in 1992, with an ATC, using a DCT – see Fig. 14.8 for speech – and later an MDCT with added simultaneous-masking principles applied within an iterative coding/masking/bit-allocation procedure. This first system, which had been developed in Germany in 1988, was called optimum coding in the frequency domain, OCF. In parallel, a similar, DFT-based system, termed PXFM, was developed in the United States of America. Both approaches were fused in 1990 [4]. As further features of the system, additional octave-band QMF banks, time-masking effects, and Bark-scale-approximating QMF trees combined with transformations were included, again in a German system, MUSICAM, 1990, but also in a US system, ASPEC, 1991, and in France, 1989. The development process finally ended in the 3-layer ISO/MPEG standard, which combines all methods which had proven successful so far.

14.4 Further Work and Results

14.4.1 Audio Coding

All the other principles which have been discussed in the speech-coding sections of this chapter could of course be also tried out for audio signals, e. g., VQ, pitch filters, linear prediction in sub-bands, or MPE. Actually, SM appeared recently in the MPEG-4 context [46], especially for really low transmission rates, such as 2.4 kbit/s. Furthermore, novel time-frequency representations became of interest, e. g., wavelets. Finer spectral resolution, window-*form* switching, and more sophisticated psychoacoustic models may show to be of advantage, as in the advanced audio codec, AAC, which has been selected for MPEG-2 and MPEG-4. The latter standards aim at, respectively, enhanced stereo-surround representations, i. e. five full channels with up to $f_B = 320$ kbit/s (!) and scalable rates and quality. As a final goal of research, transparent quality is envisaged at only $f_B = 16$ kbit/s, hopefully.

14.4.2 Speech Coding

Good telephone quality will soon be achieved at $f_B = 4$ kbit/s, while good wide-band quality is available at $f_B = \ldots12\ldots$ kbit/s. Good telephone speech with $f_B = \ldots2.4\ldots$ kbit/s is currently being studied. For instance, SM/HC ideas reappear in the MPEG-4 subsystem called harmonic vector-excited coder, HVXC, pitch-dependent waveform-interpolation concepts as proposed in [33] are investigated, models of the hearing process are used in the LP excitation [1], and also phoneme classification is being applied [8].

14.5 Further Reading

A huge variety of literature, including books, is available. The selection presented here is, of course, limited – and subjective. This holds also for the following recommendations.

A good entrance into the details of waveform-preserving coder approaches applied to speech (and video) is found in the "standard" [31], a broader overview on speech processing in [44], a less broad but more detailed and recent one in [55]. For more information on audio systems, formats, and market developments, the reader is referred to [39]. Modern DSP aspects of audio and acoustic signals are covered by a recent book [32]. For an excellent, very detailed review of coding methods, the recent report by *Painter & Spanias* [45] is recommended. All books and overview papers mentioned here are, in fact, rich sources for further references.

List of abbreviations

AAC	Advanced Audio Coding
ACELP	Algebraic CELP – with special code-book, →CELP
ADPCM	Adaptive Differential PCM – with adaptive prediction and quantization
AMR	Adaptive Multi-Rate coding
APC	Adaptive Predictive Coding – also "Block-ADPCM"
ASPEC	Audio Spectral Perceptual Entropy Coding
ATC	Adaptive Transform Coding
BB-RELP	Base-Band RELP, i. e. low-frequency band RELP →RELP
CELP	Code-(word) Excited Linear Predictor/Prediction
DCT	Discrete Cosine Transformation
DFT	Discrete Fourier Transformation
DSP	Digital Signal Processor
EDGE	Enhanced Data-rate for GSM Evolution – an add-on to GSM
EFR	Enhanced Full-Rate system – with the original GSM data rate and better quality
EMBE	Enhanced MBE →MBE
ETSI	European Telecommunication Standardization Institution
eXCELP	Extended CELP – with classified code-books →CELP
FIR	Finite (-length) Impulse Response, also "non-recursive" filter
FR	Full-Rate system . . . the first standard GSM speech coding at 13 kbit/s
GSM	Global System for Mobile communication
HC	Harmonic Coding . . . SM with a frequency grid →SM
HR	Half-Rate system – with half the original GSM data rate and less quality
HVXC	Harmonic Vector-eXcited Coding – a HC/VQ combination →HC, CELP, VQ

IMBE	Improved MBE →MBE
ISO	International Standardization Organization
ITU	International Telecommunication Union, formerly CCITT, Conseil Consultative de la Téléphonie et Télégraphie
PCM	Pulse-Code Modulation ... analog-to-digital conversion, with linear or possibly non-linear quantization
LP	Linear Prediction ... linear estimation of a new sample from n past samples
LPC	Linear Predictive Coding – used generally for the filter inside all predictive coders or specifically for the LP-based VOCODER
LSF	Line Spectral Frequencies ... transformed LPC coefficients
LTP	Long-Term Prediction – with a large lag in the order of a "pitch" period
MBE	Multi-Band Excitation – of an LP filter
MDCT	Modified DCT →DCT
Mops	Mega – i.e. million – operations per second
MOS	Mean Opinion Score – averaged-quality grades from, 1 (poor) to 5 (very good)
MPE	Multi-Pulse Excitation – with a reduced number of entries on a general time grid
MPEG	Moving-Picture Expert Group – also standardizing speech and audio systems
mp3	MPEG-1/Layer 3 audio-coding standard
MSE	Mean-Square Error
MUSICAM	Masking-pattern-adapted Universal Sub-band-Integrated Coding And Multiplexing
OCF	Optimum Coding in the Frequency Domain
PXFM	Perceptual Transform Coder
QMF	Quadrature-Mirror Filter – pair or bank
RELP	Residual (signal)-Excited Linear Predictor/Prediction
RPE	Regular-Pulse Excitation – with a reduced number of entries on a regular time grid
SBC	Sub-Band Coding
SIRP	Spherically Invariant Random Process
SM	Sinusoidal Modelling – of signals by sums of general sinusoids
SNR	Signal-to-Noise (power) Ratio – usually on a log-scale in dB
TC	Transform Coding
UMTS	Universal Mobile Telecommunication System
VOCODER	VOice CODER featuring parametric speech description and synthesis
VQ	Vector Quantization, also "block quantization"
WB	Wide-Band – speech, with 7 kHz bandwidth rather than 3.5 kHz

References

1. Ambikairajah E, Eps J, Lin L (2001) Wideband-speech and audio coding using gammatone filterbanks. Proc. IEEE Int Conf Acoustics, Speech, Sign Processing, ICASSP, Salt Lake City, II:773–776

2. Bell C G, Fujisaki H, Heinz J M, Stevens N K, House A S (1961) Reduction of speech spectra by analysis-by-synthesis techniques. J Acoust Soc Am. 33:1725–1736

3. Bessette B, Lefebvre R, Salami R, Jelinek M, Vainio J, Rotola-Pukkila J, Järvinen K, (2001) Techniques for high-quality ACELP coding of wide-band speech. Proc EUROSPEECH, Aalborg, 3:1993–1996

4. Brandenburg K, Johnston J D (1990) Second-generation audio coding: the hybrid coder. Proc 88th Conv Audio Engr Soc, AES, preprint 2937

5. Brehm H, Stammler W (1987) Description and generation of Spherically-Invariant Speech-Model Signals. EURASIP J Sign Process 12:119–141

6. Carl H (1994) Examination of different speech-coding methods and an application to band-width enhancement for narrow-band speech signals (in German). Doct Diss, Ruhr-University, Bochum.

7. Carl H, Heute U (1994) Bandwidth enhancement of narrow-band speech signals. Proc Europ Sig Process Conf, EUSIPCO, Edinburgh, 1716–1719

8. Ehnert W, Heute U. (1997) Variable-rate speech coding: replacing unvoiced excitations by linear-prediction residues of different phonemes. Proc GRETSI, Grenoble, 993–996

9. Erdmann Ch, Vary P, et al. (2001) A candidate proposal for a 3-GPP adaptive multi-rate wide-band speech codec. Proc Int Conf Acoustics, Speech, Sig Process, ICASSP, Salt Lake City, II:757–760

10. ETSI (1989) Recommendation GSM 06.10: GSM full-rate transcoding"

11. ETSI (1994) Recommendation GSM 06.20: European digital cellular telecommunications system speech codec for the half-rate speech-traffic channel

12. ETSI (1996) Recommendation GSM 06.60: Digital cellular telecommunications system: enhanced full-rate (EFR) speech transcoding

13. ETSI (1999) Recommendation AMR 06.90: Digital cellular telecommunications system: adaptive multi-rate (AMR) speech transmission

14. Fastl H (2005) Psychoacoustics and sound quality. Chap 6 this vol

15. Gluth R , Guendel L, Heute U (1986) ATC - a candidate for digital mobile-radio telephony. Proc Nordic Sem Dig Land-Mob Radio Comm, Stockholm, 230–235

16. Gray R M (1984) Vector quantization. IEEE ASSP Mag 1:4–29

17. Grill B, Edler B, et al. (1998) Information technology – very low bitrate audio-visual coding. Part 3: Audio. Subpart 1: Main Document. ISO/IEC FCD 14496-3 Subpart 1

18. Halle M, Stevens N K (1959) Analysis-by-synthesis. In: Wathen-Dunn W, Woods L (eds) AFCR–TR–59–198, Proc Sem Speech Compress Processg II:D7

19. Hammershøi D, Møller H (2005) Binaural technique: basic methods for recording, synthesis and reproduction. Chap 9 this vol

20. Hess W (1983) Pitch determination of speech signals. Springer, Berlin

21. Heute U (1988) Medium-rate speech coding - trial of a review. Speech Commm 7:125–149

22. Heute U (1994) Speech coding: approaches, trends, standards (in German). Proc ITG Conf Source & Channel Cod, Munich, 437–448

23. Hudde H (2005) A functional view on the human peripheral hearing system. Chap 2 this vol
24. Huffman D A (1952) A method for the construction of minimum-redundancy codes. Proc IRE 40:1098–1101
25. Itakura F (1975) Line-spectral representation of linear-prediction coefficients of speech signals. J. Acoust Soc Am 57:S37
26. ITU/CCITT (1988) Recommendation G.711: Coding of analogue signals by pulse-code modulation (PCM) of voice-frequencies
27. ITU-T (1988) Recommendation G.722: 7 kHz Audio Coding within 64 kbit/s. Fascicle III.4, Blue Book 269–341
28. ITU-T (1992) Recommendation G.726: 40, 32, 24, 16 kbit/s adaptive differential pulse-code modulation (ADPCM)
29. ITU-T (1992) Recommendation G.728: Coding of speech at 16 kbit/s, using low-delay code-excited linear prediction
30. ITU-T (1995) Recommendation G.729: Coding of speech at 8 kbit/s, using conjugate-structure algebraic code-excited linear prediction (CS-ACELP)
31. Jayant N S, Noll P (1984) Digital coding of waveforms. Prentice Hall, Englewood Cliffs
32. Kahrs M, Brandenburg K (1998) Application of digital signal processing to audio and acoustics. Kluwer Acad Publ
33. Kleijn B, Haagen J (1995) A speech coder based on decomposition of characteristic waveforms. Proc Int Conf Acoustics, Speech, Sig Process, ICASSP, Detroit, 508–511
34. Lacroix A (2005) Speech-production: acoustics, models, and applications. Chap 13 this vol
35. Lazzari V, Montagna R, Sereno D (1988) Comparison of two speech codecs for DMR systems. Speech Comm 7:193–207
36. Linde Y, Buzo A, Gray R M (1980) An algorithm for vector-quantizer design. IEEE Transact Comm 28:84–95
37. Max J (1960) Quantizing for minimum distortion. IRE Transact. Inf Th 6:7–12
38. Möller S (2005) Quality of transmitted speech for humans and machines. Chap 7 this vol
39. Mourjopoulos J (2005) The evolution of digital audio technology. Chap 12 this vol.
40. Noll P, Zelinski R (1977) Adaptive transform coding of speech signals. IEEE Transact Acoustics Speech Sig Process 25:299–309
41. Noll P (1993) Wideband-speech and audio coding. IEEE Comm Mag Nov:34–44
42. Noll P (1995) Digital audio coding for visual communications. Proc IEEE 83:925-943
43. Noll P (1997) MPEG digital audio coding. IEEE Sig Process Mag Sept:59–81
44. O'Shaughnessy D (1987) Speech communication - human and machine. Addison-Wesley, New York
45. Painter T, Spanias A (2000) Perceptual coding of digital audio. Proc IEEE 88:451–513
46. Painter T, Spanias A (2003) Sinusoidal analysis-synthesis of audio using perceptual criteria. EURASIP J Appl Sig Processg 15-20.
47. Paulus J (1997) Coding of wide-band speech signals at low data rates (in German). Doct Diss, RWTH, Aachen
48. Rabiner L R , Schafer R W (1978) Digital processing of speech signals. Prentice-Hall, Englewood Cliffs, N.J.

49. Spanias A S (1994) Speech coding: a tutorial review. Proc IEEE 82:1541–1582
50. Gao Y, Benyassine A, Thyssen J, Su H, Shlomot E (2001) EX-CELP: a speech-coding paradigm. Proc Int Conf Acoustics, Speech Sig Process, ICASSP, Salt Lake City, II:689–692
51. Thyssen J, Gao Y, Benyassine A, Shlomot E, Murgia C, Su H, Mano K, Hiwasaki Y, Ehara A, Yasunaga K, Lamblin C, Kovesi B, Stegmann J, Kang H (2001) A candidate for the ITU-T 4 kbit/s speech-coding standard. Proc Int Conf Acoustics, Speech Sig Process, ICASSP, Salt Lake City, II:681–684
52. Trancoso I M, Almeida L, Rodriges J, Marques J, Tribolet J (1988) Harmonic coding - state of the art and future trends. Speech Comm 7:239–245
53. Tremain T E (1982) The government standard linear-predictive coding: LPC-10. Speech Technol 1:40–49
54. Varga I (2001) Standardization of the adaptive multi-rate wideband codec. Proc ITG Conf Source & Channel Cod, Berlin, 341–346
55. Vary P, Heute U, Hess W (1998) Digitale Sprachsignalverarbeitung. Teubner, Stuttgart
56. Zwicker E, Fastl H (1990) Psychoacoustics - facts and models. Springer, Berlin Heidelberg

Index

Printing: Strauss GmbH, Mörlenbach
Binding: Schäffer, Grünstadt

Lightning Source UK Ltd.
Milton Keynes UK
UKOW031207110212

187126UK00001B/11/P